Lecture Notes in Artificial Intelligence   4457

Edited by J. G. Carbonell and J. Siekmann

Subseries of Lecture Notes in Computer Science

Gregory M.P. O'Hare   Alessandro Ricci
Michael J. O'Grady   Oğuz Dikenelli (Eds.)

Engineering Societies in the Agents World VII

7th International Workshop, ESAW 2006
Dublin, Ireland, September 6-8, 2006
Revised Selected and Invited Papers

Springer

Series Editors

Jaime G. Carbonell, Carnegie Mellon University, Pittsburgh, PA, USA
Jörg Siekmann, University of Saarland, Saarbrücken, Germany

Volume Editors

Gregory M.P. O'Hare
Michael J. O'Grady
University College Dublin
School of Computer Science and Informatics, Adaptive Information Cluster
Belfield, Dublin 4, Ireland
E-mail: {gregory.ohare, michael.j.ogrady}@ucd.ie

Alessandro Ricci
Alma Mater Studiorum - Università di Bologna
Dipartimento di Elettronica, Informatica e Sistemistica (DEIS)
via Venezia 52, 47023 Cesena, Italy
E-mail: a.ricci@unibo.it

Oğuz Dikenelli
Ege University
Department of Computer Engineering
35100 Bornova, Izmir, Turkey
E-mail: oguz.dikenelli@ege.edu.tr

Library of Congress Control Number: Applied for

CR Subject Classification (1998): I.2.11, I.2, C.2.4, D.1.3, D.2.2, D.2.7, D.2.11, I.6

LNCS Sublibrary: SL 7 – Artificial Intelligence

ISSN   0302-9743
ISBN-10   3-540-75522-5 Springer Berlin Heidelberg New York
ISBN-13   978-3-540-75522-7 Springer Berlin Heidelberg New York

This work is subject to copyright. All rights are reserved, whether the whole or part of the material is concerned, specifically the rights of translation, reprinting, re-use of illustrations, recitation, broadcasting, reproduction on microfilms or in any other way, and storage in data banks. Duplication of this publication or parts thereof is permitted only under the provisions of the German Copyright Law of September 9, 1965, in its current version, and permission for use must always be obtained from Springer. Violations are liable to prosecution under the German Copyright Law.

Springer is a part of Springer Science+Business Media

springer.com

© Springer-Verlag Berlin Heidelberg 2007
Printed in Germany

Typesetting: Camera-ready by author, data conversion by Scientific Publishing Services, Chennai, India
Printed on acid-free paper   SPIN: 12171279   06/3180   5 4 3 2 1 0

Preface

The seventh international workshop ESAW 2006 – Engineering Societies in the Agents World VII—was hosted in the School of Computer Science and Informatics, University College Dublin, Ireland in September 2006. This workshop was organized as a stand-alone event, running over three days, and continued and enhanced the high-quality conference theme that now uniquely characterizes the ESAW workshop series. ESAW VII built upon the success of prior ESAW workshops – Kuşadasi (2005), London (2004) and Toulouse (2004), going back to the inaugural workshop held in Berlin (2000). This workshop was attended by 50 participants from 13 different countries. Over 25 researchers presented their work and substantial time was allocated each day for ad-hoc interactive discussions on those presented topics. Indeed, these opportunities for the exchange of views and open discussion with fellow experts are one of the hallmarks of the ESAW series. Discussions coalesced around ESAW's main themes:

- Engineering multi-agent systems
- Methodologies for analysis, design, development and verification of agent societies
- Interaction and coordination in agent societies
- Autonomic agent societies
- Trust in agent societies

For more information about the workshop, the interested reader is referred to the ESAW 2006 WWW site[1]. The original contributions have been published as a Technical Report (UCD-CSI-2006-5) and this may be obtained freely from the Technical Report section on the WWW page of the School of Computer Science and Informatics at University College Dublin[2].

These post-proceedings continue the series published by Springer (ESAW 2000: LNAI 1972; ESAW 2001: LNAI 2203; ESAW 2002: LNAI 2577; ESAW 2003: LNAI 3071; ESAW 2004: LNAI 3451; ESAW 2005: LNAI 3963). This volume contains substantially revised and extended versions of selected papers from ESAW 2006 that both address and accommodate comments from the two rounds of reviews, as well as incorporating feedback from comments and questions that arose during the workshop. In addition to the selected papers, the Program Chairs directly invited and evaluated some papers to further improve the quality of the post-proceedings.

The workshop organizers would like to acknowledge the support of the School of Computer Science and Informatics at University College Dublin, Fáilte Ireland, and Dublin Tourism for their practical support. We would also like to

---

[1] http://esaw06.ucd.ie
[2] http://csiweb.ucd.ie/Research/TechnicalReports.html

acknowledge the generous efforts of the Steering Committee for their guidance, the Program Committee for their insightful reviews and the local Organizing Committee for arranging a thoroughly enjoyable event. We would also like to offer our thanks for the continued patronage of the large number of researchers who submitted their work for consideration. Finally, thanks to Christine Guenther and Alfred Hofmann, without whose efforts these post-proceedings would have floundered.

April 2007

Gregory O'Hare
Alessandro Ricci
Michael O'Grady
Oğuz Dikenelli

Organization

ESAW 2006 Workshop Organizers and Program Chairs

Gregory O'Hare	University College Dublin, Ireland
Alessandro Ricci	Università di Bologna, Italy
Michael O'Grady	University College Dublin, Ireland
Oğuz Dikenelli	Ege University, Izmir, Turkey

ESAW Steering Committee

Marie-Pierre Gleizes	IRIT Université Paul Sabatier, Toulouse, France
Andrea Omicini	DEIS, Alma Mater Studiorum Università di Bologna, Cesena, Italy
Paolo Petta	Austrian Research Institute for Artificial Intelligence, Austria
Jeremy Pitt	Imperial College London, UK
Robert Tolksdorf	Free University of Berlin, Germany
Franco Zambonelli	Università di Modena e Reggio Emilia, Italy

ESAW 2006 Local Organizing Committee

Gregory O'Hare	University College Dublin, Ireland
Michael O'Grady	University College Dublin, Ireland
Rem Collier	University College Dublin, Ireland
Eleni Mangina	University College Dublin, Ireland
Stephen Keegan	University College Dublin, Ireland

ESAW 2006 Program Committee

Alexander Artikis	Institute of Informatics and Telecommunications, Greece
Federico Bergenti	Università di Parma, Italy
Carole Bernon	IRIT Université Paul Sabatier, Toulouse, France
Olivier Boissier	Ecole Nationale Supérieure des Mines de Paris, France
Guido Boella	Università degli Studi di Torino, Italy
Monique Calisti	Whitestein Technologies, Switzerland
Jacques Calmet	University of Karlsruhe, Germany

Cristiano Castelfranchi — ISTC-CNR, Rome, Italy
Luca Cernuzzi — Universidad Católica, Paraguay
Helder Coelho — University of Lisbon, Portugal
Rem Collier — University College Dublin, Ireland
Dan Corkill — University of Massachusetts at Amherst, USA
Massimo Cossentino — ICAR-CNRL, Italy
Mehdi Dastani — Intelligent Systems Group, Utrecht University, The Netherlands
Paul Davidsson — Blekinge Institute of Technology, Sweden
Riza Cenk Erdur — Ege University, Izmir, Turkey
Rino Falcone — ISTC-CNR, Rome, Italy
Paul Feltovich — University of West Florida, USA
Jean-Pierre Georgé — IRIT Université Paul Sabatier, Toulouse, France
Marie-Pierre Gleizes — IRIT Université Paul Sabatier, Toulouse, France
Paolo Giorgini — Dept. of Information and Communication Technology, University of Trento, Italy
Salima Hassas — Université Claude Bernard Lyon 1, France
Anthony Karageorgos — University of Thessaly, Greece
Michael Luck — University of Southampton, UK
Eleni Mangina — University College Dublin, Ireland
Peter McBurney — University of Liverpool, UK
Fabien Michel — IUT de Reims-Châlons-Charleville, Reims, France
Pablo Noriega — IIIA, Spain
Andrea Omicini — DEIS, Alma Mater Studiorum, Università di Bologna, Cesena, Italy
Sascha Ossowski — Univesidad Rey Juan Carlos, Spain
Juan Pavon — Universidad Complutense de Madrid, Spain
Paolo Petta — The Austrian Research Institute for Artificial Intelligence, Austria
Jeremy Pitt — Imperial College, London, UK
Giovanni Rimassa — Whitestein Technologies, Switzerland
Juan A. Rodríguez Aguilar — IIIA, Spain
Maarten Sierhuis — RIACS/NASA Ames Research Center, USA
Kostas Stathis — Royal Holloway, University of London, UK
Luca Tummolini — ISTC-CNR, Rome, Italy
Paul Valckenaers — Katholieke Universiteit Leuven, Belgium
Leon Van der Torre — CWI Amsterdam, The Netherlands
Pinar Yolum — Bogazici University, Istanbul, Turkey
Mirko Viroli — DEIS, Alma Mater Studiorum Università di Bologna, Cesena, Italy
Pallapa Venkataram — IIS Bangalore, India
Danny Weyns — Katholieke Universiteit Leuven, Belgium
Franco Zambonelli — Università di Modena e Reggio Emilia, Italy

Table of Contents

Engineering of Multi-agent Systems

"It's Not Just Goals All the Way Down" – "It's Activities All the Way Down" 1
Maarten Sierhuis

The Construction of Multi-agent Systems as an Engineering Discipline 25
Jorge J. Gomez-Sanz

Current Issues in Multi-Agent Systems Development 38
Rafael H. Bordini, Mehdi Dastani, and Michael Winikoff

Architecture-Centric Software Development of Situated Multiagent Systems 62
Danny Weyns and Tom Holvoet

Organization Oriented Programming: From Closed to Open Organizations 86
Olivier Boissier, Jomi Fred Hübner, and Jaime Simão Sichman

Analysis, Design, Development and Verification of Agent Societies

Modelling and Executing Complex and Dynamic Business Processes by Reification of Agent Interactions 106
Marco Stuit and Nick B. Szirbik

Model Driven Development of Multi-Agent Systems with Repositories of Social Patterns 126
Rubén Fuentes-Fernández, Jorge J. Gómez-Sanz, and Juan Pavón

A Norm-Governed Systems Perspective of Ad Hoc Networks 143
Alexander Artikis, Lloyd Kamara, and Jeremy Pitt

Interaction and Coordination in Agent Societies

A Definition of Exceptions in Agent-Oriented Computing 161
Eric Platon, Nicolas Sabouret, and Shinichi Honiden

Toward an Ontology of Regulation: Socially-Based Support for Coordination in Human and Machine Joint Activity 175
Paul J. Feltovich, Jeffrey M. Bradshaw, William J. Clancey, and Matthew Johnson

An Algorithm for Conflict Resolution in Regulated Compound
Activities 193
   Andrés García-Camino, Pablo Noriega, and
   Juan-Antonio Rodríguez-Aguilar

Modeling the Interaction Between Semantic Agents and Semantic Web
Services Using MDA Approach 209
   Geylani Kardas, Arda Goknil, Oguz Dikenelli, and
   N. Yasemin Topaloglu

Formal Modelling of a Coordination System: From Practice to Theory,
and Back Again 229
   Eloy J. Mata, Pedro Álvarez, José A. Bañares, and Julio Rubio

Autonomic Agent Societies

Using Constraints and Process Algebra for Specification of First-Class
Agent Interaction Protocols 245
   Tim Miller and Peter McBurney

Dynamic Specifications in Norm-Governed Open Computational
Societies 265
   Dimosthenis Kaponis and Jeremy Pitt

Enhancing Self-organising Emergent Systems Design with Simulation ... 284
   Carole Bernon, Marie-Pierre Gleizes, and Gauthier Picard

Adaptation of Autonomic Electronic Institutions Through Norms and
Institutional Agents 300
   Eva Bou, Maite López-Sánchez, and J.A. Rodríguez-Aguilar

Managing Resources in Constrained Environments with Autonomous
Agents 320
   C. Muldoon, G.M.P. O'Hare, and M.J. O'Grady

Trust in Agent Societies

Towards a Computational Model of Creative Societies Using Curious
Design Agents 340
   Rob Saunders

Privacy Management in User-Centred Multi-agent Systems 354
   Guillaume Piolle, Yves Demazeau, and Jean Caelen

Effective Use of Organisational Abstractions for Confidence Models 368
   Ramón Hermoso, Holger Billhardt, Roberto Centeno, and
   Sascha Ossowski

Competence Checking for the Global E-Service Society Using Games ... 384
  Kostas Stathis, George Lekeas, and Christos Kloukinas

Author Index 401

"It's Not Just Goals All the Way Down" – "It's Activities All the Way Down"

Maarten Sierhuis

Research Institute for Advanced Computer Science
NASA Ames Research Center
M/S 269-1
Moffett Field,
CA 94035-1000, USA
Maarten.Sierhuis-1@nasa.gov

Abstract. The rational agent community uses Michael Bratman's *planning theory of intention* as its theoretical foundation for the development of its agent-oriented BDI languages. We present an alternative framework based on situated action and activity theory, combining both BDI and activity-based modeling, to provide a more general agent framework. We describe an activity-based, as opposed to a goal-based, BDI language and agent architecture, and provide an example that shows the flexibility of this language compared to a goal-based language.

1 Introduction

This chapter is written based on a talk with the same title given as an invited talk at the seventh annual international "Engineering Societies in the Agents World" (ESAW 2006) conference. The title is meant as a tongue-in-cheek provocation towards the BDI agent community being proponents of agents as goal-driven planners. A previous presentation about the Brahms multiagent simulation and development environment, at the 2006 Dagstuhl seminar on Foundations and Practice of Programming Multi-Agent Systems, had created debates with agent language researchers about the primary use of goal-driven behavior in agent languages [1]. In this talk, we intended to put forth an alternative to agents being solely goal-driven planners. We not only put forward an alternative view on human behavior (a view that relies on activities instead of goal-driven action), but we also presented our Brahms multiagent language and simulation environment that is BDI-like without the notion of goals as the driving force of agent action [2][3]. In this chapter, we try to more thoroughly explain this activity-based BDI approach and present not only the Brahms language and the workings of the architecture, but present the theoretical basis for this alternative view of human behavior. The theory of activity-based modeling is rooted in *situated action, cognition in practice, situated cognition* and *activity theory*.

It has to be noted that this paper cannot do justice to the large number of research articles that have been written on these concepts. We merely can scratch the surface, and we do not claim that we explain these concepts completely. We quote heavily from writings of other, more prominent, researchers to explain these concepts, and we advice the reader to study them in order to get a deeper understanding of the theories and research behind them. What we hope this paper contributes to is a theory, practice and tool of how to analyze, design and implement large agent systems. Our hope is also that we succeed in providing a well enough argument for activity-based rational agents as a more general alternative to goal-driven ones.

The rational agent community, over the last ten or so years, has settled on using Michael Bratman's *planning theory of intention* as its theoretical foundation for the development of its agent-oriented BDI languages [4][5]. Fundamentally, we believe that it is very good for a programming paradigm and its subsequent languages to be based on a theory. In general, programming is about modeling the world as we know it, and a specific program embeds a particular theory—the programmer's theory of the world being modeled. When a programming language itself is based on a solid theory it can help the programmer. However, this is both a good and a bad thing. The good thing is that we can have a tool that is sound—based on a theory—and thus the programs that are being developed with it will, by definition, adhere to the theory. This helps the programmer not to have to invent a new theory before implementing a model based on it. The theory is already there. For example, all computer programmers use Turing's theory of universal computing machines as their basis for developing programs as a sequence of instructions.[1] The bad thing is that believers of the theory tend to apply the theory in all sorts of ways that might, or more importantly, might not be within the theory's relevance. A second problem is that the theory might be flawed, or at least, can have alternative theories that put in question a fundamental part of the theory.

In this chapter we discuss some of the fundamental underpinnings of the theory on which the BDI programming languages of today are based. We point out that a fundamental assumption of Bratman's theory has at least one alternative theory that is based on the *inverse* assumption. This is interesting in the sense that both theories cannot be correct. After we discuss some of Bratman's theory and its fundamental assumption, we will move on to describe the alternative theory and its assumption. Then, we will merge the two theories into a theory that combines both and thus will provide a more general theory on which future BDI languages could be based. We then move on to describe an activity-based BDI language and agent architecture based on this new more general theory, and provide an example that shows the flexibility of this language compared to a goal-based BDI language.

2 Theoretical Stances

2.1 Bratman's Planning Theory

Bratman's philosophical treatise on human behavior is based on commonsense or folk psychology—a not falsifiable theory based on propositional attitude ascriptions, such as "belief", "desire", "pain", "love," "fear". The fact that the theory on which Bratman's treatise is based is not falsifiable makes it not a theory, but a theory about

---

[1] Alan Turing (1936), "On Computable Numbers, With an Application to the Entscheidungs-problem", Proceedings of the London Mathematical Society, Series 2, Volume 42 (1936). Eprint. Reprinted in The Undecidable pp.115-154.

the existence of a theory, also known as a theory-theory. Herein lays the division in camps of believers: Those who believe in the theory and those who do not, and then those who believe in some of it. It is difficult to decipher in which camp the rational agent community lays. To me there are only two possibilities. First, it is completely possible that people in the rational agent community believe in Bratman's *Planning Theory* [4]. Falling in this camp means that one agrees with the assumption that people *are* planning agents. This has a major implication, to which we will return in the next section. The other possibility is that people in this camp do not necessary believe that Bratman's theory says something in particular about people, but it says something about any kind of animal or any non-living object. People in this camp care less about the fact that Bratman's theory only talks about people and not about systems in general, but they nevertheless believe that using planning to deliberatively decide on future actions is a very attractive presupposition, and even more, it is a very useful concept for computer programming. For people in this camp every behavior, whether human or not, can be reduced to *a problem that needs to be solved*, because in that case Bratman's Planning Theory can be used to predict and execute actions to solve it. Thus, in short, Bratman's theory can easily be used to solve any behavioral "problem" with an algorithm developed by researchers of artificial intelligence and cognitive science.

Let us briefly turn to Bratman's theory, and let us start by stating that we do not have the space or the time to fully describe Bratman's theory here. We only focus on those parts that are relevant for our ultimate points. Since this paper is part of a book about engineering software agents, we assume for simplicity sake that the reader is familiar with the concepts *belief*, *desire* and *intention*, and we therefore do not explain these concepts. What does Bratman's theory say?

People are planning agents => they settle in advance on complex plans for the future and these plans guide their conduct.

This is Bratman's fundamental assumption. This is neither fully explained, nor is it proven by his theory. It is simply a supposition. The only evidence for this supposition being true is given by Bratman in the form of a reference to Simon's notion that people are *bounded rational agents* [6]. Most artificial intelligence researchers, economists and cognitive scientists agree with Simon's claim. Never mind that Simon's notion came about as a reaction on *economic models* always assuming that people are hyperrational[2]. Simon postulated that people are boundedly rational and the only way of behaving is by *satisficing*[3].

People, as planning agents, have two capacities: 1) the capacity to act purposeful, 2) the capacity to form and execute plans.

The following quote exemplifies the extent to which Bratman justifies his claim that people must be planning agents; "So we need ways for deliberation and rational

---

[2] In this context, hyperrational means that people are always rational, to a fault, and would never do anything to violate their preferences.
[3] "Satisficing is a behaviour which attempts to achieve at least some minimum level of a particular variable, but which does not necessarily maximize its value." from Wikipedia, 12/24/2006.

reflection to influence action beyond the present; and we need sources of support for coordination, both intrapersonal and interpersonal. Our capacities as planners help us meet these needs" [4, p. 3]. In here lies Bratman's basic "proof" that people are planning agents. His reasoning goes something like this; Planning is an algorithmic way to deal with deciding on what to do in the future, i.e. future action. Planning can be seen as a sort of coordination. People need to decide on future action; otherwise they could never do anything. Deliberation takes time and it is not in and of it-self useful for taking action. We need a way to influence what we do in the future. People need to coordinate their actions, as well as what they do together. Therfore, people need a planning algorithm to do this. Quod erat demonstrandum.

Goal-Based Modeling
Those who believe in Bratman's Planning Theory almost always also believe in people being general problem-solvers, and that deciding what to do next is a problem to be solved for which people use a goal-based or goal-driven reasoning method. The reason that this way of thinking is almost always synonymous with believing in the BDI agent architecture is that it fits very well with Bratman's Planning Theory. To be goal-based means that one stores information about the world and about situations that are desirable, which we refer to as the goals to be reached. Our model of current situation or current state and of goals or future state, allows us then to choose among the multiple future states we can think of and select which one should be next. This next state is then called our (sub)goal. This maps, in principle, very well onto the concepts of beliefs, desires and intentions. In a more general way, not necessarily only related to how people operate, we might change the names of these concepts into stored current state, desired future state or goal, and intent to act. It is the concept of *desire* that makes the BDI architecture inherently a goal-driven architecture and BDI agents problem-solving agents.

This model of agent behavior is compelling to people adhering to Bratman's Planning Theory, because it also fits very well with some of the most prominent theories of cognition, namely Newell's Unified Theory of Cognition and Anderson's ACT-R model of cognition [7][8]. However, there are theories and implemented architectures out there that are alternatives to goal-driven planning. One of the most prominent alternatives is the behavior-based robot architecture that shows that robots can achieve intelligent behavior that is not based on the existence of an explicit plan and of goal-driven behavior. Indeed, Brooks argues convincingly, as we are trying to do in this chapter, that "planning is just a way of avoiding figuring out what to do next." [9, Chapter 6, p. 103]. Brooks' basic argument goes something like this: The idea of planning and plan execution is based on an *intuition*. There is no evidence that it *has to be* the only way to develop intelligent behavior. Putting sensors and effectors together using a program can also create intelligent behavior.

Thus a process-way of acting is just as possible as a state way of reasoning. In the Artificial Intelligence and Robotics community Brooks is the most well known researcher that argues against Bratman's Planning Theory. However, there have also been social- and cognitive scientists that have argued against the Planning Theory, not from an engineering or computational angle, but from a social and behavioral angle. This is the subject of the next section, and together with Brooks' argument will form the basis for our thesis that an activity-based approach is a more general approach, incorporating both a goal-based and a behavior-based approach for BDI agent architectures.

2.2 The Alternative View

Let us turn to an alternative view of human behavior, namely a view of human purposeful action based on ethnomethodology[4]. Suchman, in her book "Plans and Situated Actions," takes a strong stance against the traditional view of planned purposeful action. It is this alternative view that in our opinion best expresses the problems with Bratman's Planning Theory as a theory about human purposeful action [8].

Situated Action

Suchman's view takes human purposeful action as its starting point for understanding human behavior, but it is not a theory of the brain's functioning. Suchman's view is not presented as a theory of human behavior, as it does not predict human behavior. However, Suchman investigates an alternative view to the view deeply rooted in Western cognitive science, that human action is only and essentially determined by plans. Suchman describes human activity from a socially-, culturally- and environmentally situated point of view, rather than from a cognitive point of view. This different angle on human activity opens up an entirely new take on purposeful action, namely that learning how to act is dependent on the culture in which one grows up, and thus there is variation based on the situation. Whether our actions are ad hoc or planned depends on the nature of the activity and the situation in which it occurs.

Although we might contrast goal-directed activities from creative or expressive activities, or contrast novice with expert behavior, Suchman argues convincingly that every purposeful action is a *situated action*, however planned or unanticipated. Situated action is defined as "actions taken in the context of particular, concrete circumstances." [p. viii] Suchman posits an important consequence of this definition of purposeful action that is best retold with a quote:

> "... our actions, while systematic, are never planned in the strong sense that cognitive science would have it. Rather, plans are best viewed as weak resources for what is primarily *ad hoc* activity. It is only when we are pressed to account for the rationality of our actions, given the biases of European culture, that we invoke guidance of a plan." [p. ix]

Suchman goes on to say that plans stated in advance of our action are *resources* that are necessarily vague, and they filter out the precise and necessary detail of situated actions. Plans are not only resources used in advance of situated action, they are also used as a reconstruction of our actions in an explanation of what happened. However, we filter out those particulars that define our situated actions in favor of explaining aspects of our actions that can be seen in accordance with the plan. Thus,

---

[4] Ethnomethodology (literally, 'the study of people's (folk) methods') is a sociological discipline which focuses on the ways in which people make sense of their world, display this understanding to others, and produce the mutually shared social order in which they live. [Wikipedia, 12/27/06].

according to Suchman, in strong contrast with the view of Bratman, plans are necessarily abstract representations of our actions and as such cannot serve as an account of our actions in a particular situation. In other words, our actions are not planned in advance, but are always *ad hoc*. Let us take a brief look at how cognitive science and AI are changing to a science of *situated cognition*.

Situated Cognition
Behavioral robotics is part of a new stream in AI, evolving from text-based or symbolic architectures of expert systems, descriptive cognitive models, and indeed BDI-like planning, to reactive, self-organizing situated robots and connectionist models. Many scientists have started to write about alternatives to the "wet computer" as a symbolic planning machine, such as Suchman [10], Brooks [9], Agre [11], Fodor [12], Edelman [13], Winograd and Flores [14]. We surely are leaving out others. In this section we will briefly present the work of Bill Clancey, because he is both the father of Brahms (the activity-based BDI language and multiagent architecture we describe in this paper) and a well-known father of expert tutoring systems (a purely symbolic approach), who has "gone the other way." Clancey is one of the proponents of situated cognition, and having written two books about it, one of the experts in the field [15][16].

Situated cognition goes one-step further than Suchman's situated action, namely it posits an alternative human memory architecture to the symbolic memory architecture of Newell and Anderson [7][8]. Clancey, in [16], develops an architecture of human memory at the neural level that is quite different from an architecture of a long-term memory storing productions and a short-term memory for matching and copying text (or symbols) into a "buffer." Clancey's arguments against the "memory as stored structures" hypothesis of Newell and Anderson has far reaching consequences for the Planning Theory of Bratman, namely that BDI-type planning cannot be the process for human thinking. It are exactly the concepts on which planning is based, like *search*, *retrieve*, and, *match*, that are associated with thinking as a "process of manipulating copies of structures, retrieved from memory" and "learning [as] a process of selectively writing associations back into long-term memory [p. 3]."

What is important for *situatedness* is the notion that thinking is a situated experience over time. In Newell and Anderson's model, time plays an important role in the matching of antecedents (of productions in long-term memory) and retrieval of the matched production from long-term memory and copying into short-term memory. In other words, the time it takes to think has little to do with being in the activity of doing something in the world, but rather it has to do with being in the activity of matching, retrieving and copying of symbols. In Clancey's model, however, the overall time it takes for neuronal processes to subsequently activate in the brain *is* the essence of situated action and (subconscious) conceptualization occurs as an activity. The essential part of the Newell and Anderson models of cognition that is disputed is the sequential relation of:

sensation $\Rightarrow$ *perception memory* $\Rightarrow$ *processing/thought and planning* $\Rightarrow$ *action*.

Instead, these are processes *coupled* in activation relations called *conceptual coordination*.

All else being equal, we adhere to the view that actions are situated as the starting point for developing an activity modeling approach. Not in the least, because the situated view of human activity constrains us less, by not having to commit to theories about brain function that model problem-solving tasks at the millisecond range. It frees us from having to model what is "in between our ears," and opens up our focus on modeling an agent's activities as larger time-taking situated actions within *the environment*. It finally enables us to model human behavior based on individual and social interaction with other individuals and artifacts in the world.

Activity-Based Modeling: Combining BDI with Situated Action
In the previous section, we laid the groundwork for activity-based modeling. Suchman's treatise on situated action is important, because it proofs that modeling intention to act may not only be goal-based. In fact, Suchman argues convincingly that an intention leading every-day life action is always situated in an environment, and is only goal-driven when there is a real-world problem to be solved. The overarching conclusion of this alternative view is that acting in the world is not only bounded by rationality, i.e. reasoning—or rather problem-solving—in the abstract, but is mainly situated in an environment that is historically, culturally and socially learned by its changes over time. It is thus important to not leave out the environment. In the remainder of this chapter, we define a modeling framework that combines the concepts of belief, desire and intention with that of situated action, to form an activity-based modeling approach that can be used not only to model the work practices of people, but also to model rational agent behavior in any multiagent system. By now, it should be no surprise that in this modeling framework an agent's behavior is not goal-based, but rather activity-based. This means that an agent's *desires* are *not* the possible subgoals the agent derives from its *intention* to solve a particular problem. Rather, as we will see, an agent's desires are reflected in the agent's *motives* for performing activities, and these motives are dependent on the environment and the situation, i.e. the state of the world at that moment. In the next section we describe what is meant with the word *situated*.

Modeling the Environment
The social anthropologist Jean Lave studied how people apply cognition in practice [17]. In the Adult Match Project she studied how people do math. She found that the same individuals do arithmetic completely different depending on the situation they are in. The results from her studies are significant, because solving mathematical problems has for a long time been the way scientists have studied general problem-solving in school and in the laboratory. Lave and her colleagues studied adult arithmetic practices in every-day settings, such as cooking and grocery shopping. Their work shows that solving mathematical problems is not so much constraint by how much of a math expert one is because of their math education, but because of the setting or situation in which the arithmetic activity takes place. Lave argues convincingly that, in order to understand problem solving we need to model the problem-solving process *in situ*.

Russian psychologist Lev Semyonovich Vygotsky plays an important role in developmental psychology. Vygotsky's deeply experimental work on child learning and behavior developed the notion that cognition develops at a young age through *tools* and the shared social and cultural understanding of tools in the form of *internalized signs* [18]. For example, a ten-month-old child is able to pull a string to obtain a cookie. However, it is not till the child is about 18 months that it is able to remove a ring from a post by lifting it, rather then pulling on it. Vygotsky writes; "Consequently, the child's system of activity is determined at each specific stage both by the child's degree of organic development and by his or her degree of mastery in the use of tools." [p. 21] The internalization of mimicry of tool use by young infants into a shared meaning of signs is, according to Vygotsky, intertwined with the development of speech. At the same time, speech plays an important role in the organization of a child's activities. Vygotsky writes; "Our analysis accords symbolic activity a specific organization function that penetrates the process of tool use and produces fundamentally new forms of behavior." [p. 24] While we will discuss the concept of activity later on, here it is important to see the essential importance of physical artifacts used as tools *within* an activity. Similar to the role of an artifact as a tool, the role of artifacts as the products of an activity plays an equally important role in what Vygotsky calls *practical intelligence*. Figure 1. shows the use of an artifact in the activity—its role—that transforms the artifact into a *tool* or a *product* of the activity, used or created by the subject. Outside the activity the artifact is just an object in the world. To the observer the object is necessary for the activity to be performed.

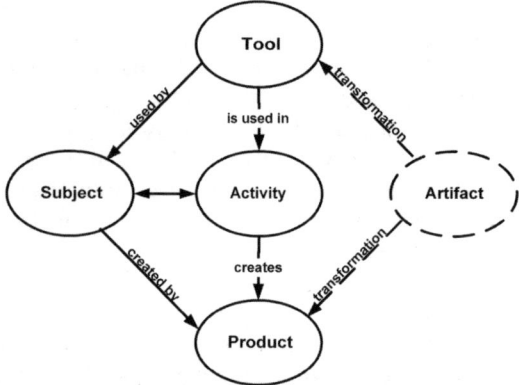

Fig. 1. Mediated relationship of artifacts in activities [3, p. 62]

Situatedness in Brahms is modeled by representing places and locations as objects in a conceptual hierarchical geography model and by representing artifacts as objects in a hierarchical object model, separate from modeling the agents and their possible activity behavior [3].

3 Modeling Places and Locations

In Brahms' geography model relevant locations are modeled conceptually as area objects in a *part-of* hierarchy. Locations are *areas*, which are *instances* of classes of areas called *areadefintions*. Areas can be part of other areas. Using these two simple concepts we are able to model any environment as a conceptual hierarchical structure of area objects. For example, we can model the geography for UC Berkeley students studying at South Hall and going for lunch on Telegraph or Bancroft Avenue in Berkeley as follows:

```
area AtmGeography instanceof World
// Berkeley
area Berkeley instanceof City partof AtmModelGeography
// inside Berkeley
area UCB instanceof UniversityCampus partof Berkeley
area SouthHall instanceof UniversityHall partof UCB
area SpraulHall instanceof UniversityHall partof UCB
area BofA_Telegraph_Av_2347 instanceof BankBranch
    partof Berkeley
area Wells_Fargo_Bancroft_Av_2460 instanceof BankBranch
    partof Berkeley
area Blondies_Pizza_Telegraph_Av_2347 instanceof
    Restaurant partof Berkeley
area Tako_Sushi_Telegraph_Av_2379 instanceof Restaurant
    part of Berkeley
```

Being situated means that Brahms agents are located within the geography model. Area objects are special kinds of objects that can inhabit agents. Agents can have an initial location. For example, agent Joe can initially be located in area SouthHall as follows:

```
agent Joe {
   intial_location: SouthHall;
}
```

People do not stay in one location, but move around. Similarly, Brahms agents can move around within a geography model. Agent (and object) movement is performed with a *move* activity. How activities are performed is described in the next section. Although a geography model and agent and object location and movement within such a model is essential for modeling situatedness, the state of the world needs to be modeled as well. Current BDI architectures do not model the state of the world and therefore do not allow for representing a factual state of the world independent of the agent's beliefs about that state. In Brahms the state of the world is modeled as *facts*[5]. For example, the location of an agent is a fact, because, regardless of the beliefs of agents, the agents have a location within the geography model.

```
Fact: (Joe.location = SouthHall)
```

To model facts about areas we give the areas attributes and relations. For example, we can model the temperature in South Hall and the fact that South Hall is part of the UC Berkeley campus. To do this, we first define these as an attribute and relation in the area definition class. This way all university hall areas we model will inherit them.

---

[5] Here we deal with solipsism, since the model is an interpretation of the world by the modeler. Facts, as such, are thus not real facts, but only the modeler's representation of world states independent of the beliefs of agents.

```
areadefinition UniversityHall {
  attributes:
    public int temperature;
  relations:
    public UniversityCampus partOf;
}
```

Having defined the attributes and relations, we can have additional facts about the world.

```
Facts:
  (SouthHall partOf UCB);
  (SouthHall.temperature = 72);
```

4 Modeling Objects and Artifacts

As discussed above, artifacts in the world are important aspects for modeling situated action. Similar to the geography model, we model artifacts as object instances of a class hierarchy. Actually, areas are special kind of objects and areadefinitions are special classes. However, artifacts are more like agents than they are like areas. Artifacts are located and some can display situated behavior. Some types of objects, for instance a book, can also store information. Then there are data objects, which can represent located information conceptually. We use a class hierarchy to model these artifacts and data objects in the world. For example, to model a book Joe the student is reading we can define the following object model:

```
class Book {
  attributes:
    public String title;
    public Writer author;
  relations:
    public Student belongsTo;
    public BookChapter hasChapter
}

object BookPlansAndSituatedAction instanceof Book {
  initial_location: SouthHall;
  initial_facts:
    (current.title = "Plans and Situated Action");
    (current.author = LucySuchman);
    (current.belongsTo = Joe);
}
```

We can model the content of the book as *conceptual* chapter data objects. For example:

```
conceptual_class BookChapter {
  attributes:
    public String title;
    public String text;
}
```

We model the information held by the object as "beliefs" of the object[6]. Thus, we can model Chapter 4 in Suchman's book as a conceptual BookChapter object the Book object "stores" information about:

```
conceptual_object Chapter4 instanceof BookChapter { }

object BookPlansAndSituatedAction instanceof Book {
   initial_beliefs:
      (current hasChapter Chapter1);
      (current hasChapter Chapter2);
      (current hasChapter Chapter3);
      (current hasChapter Chapter4);
      (current hasChapter Chapter5);
      (current hasChapter Chapter6);
      (current hasChapter Chapter7);
      (current hasChapter Chapter8);
      ...
      (Chapter4.title = "Situated Action");
      (Chapter4.text = "… I have introduced the
                        term situated action …");
}
```

The key point in modeling information is to locate the BookChapter content as information held within the Book object. To do this, we model the content of the book as the *"beliefs of the book."*

Now that we discussed how situatedness is modeled in Brahms, we turn to modeling activities. To understand what activities are, we must first understand that human action is inherently social, which means it is "outside the brain" involving the environment. The key thing is that action is meant in the broad sense of an activity, and not in the narrow sense of altering the state of the world. In the next section we discuss the concept of an activity and how to model it.

5 Modeling Activities

Like Bratman's notion of humans as planners based on commonsense psychology, the notion of human behavior in terms of activities is based on the Marxist *activity theory*, developed initially between 1920s and 1930s by Russian psychologists Vygotsky and Leont'ev [18][19]. Activity theory is also a meta-theory, as is commonsense psychology, and can better be seen as a framework with which human behavior can be analyzed as a system of activities. Activity theory has become more established in last twenty years in its further development by educational and social scientists [20][21], as well as human-computer interaction scientists [22].

The unit of behavioral analysis in Activity theory is, not surprisingly, an *activity*. Activity theory defines a relationship between an activity and the concept of *motive*. Motives are socially learned and shared with people from the same culture, organization, or more generally a *community of practice* [23]. For instance, all activities of flight controllers and directors at NASA Johnson Space Center's Mission

---

[6] We call information contained in an object also beliefs to minimize language keywords.

Control for the International Space Station are driven by the *shared motives* of keeping the space station healthy and making sure the space station crew is safe. Motives and goals, at first glance, seem to be similar, but motives, unlike goals, are situational and environmental states that we want to maintain over relatively long periods of time, such as keeping the astronauts safe and healthy onboard the space station. Motives are not internal states that drive our minute-by-minute action and deliberation, but enable a shared external understanding of our actions in the world (e.g. flight controllers are not, every minute of their time, working on tasks that are driven by the goal of keeping astronauts alive). Motives keep us in "equilibrium" with the system's environment and allow us to categorize actions as socially situated activities (e.g. everything flight controllers do is in line with their motives and have as an overall result that the ISS is safe and the astronauts are healthy).

Describing human activities as socially situated does not mean people doing things together, as in "socializing at a party" or "the social chat before a meeting." Describing human activities as social means that the tools and materials we use, and how we conceive of what we are doing, are socially constructed, based on our culture. Although an individual may be alone, as when reading a book, there is always some larger social activity in which he or she is engaged. For instance, the individual is watching television in his hotel, as relaxation, while on a business trip. Engaging in the activity of "being on a business trip," there is an even larger social activity that is being engaged in, namely "working for the company," and so on. The point is that we are always engaged in multiple social activities, which is to say that our activities, as human beings, are always shaped, constrained, and given meaning by our motives and our ongoing interactions within a business, family, and community. An activity is therefore not just something we do, but a manner of conceiving our action and our interaction with the social environment we are in. Viewing activities as a *form of engagement* emphasizes that the conception of activity constitutes a means of coordinating action, a means of deciding what to do next, what goal to pursue; In other words, a manner of being engaged with other people and artifacts in the environment. The social perspective adds the emphasis of time, rhythm, place, shared understanding and a well-defined beginning and end.

This can be contrasted with the concept of a task, which in AI is defined as being composed of only those subtasks and actions that are relevant to accomplish the goal of the task. A task is by definition goal-driven, and the conception of a task *leaves out* the inherent social, cultural and emotional aspects of why, when, where and how the task is performed within the socially and culturally bounded environment. Clancey provides a nice example that portrays the difference between an activity and a task:

> "All human activity is purposeful. But not every goal is a problem to be solved (cf. Newell & Simon 1972), and not every action is motivated by a task (cf. Kantowitz & Sorkin 1983). For example, listening to music while driving home is part of the practice of driving for many people, but it is not a subgoal [subtask] for reaching the destination. Listening to music is part of the activity of driving, with an emotional motive." [24, p. 1]

In this view of human behavior, activities are socially constructed engagements situated in the real world, taking time, effort and application of knowledge. Activities have a well-defined beginning and end, but do not have goals in the sense of problem-solving planning models. Viewing behavior as activities of individuals allows us to understand why a person is working on a particular task at a particular time, why certain tools are being used or not, and why others are participating or not. This contextual perspective helps us explain the quality of a task-oriented performance. In this sense, as is shown in Figure 2, activities are orthogonal to tasks and goals.

Fig. 2. Dimensions of behavior [3, p. 55]

While modeling an activity we might want to use a goal-plan approach to represent a specific task to solve a particular problem, but this can also be done within the activity model itself. This is to say that a goal-driven planning approach is subsumed by an activity-based approach, and a goal-oriented task can be modeled with a goal-driven activity. An activity-based approach is more general and allows the modeling of all kinds of activities, including activities that are not necessarily work-oriented and goal-driven, but are based on social and cultural norms, such as the kinds of activities described by Clancey [24, p. 4]:

- Intellectual: These include activities that represent any form of conceptual inquiry or manipulation of things or ideas. This includes work-oriented problem solving, but also activities that are less directed, such as exploration or browsing; artistic and documentation activities, such as taking photographs, writing, etc.
- Interactional: These include activities in which we interact directly with other people, such as in a fact-to-face conversation, or using a communication artifact, such as a telephone or fax machine.
- Physical/body maintenance: These include activities where we "take care" of ourselves physically, such as eating, sleeping, etc.

5.1 Activities Are Like Scripts

Engström integrates the notion of activity with the notion of practice. How do we know what to do, what actions we should take, when and how? Habitual scripts drive our practice, our learned actions in a context: "At an intermediate level, the continuity of actions is accounted for by the existence of standardized or habitual scripts that dictate the expected normal order of actions." [25, p. 964].

In cognitive science and AI, the notion of scripts was developed by Shank and Abelson for the purpose of language processing and understanding of knowledge structures. Although they did not refer to activity theory, they did refer to something that sounds very much like the notion of behavior as situated activities: "People know how to act appropriately because they have knowledge about the world they live in." [26, p. 36]. Unlike Engström, Shank and Abelson focus not on how and why people get to engage in a social activity, although they use stories of social activities (such as eating in a restaurant) as examples, but they focus on the knowledge that people need to bring to bear to understand and process a situation. They argue that known scripts, which can be based on general knowledge (i.e. knowledge we have because we are people living in the world) and specific knowledge (i.e. knowledge we get, based on being in the same situation over and over), are what people use to understand a story:

> "A script is a structure that defines appropriate sequences of events in a particular context. A script is made up of slots and requirements about what can fill those slots. The structure is an interconnected whole, and what is in one slot affects what can be in another. Scripts handle stylized everyday situations. They are not subject to much change, nor do they provide the apparatus for handling totally novel situations. Thus, a script is a predetermined, stereotyped sequence of actions that defines a well-known situation … There are scripts for eating in a restaurant, riding a bus, watching and playing a football game, participating in a birthday party, and so on." [p. 41]

Modeling activities as scripts enables us to model people's behavior as situated activities and is how we model activities in the Brahms language. This is the subject of the next section.

6 Activities in Brahms

In Brahms, an activity system is modeled by individually performed activities by agents. An agent while executing activities can be located in a geography model (see geography discussion above), representing the agent executing located behaviors. Activities are like scripts having "predetermined, stereotyped sequence of actions that defines a well-known situation." An activity abstracts a behavioral episode into a sequence of subactivities, or, if not further decomposed, it abstracts an episode into a single activity taking time, using resources—objects in the object model—or creating products—other objects in the object model—or both. In Brahms we call these *composite-* and *primitive activities* subsequently.

6.1 Primitive Activities

Activities define episodes of actions that an agent *can* perform. They are not the episodes themselves, but provide the "slot definitions" for possible instantiations of them. Following is the grammar definitions, in BNF form, for a *primitive activity* in the Brahms language:

```
primitive activity activity-name( { param-decl
[ , param-decl ]* } )
{
{ display : literal-string ; }
{ priority : [ unsigned | param-name ] ; }
```

```
{ random :      [ truth-value | param-name ] ; }
{ min duration : [ unsigned    | param-name ] ; }
{ max duration : [ unsigned    | param-name ] ; }
{ resources :   [ param-name   | object-name ] [ ,
                [param-name    | object-name ]*; }
}
```

A primitive activity can have parameters that at runtime are instantiated with parameter values, depending on the type in the parameter declaration. Parameters can be assigned to the slots in the activity definition, enabling the activity to be situated in a particular instantiation. The *display* slot provides a way to give the activity a more appropriate name when displayed. For example, the display string can include blank spaces. The *priority* slot is a positive integer that gives the activity a priority. Activity priorities are used by the agent's engine in case more than one activity can be chosen in the next event cycle. The *max_duration* slot gives the (maximum) duration of the primitive activity. Primitive activities only take an amount of time representing the time the agent performs the activity. In case the modeler wants a random time to be assigned, the modeler has to provide a *min_duration* as well, and set the *random* slot to *true*. For example, the activity bellow simulates the life of an agent in one simple primitive activity that takes a random lifetime:

```
primitive activity Being_Alive( int pri,
   int max_life_time, Body body_obj )
{
   display : "Alive and kicking" ;
   priority : pri ;
   random : true ;
   min duration : 0 ;
   max duration : max_life_time ;
   resources : body_obj ;
}
```

The above Brahms activity code is only the definition of the activity. It will not make the agent execute the activity. For this, the activity needs to be placed in a situation-action rule called a *workframe*. A workframe can have a set of preconditions that are matched against the belief-set of an agent. When there exist a set of beliefs that match the precondition, the variables in the precondition are bound to those matching the beliefs, and an instance of the workframe is created. The body of the workframe-instance is then executed. For example:

```
workframe wf_Being_Alive {
   repeat: true ;
   when (knownval(current.alive = true)) {
   do {
     Being_Alive( 100, 3600 ) ;
     conclude((current.alive = false), bc: 100, fc: 100);
   }
}
```

In this workframe, the precondition *knownval(current.active =true)* is matched against the belief-set of the agent. The *current* keyword means the agent that is executing the workframe. The *knownval* predicate returns true if the agent has a belief that matches exactly the precondition. Other possible precondition predicates are *known, unknown* and *not*, which subsequently would return true, if a) the agent has *any* belief about the agent-attribute pair *current.alive* irrespective of the value, b) the agent has *no* belief about this agent-attribute pair, and c) the agent has a belief about the agent-attribute pair, but its value is not *true* (i.e. in this specific case the value is either *false* or *unknown*).

If the preconditions are all true, a *workframe instantiation* for each set of variable bindings is created. In this above example, there are no variables and thus there will only be one workframe instantiation created. A workframe instantiation is an instance-copy of the workframe with every precondition variable bound, available to be executed (see section bellow about composite activities). If a workframe instantiation is selected as the current to be executed, the *body* of the workframe (i.e. the *do-part*) is executed. In the above workframe, the *Being_Alive* activity is subsequently executed. It should be noted that the agent's engine matches the preconditions and starts executing the first activity in the workframe all at the same time (i.e. the same simulation clock tick).

It is at the start of the execution of the activity that the activity duration time is calculated. For the *Being_Alive* activity a random duration, between the value of the min- and max-duration slot, is selected. This then becomes the duration of the of the workframe instantiation for *wf_Alive*. After the activity ends, the *conclude* statement is executed in the same clock tick. In other words, conclude statements do *not* take any time, only activities. The *conclude* statement creates a new belief for the agent, and/or a new fact in the world. In the above example there is a belief created for the agent in hundred percent of the time (*bc:100*). There is also a fact created in hundred percent of the time (*fc:100*). If the belief- or fact certainty factor were set to zero percent, no belief or fact would be created. Using these certainty factors, the modeler thus has control over the creation of facts in the world and beliefs of an agent.

6.2 Composite Activities

How can we represent complex activities that are decomposed into lower-level activities without using a goal-driven planning approach? This is an essential question in finding a way to model every-day situated activities that are not goal-driven. One of the important capabilities of people is that we can easily resume an activity that was previously interrupted by another activity. For example, while in the activity of reading e-mail your cell phone rings. Without hesitation you suspend the "reading work-related e-mail" activity and start your "talking to your child on the phone activity," switching not only your perceptual-motor context, from reading e-mail on your computer to talking on a cell phone, but also switching your situated social context from being an employee reading work-related e-mail to the role of a father. One question is how you get to change this context? It is obvious that this is not triggered by a sub-goal of the reading e-mail activity. It is detecting (i.e. hearing) your cell phone ringing that makes you interrupt your e-mail reading activity, and the social norms of today makes the answering your cell phone activity most likely be of higher priority.

The organization principle we use to enable this type of interrupt and resume behavior is Brooks' subsumption architecture [9]. Brooks developed his subsumption architecture for situated robots, with the premise that this architecture enables the development of robots without a central declarative world model and without a planning system that acts upon that model. In Brahms agents do have a declarative world model, namely the belief-set of the agent represent the agent's view of the world. However, similarly to Brooks' subsumption architecture, Brahms agents do not use a goal-directed planning system, but rather an activation system that is based upon a "computational substrate that is organized into a series of incremental layers, each, in a general case, connecting perception to action." [p.39]. In our case the substrate is a hierarchical network of situation-action rules with timing elements in the form of primitive activities with duration. This hierarchical workframe-activity network enables flexible context switching between independent activities at all levels in the hierarchy. Similar to Brooks' augmented finite state machine (AFSM) language, each individual Brahms agent engine executes activities as manageable units selectively activated and deactivated (i.e. interrupted or impassed). Each Brahms agent engine works similar to Brooks' AFSMs, namely "[(activity behaviors] are not directly specified, but rather as rule sets of real-time rules which compile into AFSMs in a one-to-one manner." [p. 40].

Activity Subsumption
It is obvious that we want to be able to decompose primitive activities into more complex activities in order to model more complex behavior. For example, we want to be able to decompose the *Being_Alive* activity into more specific activities that the agent performs while alive. In Brahms this is done with a *composite activity*. Following is the grammar definitions, in BNF form, for a *composite activity* in the Brahms language:

```
composite-activity activity-name (
{ param-decl [ , param-decl ]* } )
{
{ display : literal-string ; }
{ priority : [ unsigned | param-name ] ; }
{ end condition : [ detectable | nowork ] ; }
{ detectable-decl }
{ activities }
{ workframes }
{ thoughtframes }
}
```

The "slots" of a composite activity are different from that of a primitive activity, except for the display- and priority slots. A composite activity has sections to define the decomposed behavior of the activity: *detectable-decl, activities, workframes, thoughtframes*. First, the *end_condition* slot declares how a composite activity can end. There are two possibilities; 1) the activity ends when there is nothing more to do. This is called end when there is no work (i.e. no workframe or thoughtframe in the activity fires); 2) the activity ends, because the agent has detected some condition in the world that makes it end the activity. In this case, when the activity should be ended is defined in the *detectable-decl* slot, by declaring so-called *detectlables*. How detectables work is explained bellow in the section about reactive behavior. Here we first explain the other sections of a composite activity. The other three sections define a composite activity in terms of more specialized activities in the activities section, and workframes calling these activities in the workframes section. The thoughtframes section contains thoughtframes. Thoughtframes are simple forward-chaining production rules. Actually, thoughtframes are like workframes except they cannot call activities but only conclude beliefs, based on matching belief preconditions. Also, thoughtframes do not take any simulated time, unlike workframes that always perform activities that take time. Thus, composite activities allow the definition of detailed scripts for well-known situations (a.k.a. work practice).

Next, we provide an example of how composite activities are used. We will use the example from before about an agent who is in the activity of being alive. We expand the example to have the agent go from being alive to being in a coma. Instead of defining the *Being_Alive* activity as a primitive activity that simply takes an amount of time, let us define this activity in more detail as a composite activity of two subactivities called *PAC_1* and *PAC_2*. Both of these subactivities are primitive activities being called in two separate workframes *wf_PAC_1* and *wf_PAC_2*:

```
composite_activity Being_Alive( ) {
   priority: 0;
    detectables:
      detectable det_Impasse {
        detect((current.headTrauma = true))
          then impasse;
      }

    activities:
      primitive_activity PAC_1(int pri) {
        display: "PAC 1";
        priority: pri;
        max_duration: 900;
      }

      primitive_activity PAC_2(int pri, int dur) {
        display: "PAC 2";
        priority: pri;
        max_duration: dur;
      }

    workframes:
      workframe wf_PAC_1 {
        repeat: true;
        when (knownval(current.execute_PAC_1 = true))
        do {
          PAC_1(1);
          conclude((current.headTrauma = true), fc:50);
        } //end do
      } //end wf_PAC_1

      workframe wf_PAC_2 {
```

```
    repeat: true;
    do {
      PAC_2(0, 1800);
      conclude((current.execute_PAC_1 = true), bc:25);
      PAC_2(0, 600);
    } //end do
  } //end wf_PAC_2
} //end composite activity Being_Alive
```

The left side of Figure 3 shows the workframe-activity subsumption hierarchy for the *Being_Alive* activity. The *wf_Being_Alive* workframe from before now calls the composite activity *Being_Alive*, instead of the previous primitive activity. From the above source code you can see that, at first, the workframe *wf_PAC_2* will fire, because that workframe does not have any preconditions and can thus fire immediately. This workframe will fire forever due to the *repeat: true* statement. Thus, if nothing else changes, the agent will first execute primitive activity *PAC_2* for 1800 clock ticks, and then again for 600 clock ticks, after which the *wf_PAC_2* fires again, and again. However, the *conclude* statement, in between the two *PAC_2* activity calls concludes its specified belief 25% of the time, due to the belief-certainty of 25. This means that approximately one out of four executions of the workframe *wf_PAC_2* the agent gets the belief to execute *PAC_1*.

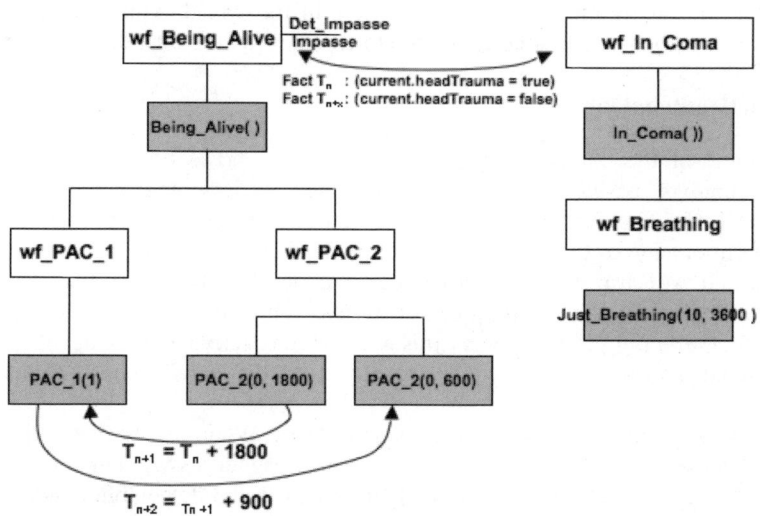

Fig. 3. Coma Model Workframe-Activity Subsumption Hierarchy

When this happens the belief immediately matches with the precondition of the *wf_PAC_1* workframe, which then becomes available to fire. Now the agent's engine needs to determine which workframe to execute in the next clock tick. This is done using the priorities of the workframes. If no priority is specified directly on the workframe, the workframe will get the priority of the highest priority activity within its body. In this example workframe *wf_PAC_2* has a priority of zero, because both

PAC_2 activity calls get a priority of zero as a parameter value. Workframe *wf_PAC_1* on the other hand will get the priority one, due to *PAC_1* having a priority parameter value of one. Thus, *wf_PAC_1* has the highest priority and the engine will pick this workframe as the next workframe to execute, with as the result that the agent will start performing *PAC_1* for its defined duration of 900 clock ticks. Workframe *wf_PAC_2* will be *interrupted* at the point of the beginning of the second *PAC_2* activity call, right after the execution of the conclude statement. Therefore, when the *wf_PAC_1* workframe finishes its execution of the *PAC_1* activity (i.e. 900 clock ticks later) the agent will switch immediately back to the execution of the interrupted *wf_PAC_2* workframe, and will continue with executing the second *PAC_2* activity in the workframe.

This example shows the selective activation and deactivation of subsumed activities via a perception-action architecture based on an activity-priority scheme. Agents can easily switch their activity context, independent of the level in the activity network hierarchy. When a workframe with activities within it becomes available, no matter where in the workframe-activity network, the workframe with the highest priority becomes the agent's current context—this is called the agent's *current work*—and thus the activities within this current workframe will execute.

Thus far the example shows how an agent can easily switch activities based on new belief creation, i.e. either through performance of activities in workframes or through pure reasoning in thoughtframes. Next, we will show how an agent can also react to changes in the environment by detecting *facts* in the world. This approach enables flexible reactive behavior by the agent due to changes in "the outside" environment.

6.3 Reactive Behavior

The example of the composite *Being_Alive* activity shows how agents can switch activity contexts based on "internally" created beliefs and activity-workframe priorities. However, we want agents be able to react to fact chances in the environment outside of the agent. This is done through the definition of detectables in activities and workframes. The above source code shows the *det_Impasse* activity detectable declared in the *Being_Alive* activity. What this means is that while the agent is "in the *Being_Alive* activity" this detectable is active and the agent will detect any *fact* changes, made by any agent or object, to the *headTrauma* attibute of the agent.

The process of firing a detectable goes as follows: When a new fact is detected 1) the fact becomes a belief for the agent that can then trigger a switch in activity context as shown in the example before, 2) the agent matches the detectable condition of all active detectables that refer to the fact and checks if the fact matches the condition, 3) in case the detectable condition matches the fact the detectable action statement is executed. There are four types of detectable actions possible, *continue*, *abort*, *complete*, and *impasse*.

In the Coma model source code for the *Being_Alive* composite activity, the *det_Impasse* detectable has an *impasse* action. An *impasse* action means that the activity will be impassed (i.e. interrupted) until the impasse condition is resolved. The impasse condition is the detect condition of the detectable. Thus, in the det_Impasse detectable, the activity is impassed when the fact *(current.headTrauma = true)* is created. The *wf_Being_Alive* will be impassed until the fact is changed.

Figure 3, on the right hand side, shows an additional workframe for the agent called *wf_In_Coma*. This workframe has a precondition that matches on the belief *(current.headTrauma = true)* created by the detection of the fact, and will activate the *In_Coma* activity. If the *In_Coma* activity, or something else, at some point in the future creates the fact *(current.headTrauma = false)*, the impasse is resolved and the *Being_Alive* activity will be available again for the agent to continue executing it, depending its priority compared to other available activities.

The above example shows how a Brahms agent's behavior is modeled as decomposed script-like activities that are executed using a perception-action subsumption architecture, enabling both rational and reactive behavior. There is one more important organizational element in the Brahms language that provides an important agent organizational modeling capability. This is briefly discussed in the next section.

7 Modeling Agent Organization

Societies consist of many different types of behaviors. As a design principle we want to be able to organize these behaviors in categories that are logically and culturally understandable, and moreover useful for the design of complex agent communities. In Brahms there is the notion of a *group* as the concept allowing the creation of agent pastiches. Not only did we develop groups based on the notion of organization in categories or classes, groups are based on the important idea of *communities of practice*:

> "Being alive as human beings means that we are constantly engaged in the pursuit of enterprises of all kinds, from ensuring our physical survival to seeking the most lofty pleasures. As we define these enterprises and engage in their pursuit together, we interact with each other and with the world and we tune our relations with each other and with the world accordingly. In other words we learn. Over time, this collective learning results in practices that reflect both the pursuit of our enterprises and the attendant social relations. These practices are thus the property of a kind of community created over time by the pursuit of a shared enterprise. It makes sense, therefore, to call these kinds of communities *communities of practice*." [23, p. 45]

Groups are thus meant to be the representation of the *practices* of communities of agents. People belong to many communities at once, blending practices from many different groups into one, so called, *work practice* [3]. We are students, parents, workers, children, engineers of a particular kind, etc. But we also are social creatures, belonging to a community of like-minded individuals playing sports, having hobbies, going to the same coffee shop every day, playing roles in an organization, etc. It is therefore that people belong to many communities of practice at once. Agents in the Brahms language can thus belong to many different groups, enabling the design of complex organization models of agents.

Groups in Brahms, just like agents, can contain *attributes, relations, initial beliefs and facts, activities, workframes* and *thoughtframes*. An agent can be *a member of* a group, but a group itself can also be a *member of* another group, enabling the design of a complex hierarchical structure of agent group-membership. An agent or group that is a member of another group will *inherit* all of the contents of that group. For example, agent *Joe*, from our first example, will inherit its behavior as a student from the *Student* group. But, we can create another group, let's call it *HumanBeing*, in which we put the *BeingAlive* and *BeingInComa* activities. We can now have agent *Joe* inherit both the behavior from the *Student* group and from the *HumanBeing* group;

group HumanBeing {…}

group Student {…}

agent Joe **memberof** HumanBeing, Student {…}

Groups and multiple group inheritance allows us to model common behavior as communities of practice, from which all group members will inherit its behavior. Using this simple group membership relation we can design any type of organization we want. Groups can represent a functional organization, such as groups representing particular roles that agents play, performing certain functions (i.e. activities) in an organization. However, groups can also represent social organizations, such the relationships and social knowledge people share about a particular community. For example, everyone who comes to drink coffee at the same coffee shop everyday knows the name of the shop's barista.

8 Conclusions

In this chapter, we discussed some of the issues and limitations of BDI agent architectures based on Bratman's Planning Theory that explicitly states that humans are goal-driven planning agents. We posited an alternative view of human behavior based on a combined notion of situated action, cognition in practice, situated cognition and activity theory. Based on this alternative view, we developed an activity-based theory of behavior that allows for the description of complex behavior that is not only based on goals. We then described the Brahms multiagent language and execution architecture that implements this activity-based theory into a BDI agent language. Brahms allows for designing and implementing complex agent societies not based on goal-based planning agents.

The Brahms multiagent language, for each agent, "groups multiple processes (each of which turns out to be usually implemented as a single [composite activity]) into behaviors. There can be message passing, suppression and inhibition between processes within a [composite activity], and there can be message passing, suppression and inhibition between [composite activities]. [Activities] act as abstraction barriers, and one [activity] cannot reach inside another." [9, p.41].

Compared to the goal-driven paradigm, the Brahms activity-based paradigm is a more flexible execution paradigm. In a goal-driven execution engine only sub-goal contexts within the task being executed can be called as the next action, unless the current task is finished and the current goal is or is not reached and thus "popped off" the goal stack. This limits an agent's ability to flexibly react to new belief "perceptions" unrelated to the current goal it is trying to reach.

With the simple examples in this paper, we hope we have convincingly shown that Brahms agents are both rational and reactive, and use composite architecture with situation-action rules to implement a perception-action approach similar to Brooks' behavioral architecture, all *without* the use of goals and goal-driven planning. However, it should not be forgotten that a forward-driven approach might just as well implement goal-directed behavior as a backward-driven goal-based approach. It is thus that in Brahms we can implement goal-driven activities without any problem, and it can be said that Brahms enables modeling of agent behaviors much more flexibly than a goal-based planning architecture. In other words, the activity approach is more general than the goal-based approach, which justifies the title of this paper and our claim that goals develop *within* an activity, but they are not the *driving force* of behavior, and are only useful in activities where problem solving is necessary. In other words: "All human behavior is activity-based, but not every activity is a problem to be solved."

References

1. Bordini, R.H., Dastani, M., Meyer, J.C. (eds.): Foundations and Practice of Programming Multi-Agent Systems. Dagstuhl Seminar Proceedings 06261, 25.06. - 30.06, Schloss Dagsthuhl (2006)
2. Clancey, W.J., et al.: Brahms: Simulating practice for work systems design. International Journal on Human-Computer Studies 49, 831–865 (1998)
3. Sierhuis, M.: Modeling and Simulating Work Practice; Brahms: A multiagent modeling and simulation language for work system analysis and design. In: Social Science Informatics (SWI), University of Amsterdam. SIKS Dissertation Series No. 2001-10, p. 350. Amsterdam, The Netherlands (2001)
4. Bratman, M.E.: Intention, Plans, and Practical Reason. The David Hume Series of Philosophy and Cognitive Science Reissues. CLSI Publications, Stanford, CA (1999)
5. Bordini, R.H., et al. (eds.): Multi-Agent Programming: Languages, Platforms and Applications, Springer Science+Business Media, Inc. (2005)
6. Simon, H.: A behavioral model of rational choice. Quarterly Journal of Economics 69, 99–118 (1955)
7. Newell, A.: Unified theories of cognition. Harvard University Press, Cambridge, MA (1990)
8. Anderson, J.R.: Rules of the mind. Lawrence Erlbaum Associates, Hillsdale, NJ (1993)
9. Brooks, R.A.: Cambrian intelligence: the early history of the new AI. MIT Press, Cambridge, MA (1999)
10. Suchman, L.A.: Plans and Situated Action: The Problem of Human Machine Communication. Cambridge University Press, Cambridge, MA (1987)
11. Agre, P.E.: Computation and Human Experience. In: Pea, R., Brown, J.S. (eds.) Learning in doing: Social, cognitive, and computational perspectives. Cambridge University Press, Cambridge, MA (1997)
12. Fodor, J.A.: The Modularity of Mind. The MIT Press, Cambridge, MA (1987)
13. Edelman, G.M.: Bright Air, Brilliant Fire: On the Matter of the Mind. BasicBooks, New York, NY (1992)

14. Winograd, T., Flores, F.: Understanding Computers and Cognition. Addison-Wesley Publsihing Corporation, Menlo Park, CA (1986)
15. Clancey, W.J.: Situated Cognition: On Human Knowledge and Computer Representations. Cambridge University Press, Cambridge (1997)
16. Clancey, W.J.: Conceptual Coordination: How the mind Orders Experiences in Time. Lawrence Erlbaum Associates, Mahwah, New Jersey (1999)
17. Lave, J.: Cognition in Practice. Cambridge University Press, Cambridge, UK (1988)
18. Vygotsky, L.S.: In: Cole, M., et al. (ed.) Mind in Society: The Development of Higher Psychological Processes. Harvard University Press, Cambridge, MA (1978)
19. Leont'ev, A.N.: Activity, Consciousness and Personality. Prentice-Hall, Englewood Cliffs, NJ (1978)
20. Engström, Y., Miettinen, R., Panamaki (eds.): Perspectives on Activity Theory. In: Pea, R., Brown, J.S. (eds.) Learning in doing: Social, Cognitive, and Computational Perspectives. Cambridge University Press, Cambridge, MA (1999)
21. Chaiklin, S., Lave, J. (eds.): Understanding practice: Perspectives on activity and context. In: Pea, R., Brown, J.S. (eds.) Learning in doing: Social, cognitive, and computational perspectives. Cambridge University Press, Cambridge, MA (1996)
22. Nardi, B.A.: Context and Consciousness: Activity Theory and Human-Computer Interaction. The MIT Press, Cambridge, MA (1996)
23. Wenger, E.: Communities of Practice; Learning, meaning, and identity (Draft). In: Pea, R., Brown, J.S. (eds.) Learning in doing: Social, cognitive and computational perspectives. Cambridge University Press, Cambridge, MA (1998)
24. Clancey, W.J.: Simulating Activities: Relating Motives, Deliberation, and Attentive Coordination. Cognitive Systems Research 3(3), 471–499 (2002)
25. Engeström, Y.: Activity theory as a framework for analyzing and redesigning work. Ergonomics 43(7), 960–974 (2000)
26. Schank, R., Abelson, R.: Scripts, Plans, Goals, and Understanding. Lawrence Erlbaum Associates, Inc. Hillsdale, NJ (1977)

The Construction of Multi-agent Systems as an Engineering Discipline

Jorge J. Gomez-Sanz

Universidad Complutense de Madrid,
Avda. Complutense s/n, 28040 Madrid, Spain
jjgomez@sip.ucm.es
http://grasia.fdi.ucm.es

Abstract. The construction of multi-agent systems is starting to become a main issue in agent research. Using the Computer Science point of view, the development of agent systems has been considered mainly a problem of elaborating theories, constructing programming languages implementing them, or formally defining agent architectures. This effort has allowed important advances, including a growing independence of Artificial Intelligence. The results have the potential to become a new paradigm, the agent paradigm. However, the acceptance of this paradigm requires its application in real industrial developments. This paper uses this need of addressing real developments to justify the use of software engineering as driving force in agent research. The paper argues that only by means of software engineering, a complex development can be completed successfully.

Keywords: agent oriented software engineering, multi-agent systems, development.

1 Introduction

Software agents have been considered as part of Artificial Intelligence (AI) for a long time. However, looking at the current literature, it seems agent researchers are focusing more on development aspects, such as debugging or modularity, paying AI issues less attention. This tendency is giving agent research an identity as a software system construction alternative, i.e., as a new paradigm. This paradigm will be named in this paper as the agent paradigm, whose existence was already pointed out by Jennings [1]. Informally, the agent paradigm is a development philosophy where main building blocks are agents.

In the elaboration of the agent paradigm, Computer Science has played a main role so far. With its aid, several agent oriented programming languages have been constructed and models of agency have been formalised. Nevertheless, it cannot be said the agent paradigm is mature, yet. Since agents are software, it is necessary proving the agent paradigm can be an alternative to known software development paradigms in a real, industrial, developments. This implies that the cost of developing a system using agents should be better than the cost of using other technologies, at least for some problems. Proving this goes beyond the scope of Computer Science and enters the domain of software engineering (it is assumed the readers accept Computer Science and Software Engineering are different disciplines).

This need of exploring the engineering side of software agents has been known in the agent community for some years. As part of this investigation, development environments have been created, debugging techniques have been elaborated, and agent oriented specification techniques have been proposed. These, together with the improvements in agent oriented programming languages, constitute important steps towards the integration with industrial practices [2]. Nevertheless, the growing incorporation of agents to industry demands a stronger effort.

The paper constributes with an argumentation in favour of applying engineering methods to achieve matureness in the agent paradigm. This argumentation starts with an introduction to the origins of software engineering in section 2. Then, section 3 introduces briefly how agents have started to focus on developments while being part of AI. These initial steps have led to results dealing with system construction that can be considered more independent of AI. These new results are introduced in section 4. Following, section 5 argues the need of engineering methods using the evolution of the agent research towards the construction of systems and drawing similitude with the origins of software engineering. This argumentation is followed, in section 6, by a suggestion of lines of work that could constribute to the acceptance of the agent paradigm. The paper finishes with some conclusions 7.

2 The Origins of Software Engineering

The claim of the paper requires understanding what is software engineering. The term *software engineering* was born in a NATO sponsored conference [3]. This conference invited reputed experts in software development to talk about the difficulties and challenges of building complex software systems. Many topics were discussed, but above all there was one called the *software crisis*. This was the name assigned to the increasing gap between what clients demanded and what was built due to problems during the development. Indeed, the smaller a development is, in the number of involved persons and in the complexity of the system to develop, the easier it is to complete successfully. Unfortunately, the software systems required in our societies are not small.

Dijkstra [4] had the opinion that a major source of problems in a development was the programming part. He remarked the importance of having proofs to ensure the correctness of the code, since much time is dedicated to debugging. Also, better code could be produced with better programming languages, like PL/I, and programming practices, like removing the goto statement from programming [5]. Dijkstra was right. Good programming and better programming languages reduce the risk of failure. Nevertheless, good code is not enough when programming large systems. For instance, De Remer [6] declares programming a large system with the languages at that time was an exercise in obscuration.

It was needed having a separated view of the system as a whole to guide the development, a view that was not provided by the structured programming approaches. In this line, Winograd [7] comments that systems are built by means of combining little programs. Hence, the problem is not building the small programs, but how the integration of packages and objects with which the system can be achieved.

Besides the programming, reasons for this gap can be found in the humans involved in a software project. Brooks [8] gives some examples of things that can go wrong in a development like being too optimistic about the effort a project requires, thinking that adding manpower makes the development going faster, or dealing with programmers wanting to exercise their creativity with the designs.

In summary, bridging the gap requires dealing with technical and human problems. Their solution implies defining a new combination of existing disciplines to build effectively software systems. This new discipline is the software engineering and their practitioners are called software engineers. The training software engineers receive is not the same of a a computer scientists [9]. This claim would be realised in the computing curricula recommendations by the ACM and IEEE for software engineering [10]. This proposal covers all aspects of a development, including programming and project management.

2.1 Lessons Learnt

There are three lessons to learn from the experience of the *software crisis* that will be applied to agent research in section 5.

The first lesson refers to the natural emergence of engineering proposals when the foundations are well established [9]. This has occurred many times with natural sciences and their application to human societies. As a result civil engineers, electrical engineers, and aerospace engineers, among others, exist.

The second is that some systems are inherently complex and require many people involved and an important amount of time. Building these systems is not a matter of sitting down in front of the computer and start working on it. It requires organising the work, monitoring the progress, and applying specific recipes at each step. Besides, there is a human component that cannot be fully controlled.

The third lesson is that the progress in programming languages is not going to eliminate all problems in a development. Three decades of progress in programming languages have made the construction of systems easier, but it has not eliminated the need of software engineering.

3 At the Beginning Everything Was Artificial Intelligence

The Artificial Intelligence discipline is a multi-disciplinary one dedicated to the creation of artificial entities capable of intelligent behaviour. The roots of agents are associated strongly to AI. Allen Newell made one of the first references to agents pointing out their relevance to AI. He distinguished agents as programs existing at the *knowledge level* that used knowledge to achieve goals [11]. Two decades after, agents continue being relevant to AI. For instance, the Artificial Intelligence textbook from Russell and Norvig [12] presents agents as the embodiment of Artificial Intelligence.

Despite these statements, considering software agents only as AI would be a mistake. AI is important if certain features are to be incorporated in the agents, like planning or reasoning. However, the role of AI is not a dominant one and needs to be reviewed depending on the kind of system developed [2]. As a consequence, it is possible having an agent paradigm that can use AI while keeping its identity as paradigm. There have been similar cases in the past. Lisp [13] and Prolog [14] programming languages were born under the umbrella of AI. They are now independent and represent an important milestone in the establishment of the functional and logic programming paradigm, respectively. Following these precedents, one could draw the conclusion that focusing on particular approaches to the construction of AI capable systems has led to the definition of new development paradigms. Hence, the efforts in showing how a multi-agent system is built would eventually take to the definition of the agent paradigm.

Looking at the agent literature, published works dealing with multi-agent systems construction can be categorised into two groups: agent oriented programming languages and agent architectures. In both, Computer Science has played an important role, specially in agent oriented programming languages. Hence, most of these steps are founded on formal descriptions and theories.

The invention of the agent oriented programming language is usually dated in 1993 with the creation of Agent0 [15]. This language proposed using agents as building blocks of a system, providing a vocabulary to describe what existed inside of the agents and some semantics of how these agents operated. This trend has been explored from a computer science perspective, mainly, trying to establish theoretical principles of agent oriented programming. This has been the case of Jason [16] an extended interpreter for AgentSpeak (L) language [17], 3APL [18] or Agent Factory - APL [19] a combination of logic and imperative programming, Claim [20] which is based on ambient calculus, ConGOLOG [21] which bases on situation calculus, or Flux [22] which is based on fluent calculus, to cite some. For a wider review about agent oriented languages, readers can consult the short survey [23], the survey of logic based languages [24], or the more detailed presentation of programming languages from [25]. The diversity of concepts and approaches evidences there is still struggle to find the best way of dealing with agent concepts.

The definition of agent architectures is the other way identified to construct agents. An agent architecture is a software architecture that explains how an agent is built. A software architecture, according to IEEE, is *the fundamental organisation of a system, embodied in its components, their relationships to each other and the environment, and the principles governing its design and evolution* [26]. This way of defining systems was applied early on AI to explain how systems were capable of intelligent behaviour. In most cases, the agent architecture has been a set of boxes with arrows representing the data flow. The computations performed inside the boxes were explained in natural language. Examples of these initial architectures are the reference architecture for adaptative intelligent systems [27], the Resource-bounded practical reasoning architecture [28], or Archon [29]. This kind of architectures were not fully detailed, but they provided important hints to a developer wanting to reuse them. These architectures provided a set of elements to look at and a way of organising them. A higher level of detail would be given by Interrap [30] and Touring Machines [31]. Interrap provided formal definitions of the inputs and outputs of the different elements of the architecture. Touring machines provided a reference architecture and detailed instructions to fill in each component of the architecture with Sicstus Prolog. Its capabilities were demonstrated in a real time traffic management scenario with dynamic changes. Perhaps, the most complete architecture described so far is DESIRE [32]. The semantics of DESIRE are formally specified using temporal logic [33] and allowed compositional verification [34]. Besides formal aspects, DESIRE differs from previous ones in how it was used. Instead of providing guidelines, it used a formal language to describe how the architecture was instantiated to produce a system. This language was processed by ad-hoc tools to produce the an instance of DESIRE for the current problem domain. For a wider review of architectures, readers can consult the surveys [35] and [36].

4 Focusing on Multi-agent Systems Construction

The examples presented in previous sections demonstrate an increasing concern about the construction of systems by means of agent architectures and agent oriented programming languages. On one hand, agent oriented programming languages implement theories of agency devising new ways of processing information. On the other hand, agent architectures explore the internal components of an agent, often letting the developer choosing how these components are implemented. These precedents have given the opportunity of focusing more on development problems and less on AI, in concrete, the experience gained allowed to propose agent standards, agent development environments, and methodologies.

The creation of a standard, FIPA, and the availability of a platform implementing it, the JADE framework [37], turned out to be an important factor that increased the number of developments using agents. Agents no more had to be created from scratch. By extending some classes from the framework, one could have a simple multi-agent system. However, JADE provided very basic support for defining the behaviour of the agent. Most of the times, developers had to build over a JADE layer other layers dedicated to implement the control of the agent. FIPA work continues under the scope of IEEE, so there is still much to say.

With respect development environments to produce agent software, the first was the Zeus tool [38]. It offered a graphic front end that allowed to define the internals of the agents (ontology, action rules, goals), communication, and deployment of the system. The information introduced served to instantiate a multi-agent system framework, which could be customised by the developer. The main contribution of Zeus was the reduction of the development effort.

Little programming was required unless the generated system shown a bug in the automatically generated part. This development environment was followed by others like JADEX, Agent Factory or AgentTool [23]. These frameworks can be considered the inheritors of the first efforts in the definition of agent architectures. They still define the system in terms of components and relationships among them, but require less information from the domain problem to produce a tentative system.

Agent oriented methodologies contributed with support tools, proposals for the specifications of agent oriented systems, steps to follow to produce these specifications as well as to implement them. One of the first was Vowel Engineering [39]. It conceived the system as the combination of five interdependent aspects: Agent, Environment, Interaction, and Organisation. Another important contribution is the adaptation of BDI theory into an iterative development process [40]. Both proposals introduced a simple idea: constructing a multi-agent system could be less an art and more a discipline by itself. This was introduced later on by [1] describing several distinguishing features of agent computing that could constitute a new approach to software engineering. The next step in methodologies was MESSAGE [41] [42]. MESSAGE provided a comprehensive collection of agent concepts, supported by tools, and covering the complete development process. It was the first in applying meta-modelling techniques in an agent oriented methodology. Later on this would be incorporated to all methodologies. For a review of agent oriented methodologies, readers can read [43]. This book has a chapter dedicated each one of ten agent oriented methodologies. It include a comparison among them using an evaluation framework. This framework evaluates the existence and quality of a set of features that are desirable in a methodology.

5 The Need of Engineering Methods

The results presented in previous sections constribute to the understanding of the agent paradigm in different ways. In general, they make a development simpler. Nevertheless, the impact of those results has not been the one expected in industry. Agentlink conducted a study of the impact and evolution of agent technology in Europe during several years. The result is the AgentLink roadmap [44]. This study shows relatively few histories of success of agent technology in real developments. There may be reasons for this like slow penetration of new technologies in the market or companies being reluctant to tell what kind of software they use. However, it seems more reasonable to find answers in the way agent research has been directed. Zambonelli and Omicini [45] tell there is little effort in generating evidences of the savings in money and resources agents can bring. Luck, McBurney, and Priest [46] provide other reasons: *most platforms are still too immature for operational environments, lack of awareness of the potential applications of agent systems, the cost of system development and implementation both in direct financial terms and in terms of required skills and timescales*, or *the absence of a migration path* from agent research to industrial applications.

Drawing a parallelism with the situation presented in section 2, there is a particular *software crisis* affecting the agent community. This crisis relates to the capability of agent research to build applications in an effective way. The crisis can be summarised as *a gap between the application a client demands and what can be actually built with existing agent results according to the constraints of a development project*. Hence, for a greater acceptance of the agent paradigm, this paper proposes the agent community a new goal consisting in the reduction of this gap. The answer to the *software crisis* was software engineering, i.e., using engineering approaches to the construction of software. Therefore, the agent community ought to focus more on engineering issues as well. The call for more engineering is not new. Concretely, readers can find a precedent in [45].

Assuming the need of software engineering, it is relevant to review the lessons from section 2.1 in the context of agent research.

The first lesson talked about the natural emergence of engineering methods when sufficient foundations exist. There exist a need of developing agent oriented systems, and the results (languages, architectures, tools, theories) to do so exist. Therefore, an engineering discipline specialised in the use of these elements to produce agent oriented systems makes sense. This discipline would propose combinations of theoretical and practical agent reseach results so that developers can solve problems effectively.

The second lesson commented on the resources needed in a development of a complex system. Developing a complex system requires investing time and people, even if software agents are used. Therefore, developers should know the particularities of using agents when there is a deadline to finish the product, a maximum available budget, and several involved workers. This covers the requirements gathering, analysis, design, implementation, testing, and maintenance stages of a development. Known development roles will have to be revisited to look for changes due to the use of the agent paradigm. For instance, it is required to assist project managers with guidelines, cost estimation methods, risk management, and other common project management tools customised with the experience of successfull application of agent results.

The third lesson was about the need of complementing programming languages with other techniques. Agent research is not going to produce any magic bullet for complex system development. The agent paradigm will provide means comparable to other existing ones in the capability of building a system. The difference will be in the balance between results obtained and the effort invested. Should this balance be positive and greater than applying other paradigms, the agent paradigm will become a valid alternative. There are many factors to consider in the equation, but it will depend ultimately on the developers. So, in order to ensure a successfull development, it is necessary to provide all kind of support to these developers. This includes development tools, training, and documentation, to cite some.

These lessons tell the agent paradigm has to produce many basic engineering results, yet. Applying engineering is not only producing a methodology. It is about, for instance, discovering patterns of recurrent configurations when using agent oriented programming languages, designing agent architectures attending to the non-functional requirements, evaluating the cost of a agent oriented development, producing testing cases for agent applications, documenting properly a development, and so on. Agent oriented methodologies, which are not so recent, do not consider most of these elements.

6 Proposed Lines of Research

This section collects several lines of work in line with the lessons from section 5. Some lines have been started already, but require further investigation. Others have not been initiated, yet.

The main line consists in performing large developments. In order to bridge the gap pointed out in section 5, agent research must experiment with large developments involving several developers and dealing with complex problems. The benefit for agent research would be twofold. Besides gaining experience and validating development techniques, these developments would produce useful prototypes demonstrating the benefits of the agent paradigm. There are some specific areas where the development of multi-agent system prototypes could attract additional attention from industry. Luck, McBurney, and Priest [46] identify the following: ambient intelligence, bioinformatics and computational biology, grid computing, electronic business, and simulation. Ideally, the prototypes should be built by developers not involved directly in the research of the techniques applied. Like in software tests [47], if the tester is the creator of the technique, the evaluation could contain some bias.

The results of these large developments would influence in the existing agent frameworks, agent development environments, and agent oriented programming languages. These should be revisited in the light of the experience from large developments. Expected improvements derived from this experience will improve the debugging techniques (it is likely to invest an important time detecting where the failure is), the quality of code (good programming practices for agent oriented programming languages), and the code reuse (large developments will force developers to find ways to save coding effort).

Also, the development of these prototypes could give opportunities to experience other aspects not documented yet in the agent literature. Project management and maintenance are two of them. Inside project management, project planning and risk management, two key activities in project management, have not been experienced when agents are applied. Cost estimation, which is relevant for planning, is almost missing in agent research. Gomez-Sanz, Pavon, and Garijo [48] present some seminal work in this direction using information extracted from European projects. With respect maintenance, the maintenability of code written with an agent oriented programaming language, agent frameworks, or an agent architecture is a big unknown at this moment. The cost of correcting defects or adding new functionality once the system is deployed is unknown as well.

With respect to project deliverables, it is a matter of discussion to determine which deliverables are needed in an agent oriented project beyond those known of software engineering. Deliverables determines the progress the project should have. Hence, defining a proper content of these deliverables is a priority for a project manager. From these deliverables, the only one being studied currently is the system specification. It is being defined with meta-models, a technique imported from software engineering. The technique has found quickly a place in agent research. In fact, most existing methodologies have a meta-model description for system specification. Agent researchers intend to unify the different existing meta-models [49], though this is a work that will take time.

The particularities of agent oriented development process are starting to be investigated. Brian Henderson-Sellers [50] [51] contributes with a metamodel to capture development methods and represent existing agent oriented development approaches. Cernuzzi, Cossentino, and Zambonelli [52] studies the different development process to identify open issues.

There is concern about the importance of debugging. Development environments like JADEX and Agent Factory provide basic introspection capabilities. Also, Botia, Hernansaez, and Skarmeta[53] study how to debug multi-agent systems using ACL messages as information source. Lam and Barber try to explain why the agent behaved in a concrete way using information extracted from traces of the system [54]. Testing is an activity less studied.

Code generation may turn out to be a key to the success of agents. Most of the works are agent frameworks which has to be instantiated sometimes with a formal language, others with a visual environment. Generating automatically the instantiation parameters should not be a hard task. There are precedents of this kind of tools in Zeus [38], AgentTool [55], and the INGENIAS Development Kit [56]. The benefits are clear, since the developer invests little effort and obtains a working system in exchange. However, there are drawbacks in the approach, like conserving the changes made in the generated code.

To conclude, there is a line of research related with software engineering that has not been considered yet. Besides structuring an agent oriented development, the effort invested in the creation of methodologies can be used to start the creation of bodies of knowledge for agent oriented software engineering. In software engineering, there exist a body of knowledge [57] that provide an agreed terminology for practitioners. This initiative was started in a joint effort by the IEEE and ACM in order to develop a profession for software engineers. In this line, readers can consult an extended review of agent research results that can be useful in each concrete stage of a development [58].

7 Conclusions

Agent research will eventually gain more independency from Artificial Intelligence. This independency will be completely realised with the establishment of the agent paradigm, the development philosophy that uses agents as main building blocks. According to recent studies, the adoption of agent research results by industry, which would be an evidence of the matureness of the agent paradigm, is not progressing as it should. A reason for this delay may be the lack of large developments that prove the benefits of choosing the agent paradigm. Agent research has been driven mainly by Computer Science focusing in concrete features of the agents and the experimentation has been limited to small developments, mainly. As a result, there is no enough experience in large developments of software systems based on agents, so the question of the capability of agent technology to deal with a real development remains. This problem has been presented as a kind of *software crisis* of the agent community.

To deal with this crisis, this paper has proposed the extensive use of software engineering principles. This requires prioritising the research on aspects that have not been considered before in an agent community, like project management issues, good practices of agent oriented programming, or maintenance costs associated to a multi-agent system. To illustrate the kind of areas to address, the paper has pointed at several lines of work, referring to some preliminary work already done in those directions.

Acknowledgements. This work has been supported by the project Methods and tools for agent-based modelling supported by Spanish Council for Science and Technology with grant TIN2005-08501-C03-01, and by the grant for Research Group 910494 by the Region of Madrid (Comunidad de Madrid) and the Universidad Complutense Madrid.

References

1. Jennings, N.: On agent-based software engineering. Artificial Intelligence 117(2), 277–296 (2000)
2. Dastani, M., Gomez-Sanz, J.J.: Programming multi-agent systems. The Knowledge Engineering Review 20(02), 151–164 (2006)
3. Naur, P., Randell, B. (eds.): Software Engineering: report on a conference sponsored by the nato science committee, Garmisch, Germany, NATO Science Committee (1968)
4. Dijkstra, E.: The humble programmer. Communications of the ACM 15(10), 859–866 (1972)
5. Dijkstra, E.: Goto statement considered harmful. Communications of the ACM 11(3), 147–148 (1968)
6. DeRemer, F., Kron, H.: Programming-in-the large versus programming-in-the-small. In: Proceedings of the international conference on Reliable software, pp. 114–121. ACM Press, New York (1975)
7. Winograd, T.: Beyond programming languages. Communications of the ACM 22(7), 391–401 (1979)
8. Brooks, F.: The mythical man-month: essays on software engineering. Addison-Wesley, London, UK (1982)
9. Parnas, D.: Software engineering programmes are not computer science programmes. Annals of Software Engineering 6(1), 19–37 (1998)

10. Diaz-Herrera, J.L., Hilburn, T.B.: SE 2004 - Curriculum Guidelines for Undergraduate Degree Programs in Software Engineering. In: The Joint Task Force on Computing Curricula IEEE Computer Society Association for Computing Machinery. IEEE Computer Society Press, Los Alamitos (2004)
11. Newell, A.: The knowledge level. Artificial Intelligence 18, 87–127 (1982)
12. Russell, S., Norvig, P.: Artificial Intelligence: A Modern Approach. Prentice-Hall, Englewood Cliffs (2003)
13. MacCarthy, J.: Lisp 1.5 Programmer's Manual. MIT Press, Cambridge (1965)
14. Roussel, P.: PROLOG: Manuel de reference et d'utilisation. Université d'Aix-Marseille II (1975)
15. Shoham, Y.: Agent-Oriented Programming. Artificial Intelligence 60(1), 51–92 (1993)
16. Bordini, R., Hübner, J., Vieira, R.: Jason and the Golden Fleece of Agent-Oriented Programming. In: Multi-Agent Programming. Multiagent Systems, Artificial Societies, and Simulated Organizations, vol. 15, pp. 3–37. Springer, Heidelberg (2005)
17. Rao, A.S.: AgentSpeak (L): BDI agents speak out in a logical computational language. In: Perram, J., Van de Velde, W. (eds.) MAAMAW 1996. LNCS, vol. 1038, pp. 42–55. Springer, Heidelberg (1996)
18. Hindriks, K., Boer, F.D., der Hoek, W.V., Meyer, J.: Agent Programming in 3APL. Autonomous Agents and Multi-Agent Systems 2(4), 357–401 (1999)
19. Ross, R.J., Collier, R.W., O'Hare, G.M.P.: Af-apl - bridging principles and practice in agent oriented languages. In: Bordini, R.H., Dastani, M., Dix, J., Seghrouchni, A.E.F. (eds.) Programming Multi-Agent Systems. LNCS (LNAI), vol. 3346, pp. 66–88. Springer, Heidelberg (2005)
20. Fallah-Seghrouchni, A.E., Suna, A.: CLAIM: A Computational Language for Autonomous, Intelligent and Mobile Agents. In: Dastani, M., Dix, J., El Fallah-Seghrouchni, A. (eds.) PROMAS 2003. LNCS (LNAI), vol. 3067, pp. 90–110. Springer, Heidelberg (2004)
21. Shapiro, S., Lesperance, Y., Levesque, H.: Specifying communicative multi-agent systems with ConGolog. In: Working Notes of the AAAI Fall 1997 Symposium on Communicative Action in Humans and Machines, vol. 1037, pp. 72–82 (1997)
22. Thielscher, M.: FLUX: A logic programming method for reasoning agents. Theory and Practice of Logic Programming 5(4-5), 533–565 (2005)
23. Bordini, R., Braubach, L., Dastani, M., Seghrouchni, A.E.F., Gomez-Sanz, J., Leite, J., O'Hare, G., Pokahr, A., Ricci, A.: A Survey of Programming Languages and Platforms for Multi-Agent Systems. Informatica 30(1), 33–44 (2006)
24. Mascardi, V., Martelli, M., Sterling, L.: Logic-based specification languages for intelligent software agents. Theory and Practice of Logic Programming Journal (2004)
25. Bordini, R.H., Dastani, M., Dix, J., Seghrouchni, A.E.F. (eds.): Multi-Agent Programming. Multiagent Systems, Artificial Societies, and Simulated Organizations, vol. 15. Springer, Heidelberg (2005)
26. Hillard, R.: Recommended practice for architectural description of software-intensive systems. Technical report, IEEE (2000)
27. Hayes-Roth, B., Pfleger, K., Lalanda, P., Morignot, P., Balabanovic, M.: A domain-specific software architecture for adaptive intelligent systems. IEEE Transactions on Software Engineering 21(4), 288–301 (1995)
28. Bratman, M.E., Israel, D.J., Pollack, M.: Plans and resource-bounded practical reasoning. Computational Intelligence Journal 4(4), 349–355 (1988)

29. Jennings, N.R., Wittig, T.: ARCHON: Theory and Practice. In: Distributed Artificial Intelligence: Theory and Praxis. Eurocourses: Computer and Information Science, vol. 5. Springer, Heidelberg (1992)
30. Müller, J., Pischel, M.: The Agent Architecture InteRRaP: Concept and Application. PhD thesis, Deutsches Forschungszentrum für Künstliche Intelligenz (1993)
31. Ferguson, I.: Touring Machines: An architecture for Dynamic, Rational Agents. PhD thesis, Ph. D. Dissertation, University of Cambridge, UK (1992)
32. Brazier, F., Dunin-Keplicz, B., Jennings, N., Treur, J.: DESIRE: Modelling Multi-Agent Systems in a Compositional Formal Framework. International Journal of Cooperative Information Systems 6(1), 67–94 (1997)
33. Brazier, F., Treur, J., Wijngaards, N., Willems, M.: Temporal semantics of complex reasoning tasks. In: Proceedings of the 10th Banff Knowledge Acquisition for Knowledge-based Systems workshop, vol. 96, pp. 1–15 (1996)
34. Cornelissen, F., Jonker, C., Treur, J.: Compositional Verification of Knowledge-based Systems: a Case Study for Diagnostic Reasoning. In: Plaza, E. (ed.) EKAW 1997. LNCS, vol. 1319, pp. 65–80. Springer, Heidelberg (1997)
35. Wooldridge, M., Jennings, N.: Agent Theories, Architectures, and Languages: A Survey. Intelligent Agents 22 (1995)
36. Wooldridge, M., Jennings, N.R.: Intelligent agents: Theory and practice. Knowledge Engineering Review 10(2), 115–152 (1995)
37. Bellifemine, F., Poggi, A., Rimassa, G.: Jade: a fipa2000 compliant agent development environment. In: AGENTS '01: Proceedings of the fifth international conference on Autonomous agents, pp. 216–217. ACM Press, New York (2001)
38. Nwana, H.: Zeus: A Toolkit for Building Distributed Multi-agent Systems. Applied Artificial Intelligence 13(1), 129–185 (1999)
39. Demazeau, Y.: From interactions to collective behaviour in agent-based systems. In: Proceedings of the 1st. European Conference on Cognitive Science.
40. Kinny, D., Georgeff, M., Rao, A.: A methodology and modelling technique for systems of BDI agents. In: Perram, J., Van de Velde, W. (eds.) MAAMAW 1996. LNCS, vol. 1038, pp. 56–71. Springer, Heidelberg (1996)
41. Caire, G., Coulier, W., Garijo, F.J., Gomez, J., Pavón, J., Leal, F., Chainho, P., Kearney, P.E., Stark, J., Evans, R., Massonet, P.: Agent oriented analysis using message/uml. In: Wooldridge, M.J., Weiß, G., Ciancarini, P. (eds.) AOSE 2001. LNCS, vol. 2222, pp. 119–135. Springer, Heidelberg (2002)
42. Evans, R., Kearney, P., Caire, G., Garijo, F., Gomez-Sanz, J.J., Pavon, J., Leal, F., Chainho, P., Massonet, P.: Message: Methodology for engineering systems of software agents (September 2001), http://www.eurescom.de/~pub-deliverables/p900-series/P907/TI1/p907ti1.pdf
43. Henderson-Sellers, B., Giorgini, P.: Agent-oriented methodologies. Idea Group Pub. USA (2005)
44. Luck, M., McBurney, P., Shehory, O., Willmott, S.: Agent Technology: Computing as Interaction (A Roadmap for Agent Based Computing). AgentLink (2005)
45. Zambonelli, F., Omicini, A.: Challenges and Research Directions in Agent-Oriented Software Engineering. Autonomous Agents and Multi-Agent Systems 9(3), 253–283 (2004)
46. Luck, M., McBurney, P., Preist, C.: A Manifesto for Agent Technology: Towards Next Generation Computing. Autonomous Agents and Multi-Agent Systems 9(3), 203–252 (2004)
47. Myers, G., Sandler, C., Thomas, T.M., Badgett, T.: The Art of Software Testing. John Wiley and Sons, West Sussex, England (2004)

48. Gomez-Sanz, J., Pavon, J., Garijo, F.: Estimating Costs for Agent Oriented Software. In: Müller, J.P., Zambonelli, F. (eds.) AOSE 2005. LNCS, vol. 3950, pp. 218–230. Springer, Heidelberg (2006)
49. Bernon, C., Cossentino, M., Pavon, J.: Agent-oriented software engineering. The Knowledge Engineering Review 20(02), 99–116 (2006)
50. Gonzalez-Perez, C., McBride, T., Henderson-Sellers, B.: A Metamodel for Assessable Software Development Methodologies. Software Quality Journal 13(2), 195–214 (2005)
51. Henderson-Sellers, B.: Creating a Comprehensive Agent-Oriented Methodology: Using Method Engineering and the OPEN Metamodel. In: Agent-Oriented Methodologies, pp. 368–397. Idea Group, USA (2005)
52. Cernuzzi, L., Cossentino, M., Zambonelli, F.: Process models for agent-based development. Engineering Applications of Artificial Intelligence 18(2), 205–222 (2005)
53. Botía, J.A., Hernansaez, J.M., Skarmeta, F.G.: Towards an approach for debugging mas through the analysis of acl messages. In: Lindemann, G., Denzinger, J., Timm, I.J., Unland, R. (eds.) MATES 2004. LNCS (LNAI), vol. 3187, pp. 301–312. Springer, Heidelberg (2004)
54. Lam, D.N., Barber, K.S.: Comprehending agent software. In: Proceedings of the fourth international joint conference on Autonomous agents and multiagent systems, pp. 586–593. ACM Press, New York (2005)
55. DeLoach, S., Wood, M.F.: Developing multiagent systems with agenttool. In: Castelfranchi, C., Lespérance, Y. (eds.) ATAL 2000. LNCS (LNAI), vol. 1986, pp. 46–60. Springer, Heidelberg (2001)
56. Pavon, J., Gomez-Sanz, J.J., Fuentes, R.: The INGENIAS Methodology and Tools. In: Agent-Oriented Methodologies, pp. 236–276. Idea Group Publishing, USA (2005)
57. Bourque, P., Dupuis, R., Abran, A., Moore, J., Tripp, L.: The guide to the Software Engineering Body of Knowledge. Software 16(6), 35–44 (1999)
58. Gomez-Sanz, J.J., Gervais, M.P., Weiss, G.: A Survey on Agent-Oriented Software Engineering Research. In: Methodologies and Software Engineering for Agent Systems, pp. 33–62. Kluwer Academic Publishers, Dordrecht (2004)

Current Issues in Multi-Agent Systems Development

Rafael H. Bordini[1,*], Mehdi Dastani[2], and Michael Winikoff[3,**]

[1] University of Durham, U.K.
R.Bordini@durham.ac.uk
[2] Utrecht University, The Netherlands
mehdi@cs.uu.nl
[3] RMIT University, Australia
michael.winikoff@rmit.edu.au

Abstract. This paper surveys the state-of-the-art in developing multi-agent systems, and sets out to answer the questions: "what are the key current issues in developing multi-agent systems?" and "what should we, as a research community, be paying particular attention to, over the next few years?". Based on our characterisation of the current state-of-the-art in developing MAS, we identify three key areas for future work: techniques for integrating design and code; extending agent-oriented programming languages to cover certain aspects that are currently weak or missing (e.g., social concepts, and modelling the environment); and development of debugging and verification techniques, with a particular focus on using model checking also in testing and debugging, and applying model checking to design artefacts.

1 Introduction

In this paper we survey the current state-of-the-art in multi-agent system development and identify current issues. These issues are areas where we believe the research community should concentrate future research efforts, since they are, in the authors' opinion, crucial to practical adoption and deployment of agent technology.

This paper was based on an invited talk at ESAW 2006. The talk was presented by Rafael Bordini, but the contents of the talk grew out of each of the author's opinions, as presented and discussed at a Dagstuhl Seminar[1], and incorporating ideas from the AgentLink ProMAS technical forum group[2].

This paper is structured as follows. We begin by reviewing the state-of-the-art in MAS development (Section 2), focussing in particular on Agent Oriented Software Engineering (AOSE), Agent Oriented Programming Languages (AOPLs), and on verification of agent systems. We then identify a number of key issues in those areas (Section 3). For each issue, we discuss what we believe is the way forward. Since this paper is all

---

[*] Rafael Bordini gratefully acknowledges the support of the Nuffield Foundation (grant number NAL/01065/G).
[**] Michael Winikoff acknowledges the support of the Australian Research Council & Agent Oriented Software (grant LP0453486).
[1] http://www.dagstuhl.de/de/program/calendar/semhp/?semnr=06261
[2] http://www.cs.uu.nl/~mehdi/al3promas.html

about future work, our conclusions (in Section 4) merely summarises the key points of the paper.

2 State of the Art

There are a number of methodologies that provide developers with a process for doing software engineering of multi-agent systems, and there is a range of programming languages especially designed to facilitate the programming of agent systems. The field of verification and validation of agent systems is comparatively less well-developed, but there has been work on both formal verification using model checking, and on approaches for debugging and testing agent systems.

In current practice, the way in which a multi-agent system is typically developed is that the developer designs the agent organisation and the individual agents (perhaps using an AOSE methodology), then takes the detailed design and manually codes the agents in some programming language, perhaps agent-oriented, but more often using traditional programming languages. The resulting system is debugged (at best) using a combination of tracing messages and agent inspectors.

Thus developing a multi-agent system, like developing any software system, encompasses activities that are traditionally classified into three broad areas: software engineering (e.g., requirements elicitation, analysis, design[3]), implementation (using some *suitable* programming language), and verification/validation. To help structure this paper, which has also a focus on agent programming languages and verification, we have separated the last two types of activities from general software engineering (specifically analysis and design). Therefore, in the following subsections we briefly review the state-of-the-art in AOSE, programming languages for multi-agent systems, and verification of multi-agent systems (including debugging).

2.1 Agent Oriented Software Engineering

Agent Oriented Software Engineering is concerned with how to do software engineering of agent-oriented systems. It is a relatively youthful field, with the first AOSE workshop held in the year 2000. Nevertheless, over the past decade or so there has been considerable work by many people, resulting in quite a number of methodologies in the literature. These methodologies vary considerably in terms of the level of detail that is provided, the maturity of the methodology, and the availability of both descriptions that are accessible to developers (i.e., not researchers) and of tool support. Of the many methodologies available (e.g., see [6,32]), key methodologies that are widely regarded as mature include Gaia [64,67], MaSE [24], Tropos [14] and Prometheus [44].

It is important to clarify what is meant by a *methodology*. From a pragmatic point of view, a methodology needs to include all that a software engineer requires to do analysis and design, namely:

Concepts: While for object-oriented design the concepts used — classes, objects, inheritance, etc. — are so commonly understood as to be taken for granted, for agents

---
[3] Design includes various sorts of design activities, such as architectural design, social design, detailed design.

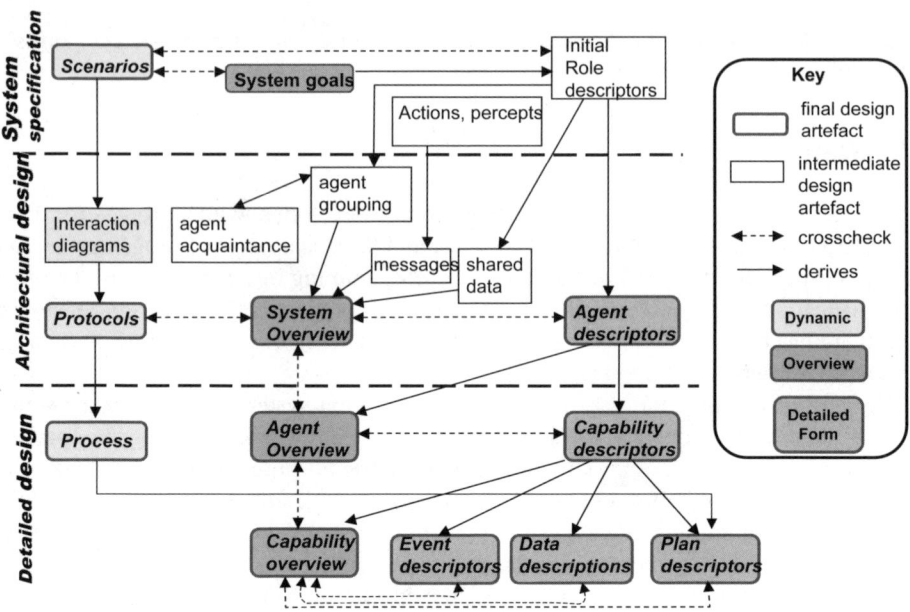

Fig. 1. Prometheus

a single set of basic concepts is not (yet) universally accepted or known, so a methodology needs, for completeness, to define a set of basic concepts that are used. For example, the Prometheus[4] methodology uses a set of concepts that are derived from the definition of an agent: action, percept, goal, event, plan, and belief [61].

Process: A methodology needs to provide an overall process that specifies what is done after what. For example, in Prometheus there are three phases — system specification, architectural design, and detailed design — where each phase consists of steps. For example, system specification includes the steps of identifying the system's goals, and of defining the interface between the system and its environment. Although these steps are often most easily described as being sequential, it is usual to recognise that iteration is the norm when doing software analysis/design. The process used by Prometheus is summarised in Figure 1.

Models and Notations: The results of analysis and design are a collection of models, for example a goal overview model or a system overview model. These models are depicted using some notation, often graphical, typically some variation on "boxes and arrows". Figure 2 shows an example model.

Techniques: It is not enough to merely say that, for example, the second step of the methodology is to develop a goal overview diagram. The software designer, especially if not particularly experienced in designing agent systems, benefits from a collection of specific techniques — usually formulated as heuristics — that guide them in *how* that step is carried out. For example, a goal overview diagram can

---

[4] We shall use Prometheus as an example because we are most familiar with it.

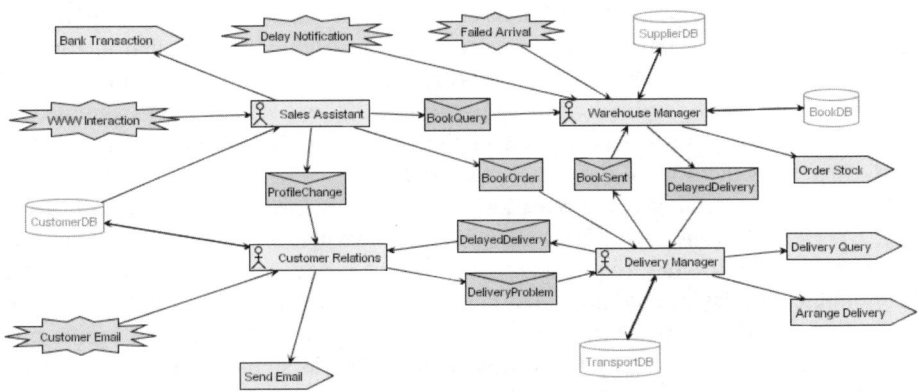

Fig. 2. System Overview Diagram (using Prometheus)

be refined by asking for each goal "how?" and "why?", yielding respectively new child and new parent goals [54].

Tool Support: Whilst arguably not essential, for any but the smallest design, and for any design that is iteratively refined, having tool support is of enormous benefit. Tools can range from simple drawing packages, to more sophisticated design environments that provide various forms of consistency checking.

Although each methodology has its own processes and notations, there are some common aspects. Many methodologies are broken down into some sort of requirements phase (e.g., "system specification" in Prometheus), some sort of system design (e.g., "architectural design" in Prometheus), and detailed design. The requirements phase specifies what it is that the system should do, with one of the commonly used models being some sort of goal hierarchy. The system design phase determines what agent types exist in the system, and how they interact (i.e., interaction design). Finally, the detailed design phase determines how each agent operates. It is also important to note that many methodologies capture in some way the environment that the agent system will inhabit. In Prometheus this is done, fairly simply, by specifying the interface to the environment in terms of actions and percepts.

2.2 Agent Oriented Programming Languages

There exist many Agent Oriented Programming Languages (AOPLs). In addition, there also exist platforms that focus on providing certain functionalities — such as communication and general infrastructure (e.g., white/yellow pages) — but which do not provide a programming language for developing agents. Such so-called platforms, such as OAA [16], JADE [5] and FipaOS [48], are not in the scope of this paper as they are not AOPLs.

In this section we shall, instead, focus on agent oriented programming languages for defining the behaviour of individual agents in a multi-agent system. In general, in so-called "cognitive agent programming languages", the focus is on how to describe the behaviour of an agent in terms of constructs such as plans, events, beliefs, goals, and messages. Although there are differences between various proposed AOPLs, they also have significant common characteristics. It is instructive to note that in some of these languages the environment is not captured, that agent interaction is implemented at the level of sending individual message — interaction protocols, for example, are not represented — and that in many languages goals are not provided, but they are approximated by the notion of events instead [62].

Most cognitive agent programming languages such as *Jason* [12], Jadex [12], JACK [46], 3APL [33,22], and 2APL[20] come with their own platforms. These platforms provide general infrastructure (e.g., agent management system, white and yellow pages), a communication layer, and integrated development environment (IDE). The existing IDE's provide editors with syntax highlighting facilities, enable a set of agents programs to be executed in parallel in various modes such as step-by-step or continuous, provide tools to inspect the internal states of the agents during execution, and examine messages that are exchanged by the agents. Further, some of the platforms that come with a cognitive agent programming languages, such as Jadex [47] and 2APL [20], are built on an existing agent platform, such as JADE [5]. The resulting platforms use the functionalities of the existing platforms such as the general infrastructure, communication facilities, and inspection tools.

Loosely speaking, agent-oriented programming languages can be classified as being either "theoretical" (i.e., having formal semantics, but arguably being impractical for serious development) or "practical" (i.e., being practical for serious development, but lacking formal semantics). However, it must be noted that this classification is simplistic in that languages with formal semantics are not precluded from being practical. Additionally, there are a number of more recent languages (or extensions of existing languages, such as *Jason*, which extends AgentSpeak(L)) that aim to be practical, yet have formal semantics. In particular, we mention 2APL, *Jason*, and SPARK [20,12,41] as examples of languages that aim to be practical yet have formal semantics. Below we briefly present some relevant features of *Jason* and 2APL. For more information on AOPLs, including various agent programming languages and platforms not mentioned here, see [8,7].

AgentSpeak(L) and *Jason*

AgentSpeak(L)[5] was proposed by Anand Rao in the mid 90s [51]. Perhaps surprisingly, the proposal received limited attention for a number of years before being "revived": in recent years there have been a number of implementations of AgentSpeak. Of these, the best developed is *Jason*, which extends AgentSpeak with a range of additional features.

The AgentSpeak language was intended to capture, in an abstract form, the key execution mechanism of existing Belief-Desire-Intention (BDI) platforms such as PRS [29,37] and dMARS[25]. This execution cycle links events and plans. Each event has a number of plans that it triggers. These plans have so-called context conditions, that specify under what conditions a plan should be considered to be applicable to handle an event. A given event instance is handled by determining which of the plans that it triggers are currently applicable, and then selecting one of these plans and executing it.

---

[5] In the remainder of this paper we shall simply refer to this language and its variants as "AgentSpeak".

Execution of plans is done step-by-step in an interleaved manner; plan instances form *intentions*, each representing one of the various foci of attention for the agent. The execution switches to the plan in the focus of attention of currently greatest importance for the agent.

Figure 3 shows examples of AgentSpeak plans for an abstract scenario of a planetary exploration robot. The first plan is triggered when the robot perceives a green patch on a rock, which indicates that the rock should be examined first as its analysis is likely to contain important data for a scientist working on the mission. In particular, the triggering event is green_patch(Rock) which occurs when the agent has a new belief of that form. However, the plan is only to be used when the battery charge is not low (this is the plan *context*, in this case battery_charge(low) must not be believed by the agent). The course of action prescribed by this plan, for when the relevant event happens and the context condition is satisfied is as follows: the location of the rock with the perceived green patch is to be retrieved from the belief base (?location(Rock,Coordinates)), then the agent should have a new goal to traverse to those coordinates (!traverse(Coordinates)), and finally, after the traverse has been successfully achieved, having a new goal to examine that particular rock (!examine(Rock)). The other two plans give alternative courses of action for when the robot comes to have a new goal of traversing to a certain coordinate. When the robot believes there is a safe path towards there, all it has to do is to execute the action (that the robot is hardwired to do) of physically moving towards those coordinates. The figure omits for the sake of space the alternative course of action (which would be to move around to try and find a different location from which a safe path can be perceived, probably by using some image processing software).

```
+green_patch(Rock) :
   not battery_charge(low) <-
      ?location(Rock,Coordinates);
      !traverse(Coordinates);
      !examine(Rock).
+!traverse(Coords) :
   safe_path(Coords) <-
      move_towards(Coords).
+!traverse(Coords) :
   not safe_path(Coords) <-
      ...
```

Fig. 3. Examples of AgentSpeak Plans

Jason[6] is a Java-based platform that implements the operational semantics of an extended version of AgentSpeak. The purpose of the language extensions was to turn the abstract AgentSpeak(L) language originally defined by Rao into a practical programming language. The language extensions and the platform have the following features:

---

[6] ***Jason*** is jointly developed by Rafael Bordini and Jomi F. Hübner (FURB, Brazil) and available *open source* at http://jason.sf.net

Strong negation: as agents typically operate under uncertainty and in dynamic environments, it helps the modelling of such applications if we are able to refer to things agents believe to be true, believe to be false, or are ignorant about.

Handling of plan failures: as multi-agent systems operate in unpredictable environments, plans can fail to achieve the goals they were written to achieve. It is therefore important that agent languages provide mechanisms to handle plan failure.

Speech-act based communication: as the mental attitudes that are classically used to give semantics for speech-act based communication are formally defined for AgentSpeak we can give precise semantics for how agents should interpret the basic illocutionary forces (e.g., inform, request), and this has been implemented in *Jason*. An interesting extension of the language is that beliefs can have "annotations" which can be useful for application-specific tasks, but there is one standard annotations that is done automatically by *Jason*, which is on the *source* of each particular belief.

Plan annotations: in the same way that beliefs can have annotations, programmers can add annotations to plan labels, which can be used by elaborate (e.g., using decision-theoretic techniques) selection functions. Selection functions are user-defined functions which are used by the interpreter, including which plan should be given preference in case various different plans happen to be considered applicable for a particular event.

Distribution: the platform makes it easy to define the agents that will take part in the system and also determine in which machines each will run, if actual distribution is necessary. The infrastructure for actual distribution can be changed (e.g., if a particular application needs to use a particular distribution platform such as JADE).

Environments: multi-agent systems will normally be deployed in some real-world environment. Even in that case, during development a simulation of the environment will be needed. *Jason* provides support for developing environments, which are programmed in Java rather than an agent language.

Customisation: programmers can customise two important parts of the agent platform by providing application-specific Java methods: the agent class and the agent architecture (note that the AgentSpeak interpreter provides only the reasoning component of the overall agent architecture). For more details, see [11].

Language extensibility and legacy code: the AgentSpeak extension available with *Jason* has a construct called "internal actions". These are then implemented in Java (or indeed any other language using JNI). This provides for straightforward language extensibility, which is also a straightforward way of invoking legacy code from within the high-level agent reasoning in an elegant manner. *Jason* comes with a library of essential standard internal actions. These implement a variety of useful operations for practical programming, but most importantly, they provide the means for programmers to do important things for BDI-inspired programming that were not possible in the original AgentSpeak language, such as checking and dropping the agent's own desires/intentions.

Integrated Development Environment: *Jason* is distributed with an IDE which provides a GUI for managing the system's project (the multi-agent system), editing the source code of individual agents, and running the system. Another tool provided as part of the IDE allows the user to inspect agents' internal (i.e., "mental") states when the system is running in debugging mode. The IDE is a plug-in to jEdit (http://www.jedit.org/), and an Eclipse plug-in is likely to be available in the future.

There is also much ongoing research to extend *Jason* is various ways, including: plan patterns for declarative goals [35], combination with the Moise+ [36] organisational model (http://moise.sf.net), automated belief revision [1], and combination with a high-level environment language aimed at social simulation, which in recent work aims to allow normative and organisational aspects to be associated with, for example, certain environment locations [42].

2APL: A Practical Agent Programming Language

One of the challenges of practical cognitive agent programming languages is an effective integration of declarative and imperative style programming. The declarative style programming should facilitate the implementation of the mental state of agents allowing agents to reason about their beliefs and goals and update them accordingly. An important issue here is the expressivity of the beliefs and goals, the expressions with which they can be updated, interface to existing declarative languages, and the relation between beliefs and goals (e.g., is it possible for an agent to have an expression as belief and goal at the same time? That is, can an agent desire a state which is believed to be achieved?) [63,21]. The imperative style programming should facilitate the implementation of processes, their execution modes, the flow of control, interface to existing imperative programming languages, and processing of events and exceptions. The question is how to integrate these declarative and imperative programming aspects in an effective way. This design objective is the main motivation for introducing a new agent programming language called 2APL (A Practical Agent Programming Language) [20].

Agents that are implemented by 2APL programs can generate plans by reasoning about their goals and beliefs, which are implemented in a declarative way. Plans can consist of actions of different types. Like most BDI-based programming languages, 2APL provides different types of actions such as belief and goal update actions, belief and goal test actions, external actions (to be performed in the agents' shared environment), and communication actions. As agents may operate in dynamic environments, they have to observe (be notified about) their environmental changes. In 2APL such environmental changes will be notified to the agents by means of *events*.

A characterising feature of 2APL is its distinction between events and goals. In some agent programming languages events are used for various purposes, e.g., for modelling an agent's goals or for notifying an agent about internal changes. Although both goals and events cause a 2APL agent to execute actions, they differ from each other in a principle way. For example, an agent's goal denotes a desirable state that the agent performs actions to achieve (goals are directly related with beliefs such that a goal is automatically dropped as soon as it is achieved), while an event carries information about (environmental) changes which may cause an agent to react and execute certain actions. After the execution of actions, an agent's goal may be dropped if the state denoted by it is believed to be achieved, while an event can be dropped just before

```
Beliefs:
   post(5,5).
   dirt(3,6).
   dirt(5,4).
   clean(world) :- not dirt(X,Y).
Goals:
   hasGold(2) and clean(world) ,
   hasGold(5)
PG-rules:
   clean(world) <- dirt(X,Y) |
   {  goto(X,Y);
      PickUpDirt();
      goto(2,2);
      DropDirt() }
PC-rules:
   goldAt(X,Y) <- true | { [goto(X,Y); PickUpGold()] }
PR-rules:
   PickUpDirt();R <- dirt(X,Y) | { goto(X,Y);PickUpDirt();R }
```

Fig. 4. Examples of 2APL Program

executing the actions that are triggered by it. Moreover, because of the declarative nature of goals (logical expressions), an agent can reason about its goals while an event only carries information which is not necessarily the subject of reasoning.

For example, consider the 2APL program as illustrated in Figure 4. This program, which for simplicity reasons does not include all details, indicates that the agent starts with the beliefs that it is on position (5,5), that there is dirt at positions (3,6) and (5,4), and that the world is clean if there is no dirt at any position. The agent wants to achieve two states (the goals are separated by a comma): one state in which he has two pieces of gold *and* the world is clean of dirt, and another state in which he has five pieces of gold. Note that these two states do not need to be achieved simultaneously. The planning goals rule (PG-rules) indicates that the state in which the world is clean can be achieved by going to the dirts' positions, picking them up, bringing them to the depot position (2,2), and dropping them in the depot. Note that the application of this rule can only achieve the subgoal clean(world), not the desired state hasGold(2) and clean(world). The ability to achieve subgoals requires reasoning about goals. Different notions of reasoning about goals are discussed in [21,55]. Note also that if all the dirt is picked up and dropped in the depot position, then the agent will believe that the world is clean. If the agent also believes that it has two pieces of gold, then it automatically drops the goal hasGold(2) and clean(world).

The difference between goals and events can be illustrated by the procedural rules (PC-rules). This rule indicates that if the agent is notified by an event that there is a gold piece at a certain position, then the agent should go to that position and pick up the gold piece. Note that both goals and event can cause the agent to perform actions.

Other characterising 2APL features are related to the constructs designed with respect to an agent's plans. The first construct is a part of an exception handling mechanism allowing a programmer to specify how an agent should repair its plans when the execution of its plans fail. This construct has the form of a rule which indicates that a plan should be replaced by another one. For example, consider again the agent program illustrated in Figure 4. The plan repair rule (PR-rules) indicates that if the execution of a plan that starts with the action `PickUpDirt()` (followed by the rest R of the plan) fails (for example because the dirt was already removed by another agent), then the plan should be replaced by the `goto(X,Y); PickUpDirt(); R` plan if the agent believes that there is dirt at another position (X,Y). Note the use of variable R which stands for the rest of the original plan. The second 2APL programming construct related to plans is the so-called non-interleaving (region of) plans. In most agent-oriented programming languages, an agent can have a set of plans whose executions can be interleaved. The arbitrary interleaving of plans may be problematic in some cases such that a programmer may want to indicate that a certain part of a plan should be executed at once without being interleaved with the actions of other plans. A non-interleaving plan region can be marked by putting the region of the plan within [] brackets. For example, in Figure 4 the plan for picking up a gold piece when notified by an event is a non-interleaving plan.

In addition to these features, 2APL provides specific programming mechanisms such as procedures, recursion and encapsulation. A procedure can be implemented by means of a specific rule that relates an abstract action (procedure call) to a concrete plan (procedure body). A recursion can be implemented simply by including the procedure call in the procedure body. Although these mechanisms can be implemented in other cognitive agent programming languages, 2APL follows the separation of concerns principle and provides specific constructs for the purpose of implementing procedures and recursions. In comparable cognitive agent programming languages (programming languages with formal semantics), procedures can be implemented by means of rules that relate events (or goals) to plans. In 2APL, procedures and recursion are considered as inherently different concepts from goals and events such that their implementation is independent of these concepts.

2.3 Verification and Validation

Multi-agent systems are distributed and concurrent, and the agents that make up a MAS are able to exhibit complex flexible behaviour in order to achieve its objectives in the face of a dynamic and uncertain environment. This flexible behaviour is key in making agent technology useful, but it makes it difficult to debug agent systems, and, once the system is (supposedly) debugged and ready for deployment, makes it hard to obtain confidence that the system will work as desired.

Debugging is an essential part of the process of developing software, and so good support for debugging is important. In the case of agent systems, there has been some work on debugging (e.g. [40,13,27,45]), but debugging techniques used in practice still rely on tracing and state inspection. The better agent development environments provide facilities to view, browse, and analyse the messages that are being exchanged, and facilities to examine the internal state of the agents. As examples, Figures 5 and 6 show the *Mind Inspector* tool provided by ***Jason*** and 2APL's State Trace, respectively; other platforms, such as JACK, provide similar functionality.

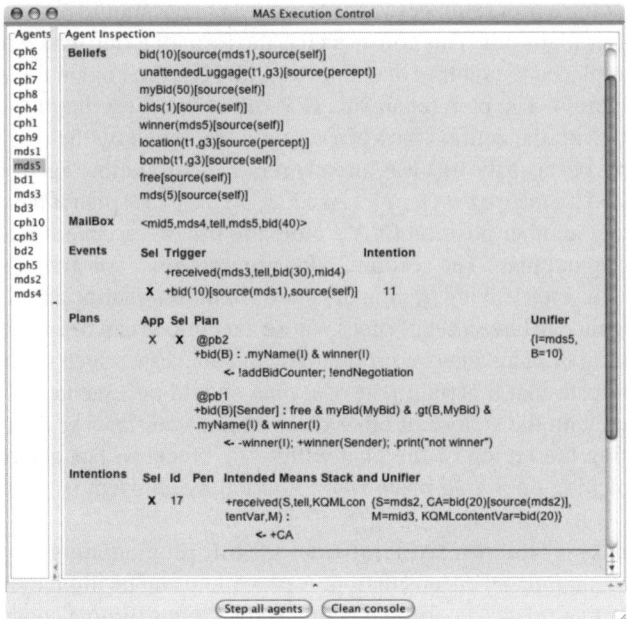

Fig. 5. *Jason* Debugging: Agent Mind Inspector

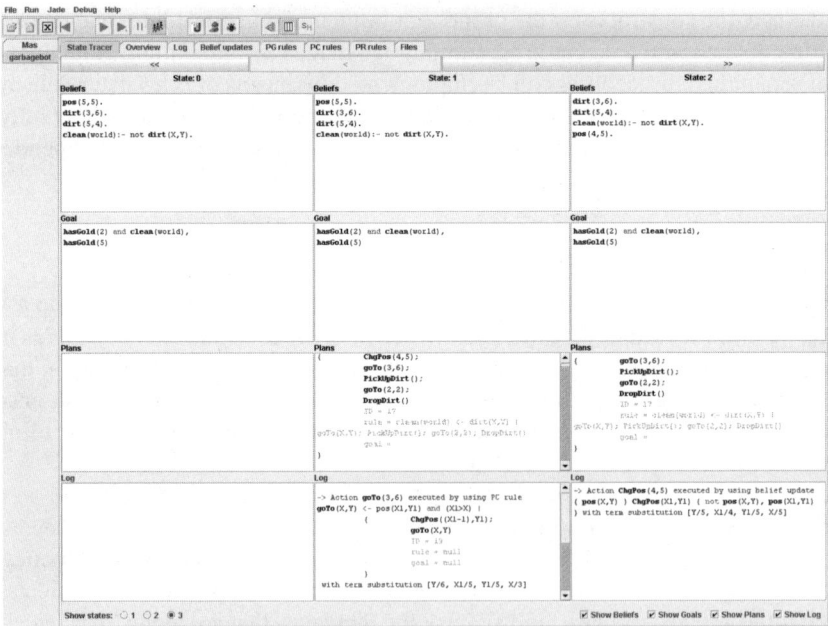

Fig. 6. 2APL Debugging: Agent State Tracer

In addition to using standard debugging techniques, there has been some work that aims to provide semi-automatic bug detection. For example, the work of Poutakidis et al. [49,50,45] automatically detects bugs in agent interactions by comparing an execution trace with the interaction protocol that is supposed to describe the valid message sequences. Any sequence of messages that occurs in the system's execution but that is illegal according to the protocol is automatically identified and flagged as an error. The general principle is that design artefacts can be used to assist in debugging.

In any software system it is essential that when the system is deployed and used, there is confidence that it will do what it is supposed to do. Typically, this confidence is achieved through testing. However, for agents that are able to exhibit flexible behaviour, achieving their goals in a range of ways depending on the situation, it is harder to achieve confidence in the system through testing. Hence, there has been a rather limited amount of work on testing agent systems, but there has been interest in using formal methods, especially model checking, to verify agent systems.

Work has focussed on model checking because it is easier to use than theorem proving, and, more importantly, because it can provide counter examples when the system fails to satisfy a desired property. Further, much work is devoted to state-space reduction techniques which can make model checking practical even for very large systems. However, although the technology is promising, at present it is fairly preliminary: the languages handled are limited, and the techniques have not been applied to industrial-scale case studies in multi-agent system.

Another type of work related to the correct behaviour of agent programs aims at specifying the semantics of the agent programming language is such a way to guarantee the satisfaction of certain behaviour. For example, in [23] it was shown that the semantics of an agent programming language can be defined in such a way that any agent implemented in that agent programming language will drop its goals if the goal is either achieved or believed not to be achievable anymore.

Most of the work done on model checking within the multi-agent systems research area is quite theoretical, although there are approaches that use existing model checkers, typically to check properties of particular aspects of a multi-agent system. A survey paper on the use of logic-based techniques for specifying but particular for *verifying* multi-agent systems is to appear in print around the same time as this paper, so instead of giving references here, we refer the interested reader to [26]. When it comes to model checking software (i.e., a complete running system) there is little work that applies to multi-agent systems in particular. More specifically on model checking agent *programs* written in an agent-oriented programming language, to our knowledge the only existing approach is the one presented in [10,56].

3 Problems with the Current State of the Art

Let us briefly summarise the current state-of-the-art in developing MAS. Not all current multi-agent system development projects use all of these, but rather we describe what is already available and in our opinion is likely to be used. Indeed adoption of these techniques will make short-term future development of agent-based system far more successful than previous attempts, in our opinion. We then discuss how this process should be improved by future research. We consider the state-of-the-art development process to be as follows:

1. Designing organisation and individual agents *using an AOSE methodology*
2. Taking the resulting design and (manually) coding the agents *in some AOPL*, based on the design
3. Debugging the system using message tracing and agent inspectors
4. Possibly using model checking on agent code (but unlikely)

Even though we believe that adoption of this development process would already improve significantly the development of multi-agent systems, the above summary highlights a number of areas where the current state-of-the-art is, in our opinion, seriously lacking, and where future work is sorely needed. There are three key issues:

- The implementation is developed completely *manually* from the design. This creates the possibility for the design and implementation to diverge, which tends to make the design less useful for further work in maintenance and comprehension of the system.
- Although present AOPLs provide powerful features for specifying the internals of a single agent, they mostly[7] only provide messages as the mechanism for agent interaction. Messages are really just the least common denominator for interaction, and, especially if flexible and robust agent interactions are desired, it is important to design and implement agent interactions in terms of higher-level concepts such as social commitments [65,28], delegation of goasl/tasks, responsibility [30,31], or interaction goals [15]. Additionally, AOPLs are weak in allowing the developer to model the environment within which the agents will execute.
- In most of the practical approaches for verification of multi-agent systems, verification is done on code. While this has the advantage of proving properties of the system that will be actually deployed, it is also often useful to check properties during the system design, so more work is required in verification of agent design artefacts. In fact, all the work on model checking for multi-agent systems is still in early stages so not really suitable for use on large and realistic systems.

In the remainder of this section we tackle each of these issues, and describe where we believe we should be heading, and what we believe needs to be done to address each of these issues. In brief, we believe that the key things we, as a research community, should be doing with respect to these issues are:

- Working on developing techniques and tools that allow for designs and code to be strongly integrated with consistency checking and change propagation.
- Developing better integrated designs and code would be facilitated by AOPLs being closer to the design in terms of covered concepts — while this is already true for individual agent abstractions, that is not the case for social abstractions. Thus, we believe that research effort in AOPLs should in the short-term concentrate on

---

[7] Although there has been work on AOPL support for programming *teams* of agents (e.g. [18,34,53]), this approach only applies to certain problem domains, where agents are cooperative.

extending AOPLs so they cover design concepts that are presently either missing or not covered well. Such concepts include interaction concepts at a higher level than messages (e.g., interaction protocols, social commitments, norms, obligations, responsibility, trust), and the environment (e.g., resources, services, actions), although further work on certain types of declarative goals is still required [19].
- Develop better techniques and tools for debugging and verification. One approach that is enabled by the existence of design that is reliably consistent with the code[8] is to use design artefacts to assist with debugging (e.g., [45]). However, debugging alone cannot assure us of the correctness of a system, and so formal verification techniques are also important. Interestingly, formal verification techniques such as model checking can be used to help validation when formal verification turns out not to be possible in practice (e.g., [58]).

3.1 Integrating Code and Design

There is an unfortunate tendency in the computing world to regard design and code as being completely different beasts. There are some clear differences between them: for instance, code is usually textual and detail-rich, whereas design is usually graphical and high-level. However, as was lucidly argued as far back as 1992 "Programming is a design activity" [52]. That is, the programming process, which is often (incorrectly) related by analogy to manufacturing a design in other engineering disciplines, is in fact a design activity, which is why it involves considerable rework in the form of debugging. Thus, it is highly desirable to have code and design being seen as different views on what is really a single conceptual activity.

Unfortunately, the current state-of-the-art in linking design and code is surprisingly primitive: "In most cases, the reverse-engineering facilities provided by CASE-tools supporting the Unified Modelling Language (UML) are limited to class diagram extraction" [38].

In an attempt to be systematic, we briefly present a taxonomy of the possible approaches for eliminating the "gap" between code and design. We have identified eight possible approaches:

Eliminate design: one way of avoiding discrepancies between two entities is to eliminate one of them! By "eliminating design" we do not mean that design activities are not performed, but that the results of these activities (in the form of design artefacts) are not retained and maintained. This approach, which may sound impossibly naive, is in fact what agile methodologies such as XP [4] propose. This approach is feasible if the application's design is relatively simple and/or is familiar to the system's developers.

Eliminate code: instead of eliminating design, we could eliminate code. Clearly, in order to have running software we need to augment the design with additional details. This approach corresponds to Model-Driven Development (MDD). This has been shown to be practical in certain cases, but has the drawback that the design can become cluttered with the additional details needed to make it executable.

---
[8] In fact, using design artefacts for debugging can also assist in detecting inconsistencies between design and code.

Generate code from design: a third approach is to generate the code from the design. This can be done fairly easily (although usually there is not enough information in the design to generate more than skeleton code). However, without additional techniques to then ensure the continued consistency of design and code as one or the other is changed, this is not a useful solution.

Extract design from code (reverse engineering): this automation possibility is intriguing, but not practical yet. Also, code typically does not contain all desired design information. However, the code can be extended to encompass such information.

Extract changes from design and apply to code: there is an issue here with language-dependence; that is, tools need to be developed for the particular design notation and the target programming language so that changes in the design can be reflected in the right way for that programming language. Also, it does not actually solve the problem (in case the code is changed directly)!

Extract changes from code and apply to design: the same issue with language-dependence as in the item above exists here. Also, it does not actually solve the problem (in case the design gets changed)!

Extract changes in design/code and apply to the other: although this approach completely solves the problem, the issue of language dependence still remains.

Integrate code and design into a single model: in this approach design and code become just different views on an underlying model which encompasses both. This avoids problem with language dependence by committing to a given programming language for the methodology, but requires integration between design and programming tools.

An issue in integrating code and design is that there are many design notations (and associated tools), and many AOPLs. Having to develop a link between each possible design tool and each possible language is clearly undesirable. A naive solution to this problem would be firstly to get the AOSE research community to agree on a single methodology and come together to develop a single support tool, and then secondly to get the AOPL research community to agree on a single AOPL. Clearly, this is not something that is likely to happen any time soon!

A more complex, but far more practical approach is to standardise interchange formats and APIs, while allowing the underlying notations/languages/tools to remain diverse. This is the approach we believe is most suitable for the multi-agent systems community and therefore we propose that the research community:

- Develop a standard abstraction for AOPLs
- Develop a standard API for making changes to an agent program
- For each AOPL's implementation, an implementation of the API is created
- Each design tool is extended with the ability to push changes into code via the API.

and, symmetrically, it is also required that the community:

- Agree on a common set of design abstractions
- Develop a standard API for making changes to a multi-agent system design
- For each AOSE methodology, an implementation of the API is created
- Each programming tool is extended with the ability to push changes into design via the API.

Clearly this will require major research effort and collaboration within (and between) the AOSE and ProMAS communities, but we believe this will have a significant impact in future MAS development, and that this line of research is worth pursuing.

3.2 Extending Agent-Oriented Programming Languages with Organisation and Interaction Aspects

Organisations are useful because they allow us to address at design and runtime how a complex multi-agent system should behave. Concepts such as responsibility, power, task delegation, norms, role enactment, workflows, shared goals, access control, groups and social structure help a software developer to understand and implement large multi-agent systems. Agent development methodologies need to provide concepts to specify, design, and implement static and dynamic aspects of such organisations. Though concepts for static views of organisations appear in almost all methodologies, the dynamics are not widely and thoroughly considered yet.

Historically, agent-oriented programming languages have focused on the *internals* of agents and have somewhat neglected *social* and organisational aspects. Most existing agent programming languages do not provide programming constructs to implement such multi-agent aspects so that programmers have to translate and incorporate these features at the level of individual agents' internals. However, some existing programming languages allow the implementation of these aspects, although to a very limited extent. For example, **Jason** provides programming constructs to indicate the infrastructure to be used by the agents, the environment the agent will share, and the agents and their numbers to be created and executed. Also, 2APL provides programming constructs to indicate which individual agents and how many of them should be created, and which agent has access to which environment.

One reason for neglecting these issues is the lack of clear and computational semantics for these social and organisational concepts. A starting point to tackle this issue is to develop formal and computational semantics for social and organisational concepts based on theories of concurrency and coordination, and possibly inspiration from theories of human organisations. It should be noted that work on electronic institutions, such as Islander, which maps to an implementation platform called Ameli [2], regulates agent interactions and ensures that the laws of the institution are obeyed. Although work in that area does not focus on designing agent programming languages, they can be a source of inspiration for designing agent programming languages with specific programming constructs that allow the implementation of multi-agent organisations and interactions.

Extending AOPLs: Interaction

Current agent-oriented programming languages allow the implementation of agent interactions at the level of messages. It is, however, desirable to move beyond messages because designing and implementing at the message level gives "brittle" interactions. Also, designing and implementing at this level makes it very hard to verify/debug and modify the interaction between agents. In order to overcome these problems, the following options can be considered.

One can extend agent-oriented programming languages with programming constructs that enable the implementation of interaction protocols. The execution of implemented protocols should enable individual agents to perform appropriate actions to achieve desirable interactions when they so choose. Alternatively, the execution of these programming constructs could extend the individual agent programs with the appropriate actions such that the execution of extended agent programs guarantees the desirable interactions between the agents.

Other options include using alternatives to message-centric interaction protocols, such as specifying interactions in terms of social commitments (e.g. [28,66]), landmarks [39] or interaction goals [15]; and extending AOPLs with support for these concepts [60].

Extending AOPLs: Environments

The environment shared by agents can be seen as a first-class abstraction which is as important as agents [59]. It provides the surrounding conditions for agents to exist and contains elements that are not present in agents, which are often important means for agent interaction. The environment can be used to help build a solution (coordination marks, such as pheromones, are a typical example). Agents can influence the environment to make it change or to extract meaningful information (perception). Agents can also communicate indirectly via the environment by adding and reading information from the environment. Finally, from the decision theory point of view, an agent decides which action to perform in an environment while the environment determines the actual effects of the action.

The environment benefits agent technology because it contributes to the separation of concerns and forces designers to incorporate appropriate agent features. Environments can be considered as a set of artefacts [43] used by agents to achieve goals and that regulate agent interaction. More elaborate approaches try to give more explicit definitions of environments by defining a framework. This framework would be responsible for executing agent actions and determining the effects of such actions.

Some agent development methodologies such as Prometheus have agent system designs that include a primitive environment model. The environment model is captured as *actions* and *percepts*. Also, some of the existing agent programming languages such as 3APL, 2APL, and **Jason** support the implementation of external shared environments. These environment are implemented as Java classes, for example so that their methods correspond with the actions that agents can perform in the environment. The state of the environment is then implemented by class variables which will be changed by the agents' actions (method calls). The modification of the state of the environment is implemented by the methods of the class. It is important to emphasis that these agent programming languages use Java to implement the agents' shared environment. Future work should consider extending the existing agent programming languages with specific abstract programming constructs that facilitate implementation of the environment in terms of high-level concepts such as resource, service, actions, and action effects.

3.3 Verification and Validation

Some of the reasons why debugging MAS is so hard are: agents exhibit *flexible* behaviour so it may turn out to be difficult to detect the circumstance that led to the faulty behaviour and even more so to fix the problems in a way that is consistent in all behaviours; the inherent *concurrency* in the system is an obvious complication as concurrent systems are notoriously difficult to debug; the environment is typically failure-prone so it may again be difficult to detect/reproduce the exact circumstances that cause problems and ensure that it will work correctly in the future; there will be typically a large number of agents which clearly makes things more difficult; systems might be open, so possibly difficult to consider the consequence of changes for various combinations of participating agents; each agent has a complex mental structure which needs to be not only inspected, but also understood; there will typically be a large number of communication messages that might need to be analysed in conjunction with agents' mental states. More importantly, the whole process needs to be tailored to account for the high-level notions used in MAS, such as beliefs, goals, plans, norms, roles, groups, etc. We strongly expect a lot of research to be done in this area to produce debugging approaches and tools which address these and many other specific issues in debugging multi-agent systems.

Whilst debugging and testing are fundamental, some applications require more than that. Many applications of multi-agent systems need to be *dependable systems*. Ideally, we would like to be able to fully *verify*, using formal methods, a multi-agent system which is safety/business/mission-critical. A popular current approach for formal verification of software is *model checking* [17]. Unfortunately, model checking techniques for verification of agent systems are still in their infancy (particularly in regards to practical tools). We expect to see a lot of work being done also in this area, and indeed there is already an active research community with ongoing projects in this direction.

In summary, in order to provide good support for ensuring that MAS run correctly, much work is needed in testing, debugging, and verification approaches and tools. More interestingly, approaches that combine these three activities are also likely to emerge for MAS, as they have in the automated software verification literature with approaches for traditional systems/languages. For example, when full verification is not possible because the system's state space is too large even after the use of state-space reduction techniques, practical model checking tools can still be used, for example, to help find the required input leading to special cases that can be potentially useful in testing a system [58].

One possible direction for research on model-checking multi-agent systems is as follows. In the same way that there is work being done for model checking to be applied to systems programmed in agent-oriented programming languages, it would be interesting to see approaches that apply directly to design documents of AOSE methodologies. This would, however, require that the design notations are given formal semantics, which is another interesting research strand. Both approaches exist in model checking traditional software. The idea in model checking programs [57] is to verify the system as it will be run by the users. Because code is much more detailed than design, the state-space explosion problems is normally much worse for programs than for design. In this approach, the use of state-space reduction techniques is particularly important. In fact, much work is done by the automated verification community on that topic, and a variety of different techniques exists for various types of state-space reduction used in model checking (e.g., data abstraction, partial-order reduction, property-based slicing, and compositional reasoning). The advantage of model checking programs is that if we succeed in the verification exercise, we know that the system *as actually run* satisfies the checked properties. When model-checking a high-level description of the system, we need to ensure that no errors are introduced in the implementation, which is typically done by a process of "refinement".

In general terms, what we would like to see in the future, ideally, are model checking techniques that are tailored particularly for MAS; that is, taking into account important agent abstractions such as goals, coalitions, etc. It must be noted though, that certain characteristics of MAS might prove to be particularly difficult to deal with for model checking, such as openness, emergence, etc. On the other hand, it is also possible that characteristics that are typical of MAS specifically can be explored for more efficient verification than normally expected in traditional software (e.g., compositional reasoning may turn out to work particularly well for agent organisations), but at the moment this is at best a conjecture.

One issue is that work on verifying agent programs has been done in the context of a given agent-oriented programming language. Clearly, it is desirable to have model checking tools that can be used on programs written in a range of languages. One approach to doing this is currently being pursued by Fisher, Wooldridge, Bordini, and colleagues. They aim to develop an "Agent Infrastructure Layer" (AIL) in the form of a Java library. There will then be provably correct translations of various programming languages into that Java library, so that JPF [57](http://javapathfinder.sf.net) can be used as a model checker. This would extend the previous approach so that it would apply to a variety of agent programming languages rather than only AgentSpeak. The development of AIL itself should be of interest as it would highlight common aspects of existing programming languages for multi-agent systems. For up-to-date information on that project, see http://www.csc.liv.ac.uk/~michael/mcapl06.html.

Another important area for future research is devising *state-space reduction techniques* specifically created for MAS. As mentioned earlier, much work in the automated verification community centres on state-space reduction techniques, and indeed they are responsible for the (relatively recent) success and popularity of model checking techniques. The availability of such techniques would have a major impact in the scale of multi-agent systems that will be verifiable in practice. Unfortunately, almost no work has yet been done in this direction; some (initial) work in this direction was done in [9], where a property-based slicing technique for an agent language was presented.

Still, we cannot be sure to be able to verify all multi-agent systems, but we would like to emphasise that there is much to be explored in the use of model checkers besides verification (e.g., for debugging and testing). We expect to see work in that direction and we would like to see, eventually, off-the-shelf MAS algorithms with verified properties available for MAS practitioners. That way, even when verification of the whole system is not possible, we would be able to know properties of particulars techniques used in parts of the system, which may turn out to be of great usefulness in many applications, and have such techniques immediately available for reuse by MAS developers.

4 Conclusion

We have surveyed the state-of-the-art in developing multi-agent systems, focusing on the three core areas: software engineering (analysis, design), programming, and techniques for verification and validation. Based on this, we have identified three key areas where we feel the state-of-the-art is lacking, and could be improved. Specifically we believe that there is a need to:

- integrate design and code more systematically;
- extend AOPLs with the ability to represent social aspects and the environment; and
- develop practical tools for verification and validation that are tailored specifically for multi-agent systems.

For each of these three key topics we discussed ways of meeting the challenges, and suggested some possible directions for the Agents research community.

References

1. Alechina, N., Bordini, R.H., Hübner, J.F., Jago, M., Logan, B.: Automating belief revision for agentspeak. In: Baldoni, M., Endriss, U. (eds.) DALT 2006. LNCS (LNAI), vol. 4327, pp. 1–16. Springer, Heidelberg (2006)
2. Arcos, J.L., Esteva, M., Noriega, P., Rodríguez, J.A., Sierra, C.: Engineering open environments with electronic institutions. Journal on Engineering Applications of Artificial Intelligence 18(2), 191–204 (2005)
3. Avrunin, G.S., Rothermel, G. (eds.): In: ISSTA 2004. Proceedings of the ACM/SIGSOFT International Symposium on Software Testing and Analysis, Boston, Massachusetts, USA, July 11-14, 2004. ACM Press, New York (2004)
4. Beck, K.: Extreme Programming Explained: Embrace Change. Addison-Wesley, London, UK (2000)
5. Bellifemine, F., Bergenti, F., Caire, G., Poggi, A.: Jade - a java agent development framework. In: Bordini, R.H., Dastani, M., Dix, J., El Fallah Seghrouchni, A. (eds.) Multi-Agent Programming: Languages, Platforms and Applications, ch. 5. Springer, Heidelberg (2005)
6. Bergenti, F., Gleizes, M.-P., Zambonelli, F. (eds.): Methodologies and Software Engineering for Agent Systems. Kluwer Academic Publishers, New York (2004)
7. Bordini, R., Braubach, L., Dastani, M., Seghrouchni, A., Gomez-Sanz, J., Leite, J., O'Hare, G., Pokahr, A., Ricci, A.: A survey of programming languages and platforms for multi-agent systems. Informatica 30(1), 33–44 (2006)
8. Bordini, R., Dastani, M., Dix, J., El Fallah Seghrouchni, A. (eds.): Multi-agent Programming: Languages, Platforms, and Applications. Springer, Heidelberg (2005)
9. Bordini, R.H., Fisher, M., Visser, W., Wooldridge, M.: State-space reduction techniques in agent verification. In: Kudenko, D., Kazakov, D., Alonso, E. (eds.) Adaptive Agents and Multi-Agent Systems II. LNCS (LNAI), vol. 3394, pp. 896–903. Springer, Heidelberg (2005)
10. Bordini, R.H., Fisher, M., Visser, W., Wooldridge, M.: Verifying multi-agent programs by model checking. Journal of Autonomous Agents and Multi-Agent Systems 12(2), 239–256 (2006)

11. Bordini, R.H., Hübner, J.F., et al.: Jason: A Java-based interpreter for an extended version of AgentSpeak, manual, release 0.9 edition (July 2006), http://jason.sourceforge.net/
12. Bordini, R.H., Hübner, J.F., Vieira, R.: Jason and the Golden Fleece of agent-oriented programming. In: Bordini, R.H., Dastani, M., Dix, J., El Fallah Seghrouchni, A. (eds.) Multi-Agent Programming: Languages, Platforms and Applications, ch. 1, pp. 3–37. Springer, Heidelberg (2005)
13. Botía, J.A., Hernansaez, J.M., Skarmeta, F.G.: Towards an approach for debugging MAS through the analysis of ACL messages. In: Lindemann, G., Denzinger, J., Timm, I.J., Unland, R. (eds.) MATES 2004. LNCS (LNAI), vol. 3187, pp. 301–312. Springer, Heidelberg (2004)
14. Bresciani, P., Giorgini, P., Giunchiglia, F., Mylopoulos, J., Perini, A.: Tropos: An agent-oriented software development methodology. Journal of Autonomous Agents and Multi-Agent Systems 8, 203–236 (2004)
15. Cheong, C., Winikoff, M.: Hermes: Designing goal-oriented agent interactions. In: Müller, J.P., Zambonelli, F. (eds.) AOSE 2005. LNCS, vol. 3950. Springer, Heidelberg (2006)
16. Cheyer, A., Martin, D.: The open agent architecture. Journal of Autonomous Agents and Multi-Agent Systems 4(1), 143–148 (2001)
17. Clarke Jr., E.M., Grumberg, O., Peled, D.A.: Model Checking. The MIT Press, Cambridge (1999)
18. Cohen, P.R., Levesque, H.J.: Teamwork. Nous 25(4), 487–512 (1991)
19. Dastani, M., Gomez-Sanz, J.: Programming multi-agent systems. The Knowledge Engineering Review 20(2), 151–164 (2006)
20. Dastani, M., Hobo, D., Meyer, J.-J.C.: Practical extensions in agent programming languages. In: Proceedings of the sixth International Joint Conference on Autonomous Agents and Multi-agent Systems (AAMAS'07). ACM Press, New York (2007)
21. Dastani, M., van Riemsdijk, M., Meyer, J.-J.: Goal types in agent programming. In: Proceedings of the 17th European Conference on Artificial Intelligence (ECAI'06) (2006)
22. Dastani, M., van Riemsdijk, M.B., Dignum, F., Meyer, J.-J.C.: A programming language for cognitive agents: goal directed 3APL. In: Dastani, M., Dix, J., El Fallah-Seghrouchni, A. (eds.) PROMAS 2003. LNCS (LNAI), vol. 3067, pp. 111–130. Springer, Heidelberg (2004)
23. Dastani, M., van Riemsdijk, M.B., Meyer, J.-J.C.: On the relation between agent specification and agent programming languages. In: Proceedings of the sixth International Joint Conference on Autonomous Agents and Multi-agent Systems (AAMAS'07). ACM Press, New York (2007)
24. DeLoach, S.A.: Analysis and design using MaSE and agentTool. In: Proceedings of the 12th Midwest Artificial Intelligence and Cognitive Science Conference (MAICS 2001) (2001)
25. d'Inverno, M., Kinny, D., Luck, M., Wooldridge, M.: A formal specification of dMARS. Technical Note 72, Australian Artificial Intelligence Institute (1997)
26. Fisher, M., Bordini, R.H., Hirsch, B., Torroni, P.: Computational logics and agents: a roadmap of current technologies and future trends. Computational Intelligence Journal 2007 (to appear)
27. Flater, D.: Debugging agent interactions: a case study. In: Proceedings of the 16th ACM Symposium on Applied Computing (SAC2001), pp. 107–114. ACM Press, New York (2001)
28. Flores, R.A., Kremer, R.C.: A pragmatic approach to build conversation protocols using social commitments. In: Jennings, N.R., Sierra, C., Sonenberg, L., Tambe, M. (eds.) Autonomous Agents and Multi-Agent Systems (AAMAS), pp. 1242–1243. ACM Press, New York (2004)

29. Georgeff, M.P., Lansky, A.L.: Procedural knowledge. Proceedings of the IEEE Special Issue on Knowledge Representation 74, 1383–1398 (1986)
30. Grossi, D., Dignum, F., Dastani, M., Royakkers, L.: Foundations of organizational structures in multi-agent systems. In: Proceedings of the Fourth International Joint Conference on Autonomous Agents and Multi-agent Systems (AAMAS'05). ACM Press, New York (2005)
31. Grossi, D., Dignum, F., Dignum, V., Dastani, M., Royakkers, L.: Structural aspects of the evaluation of agent organizations. In: Pre-proceedings of COIN@ECAI'06 (2006)
32. Henderson-Sellers, B., Giorgini, P. (eds.): Agent-Oriented Methodologies. Idea Group Publishing, USA (2005)
33. Hindriks, K.V., Boer, F.S.D., der Hoek, W.V., Meyer, J.-J.C.: Agent programming in 3APL. Autonomous Agents and Multi-Agent Systems 2(4), 357–401 (1999)
34. Hodgson, A., Rönnquist, R., Busetta, P.: Specification of coordinated agent behaviour (the simple team approach). Technical Report 5, Agent Oriented Software, Pty. Ltd. (1999), Available from http://www.agent-software.com
35. Hübner, J.F., Bordini, R.H., Wooldridge, M.: Programming declarative goals using plan patterns. In: Baldoni, M., Endriss, U. (eds.) DALT 2006. LNCS (LNAI), vol. 4327, pp. 65–81. Springer, Heidelberg (2006)
36. Hübner, J.F., Sichman, J.S., Boissier, O.: Using the $\mathcal{M}$oise+ for a cooperative framework of MAS reorganisation. In: Bazzan, A.L.C., Labidi, S. (eds.) SBIA 2004. LNCS (LNAI), vol. 3171, pp. 506–515. Springer, Heidelberg (2004)
37. Ingrand, F.F., Georgeff, M.P., Rao, A.S.: An architecture for real-time reasoning and system control. IEEE Expert 7(6) (1992)
38. Kollman, R., Selonen, P., Stroulia, E., Systa, T., Zundorf, A.: A study on the current state of the art in tool-supported UML-based static reverse engineering. In: Ninth Working Conference on Reverse Engineering (WCRE'02) (2002)
39. Kumar, S., Huber, M.J., Cohen, P.R.: Representing and executing protocols as joint actions. In: Proceedings of the First International Joint Conference on Autonomous Agents and Multi-Agent Systems, Bologna, Italy, July 15–19, 2002, pp. 543–550. ACM Press, New York (2002)
40. Lam, D.N., Barber, K.S.: Debugging agent behavior in an implemented agent system. In: Bordini, R.H., Dastani, M., Dix, J., Seghrouchni, A.E.F. (eds.) Programming Multi-Agent Systems. LNCS (LNAI), vol. 3346, pp. 104–125. Springer, Heidelberg (2005)
41. Morley, D., Myers, K.L.: The spark agent framework. In: Jennings, N., Sierra, C., Sonenberg, L., Tambe, M. (eds.) 3rd International Joint Conference on Autonomous Agents and Multiagent Systems (AAMAS 2004), pp. 714–721. IEEE Computer Society, Los Alamitos (2004)
42. Okuyama, F.Y., Bordini, R.H., da Rocha Costa, A.C.: Spatially distributed normative objects. In: Boella, G., Boissier, O., Matson, E., Vázquez-Salceda, J. (eds.) Proceedings of the Workshop on Coordination, Organization, Institutions and Norms in Agent Systems (COIN), held with ECAI 2006, Riva del Garda, Italy (August 28, 2006)
43. Omicini, A., Ricci, A., Viroli, M.: Coordination artifacts as first-class abstractions for MAS engineering: State of the research. In: Garcia, A., Choren, R., Lucena, C., Giorgini, P., Holvoet, T., Romanovsky, A. (eds.) Software Engineering for Multi-Agent Systems IV. LNCS, vol. 3914, pp. 71–90. Springer, Heidelberg (2006)
44. Padgham, L., Winikoff, M.: Developing Intelligent Agent Systems: A Practical Guide. John Wiley and Sons, West Sussex, England (2004)
45. Padgham, L., Winikoff, M., Poutakidis, D.: Adding debugging support to the Prometheus methodology. Engineering Applications of Artificial Intelligence 18(2), 173–190 (2005)

46. Papasimeon, M., Heinze, C.: Extending the UML for designing JACK agents. In: Proceedings of the Australian Software Engineering Conference (ASWEC 01) (2001)
47. Pokahr, A., Braubach, L., Lamersdorf, W.: Jadex: A BDI reasoning engine. In: Bordini, R.H., Dastani, M., Dix, J., Seghrouchni, A.E.F. (eds.) Multi-Agent Programming: Languages, Platforms and Applications, ch. 6, pp. 149–174. Springer, Heidelberg (2005)
48. Poslad, S., Buckle, P., Hadingham, R.: The fipa-os agent platform: Open source for open standards. In: Proceedings of the 5th International Conference and Exhibition on the Practical Application of Intelligent Agents and Multi-Agents, pp. 355–368 (2000)
49. Poutakidis, D., Padgham, L., Winikoff, M.: Debugging multi-agent systems using design artifacts: The case of interaction protocols. In: Proceedings of the First International Joint Conference on Autonomous Agents and Multi Agent Systems (AAMAS'02) (2002)
50. Poutakidis, D., Padgham, L., Winikoff, M.: An exploration of bugs and debugging in multi-agent systems. In: Zhong, N., Raś, Z.W., Tsumoto, S., Suzuki, E. (eds.) ISMIS 2003. LNCS (LNAI), vol. 2871, pp. 628–632. Springer, Heidelberg (2003)
51. Rao, A.S.: AgentSpeak(L): BDI agents speak out in a logical computable language. In: Perram, J., Van de Velde, W. (eds.) MAAMAW 1996. LNCS, vol. 1038, pp. 42–55. Springer, Heidelberg (1996)
52. Reeves, J.: What is software design? C++ Journal (1992)
53. Tambe, M.: Agent architectures for flexible, practical teamwork. In: National Conference on Artificial Intelligence (AAAI-97) (1997)
54. van Lamsweerde, A.: Goal-oriented requirements engineering: A guided tour. In: Proceedings of the 5th IEEE International Symposium on Requirements Engineering (RE'01), Toronto, pp. 249–263. IEEE Computer Society Press, Los Alamitos (2001)
55. van Riemsdijk, M.B., Dastani, M., Meyer, J.-J.C.: Subgoal semantics in agent programming. In: Bento, C., Cardoso, A., Dias, G. (eds.) EPIA 2005. LNCS (LNAI), vol. 3808, pp. 548–559. Springer, Heidelberg (2005)
56. van Riemsdijk, M.B., de Boer, F., Dastani, M., Meyer, J.-J.C.: Prototyping 3APL in the maude term rewriting language. In: Inoue, K., Satoh, K., Toni, F. (eds.) Computational Logic in Multi-Agent Systems. LNCS (LNAI), vol. 4371. Springer, Heidelberg (2007)
57. Visser, W., Havelund, K., Brat, G., Park, S.: Model checking programs. In: Proceedings of the Fifteenth International Conference on Automated Software Engineering (ASE'00), Grenoble, France, 11-15 September, pp. 3–12. IEEE Computer Society Press, Los Alamitos (2000)
58. Visser, W., Pasareanu, C.S., Khurshid, S.: Test input generation with Java PathFinder. In: Avrunin and Rothermel [3], pp. 97–107.
59. Weyns, D., Parunak, H.V.D., Michel, F.: Environments for Multi-Agent Systems II. In: E4MAS 2005. LNCS (LNAI), vol. 3830. Springer, Heidelberg (2006)
60. Winikoff, M.: Implementing commitment-based interactions. In: Autonomous Agents and Multi-Agent Systems (AAMAS) (2007)
61. Winikoff, M., Padgham, L., Harland, J.: Simplifying the development of intelligent agents. In: Stumptner, M., Corbett, D.R., Brooks, M. (eds.) AI 2001: Advances in Artificial Intelligence. LNCS (LNAI), vol. 2256, pp. 555–568. Springer, Heidelberg (2001)
62. Winikoff, M., Padgham, L., Harland, J., Thangarajah, J.: Declarative & procedural goals in intelligent agent systems. In: Proceedings of the Eighth International Conference on Principles of Knowledge Representation and Reasoning (KR2002), Toulouse, France (2002)
63. Winkelhagen, L., Dastani, M., Broersen, J.: Beliefs in agent implementation. In: Baldoni, M., Endriss, U., Omicini, A., Torroni, P. (eds.) DALT 2005. LNCS (LNAI), vol. 3904. Springer, Heidelberg (2006)
64. Wooldridge, M., Jennings, N., Kinny, D.: The Gaia methodology for agent-oriented analysis and design. Autonomous Agents and Multi-Agent Systems 3(3) (2000)

65. Yolum, P., Singh, M.P.: Flexible protocol specification and execution: Applying event calculus planning using commitments. In: Proceedings of the 1st Joint Conference on Autonomous Agents and MultiAgent Systems (AAMAS), pp. 527–534 (2002)
66. Yolum, P., Singh, M.P.: Reasoning about commitments in the event calculus: An approach for specifying and executing protocols. Annals of Mathematics and Artificial Intelligence (AMAI), Special Issue on Computational Logic in Multi-Agent Systems (2004)
67. Zambonelli, F., Jennings, N., Wooldridge, M.: Developing multiagent systems: the Gaia methodology. ACM Transactions on Software Engineering and Methodology 12(3), 317–370 (2003)

Architecture-Centric Software Development of Situated Multiagent Systems

Danny Weyns and Tom Holvoet

DistriNet, Katholieke Universiteit Leuven
Celestijnenlaan 200 A, B-3001 Leuven, Belgium
{danny.weyns,tom.holvoet}@cs.kuleuven.be

Abstract. A multiagent system (MAS) structures a software system as a set of autonomous agents that interact through a shared environment. Software architecture is generally considered as the structures of a system which comprise software elements and the relationships among the elements. So there is a clear connection between MAS and software architecture. In our research, we study situated MAS, i.e. systems in which agents have an explicit position in the environment. We apply situated MAS to domains that are characterized by highly dynamic operating conditions and an inherent distribution of resources. We use an architecture-centric approach for developing such MAS. From our experiences with building various applications, we have developed a reference architecture for situated MAS. The reference architecture provides an asset base architects can draw from when developing new systems that share the common base of the reference architecture. In this paper, we explain our perspective on architecture-centric software development of MAS. We give an overview of the reference architecture and we show an excerpt of the software architecture of an industrial application in which we have used the reference architecture. The reference architecture shows how knowledge and experience with MAS can be documented and matured in a form that has proven its value in mainstream software engineering. We believe that this integration is a key to industrial adoption of MAS.

1 Introduction

Five years of application-driven research taught us that there is a close connection between multiagent systems (MAS) and software architecture. Our perspective on the essential purpose of MAS is as follows:

> *A multiagent system provides the software to solve a problem by structuring the system as a number of interacting autonomous entities (agents) embedded in an environment in order to achieve the functional and quality requirements of the system.*

This perspective states that a MAS provides the software to *solve* a problem. In particular, a MAS *structures the system* as a number of interacting agents embedded in an environment. The purpose of the system is to achieve the *requirements* of the system. This is exactly what *software architecture* is about. [6] defines software architecture as: "the structure or structures of the system, which comprise software elements, the externally visible properties of those elements, and the relationships among them." Software elements (or in general architectural elements) provide the functionality of the system, while the required quality attributes are primarily achieved through the structures of the software architecture.

As such, MAS are in essence a family—yet a large family—of software architectures. Based on the problem analysis that yields the functional and quality attribute requirements of the system, the architect may or may not choose for a MAS-based solution. Quality attribute requirements such as flexibility, openness, and robustness may be arguments for the designer to choose for a MAS software architecture. As such, we consider MAS as one valuable family of approaches to solve software problems in a large spectrum of possible ways to solve problems. Typical architectural elements of MAS software architectures are agents, coordination infrastructure, resources, services, etc. The relationships between the elements are very diverse, ranging from environment mediated interaction between cooperative agents via digital pheromone trails to complex negotiation protocols in a society of self-interested agents. In short, MAS are a rich family of architectural approaches with specific characteristics, useful for a diversity of challenging application domains. By considering MAS essentially as software architecture, MAS gets a clear and prominent role in the software development process paving a way to integrate MAS with mainstream software engineering.

Architecture-Centric Software Development of Situated MAS. In our research, we study situated MAS, i.e. systems in which agents have an explicit position in the environment. We apply situated MAS to domains that are characterized by highly dynamic operating conditions and an inherent distribution of resources. We use an architecture-centric approach for developing such MAS. From our experiences with building various applications, we have developed a reference architecture for situated MAS. The reference architecture provides an asset base architects can draw from when developing new systems that share the common base of the reference architecture. In this paper, we explain our perspective on architecture-centric software development of MAS. We give an overview of the reference architecture for situated MAS and we show an excerpt of the software architecture of an industrial application in which we have used the reference architecture.

Overview. The paper is structured as follows. We start with a brief introduction of architecture-centric software development in general. Next, in Sect. 3 we give a high-level overview of the reference architecture for situated MAS. Section 4 shows an excerpt of the software architecture of an industrial AGV transportation system in which we have used the reference architecture for architectural design. Section 5 discusses related work, and in Sect. 6 we draw conclusions.

2 Architecture-Centric Software Development

To understand our perspective on software engineering of MAS, we give a brief overview of architecture-centric software development in general. We use the evolutionary delivering life cycle [26,6], see Fig. 1. This life cycle model situates architectural design in the centre of the development activities. The main idea of the model is to support incremental software development and to incorporate early feedback from the stakeholders. The life-cycle consists of two main phases: developing the core system and delivering the final software product.

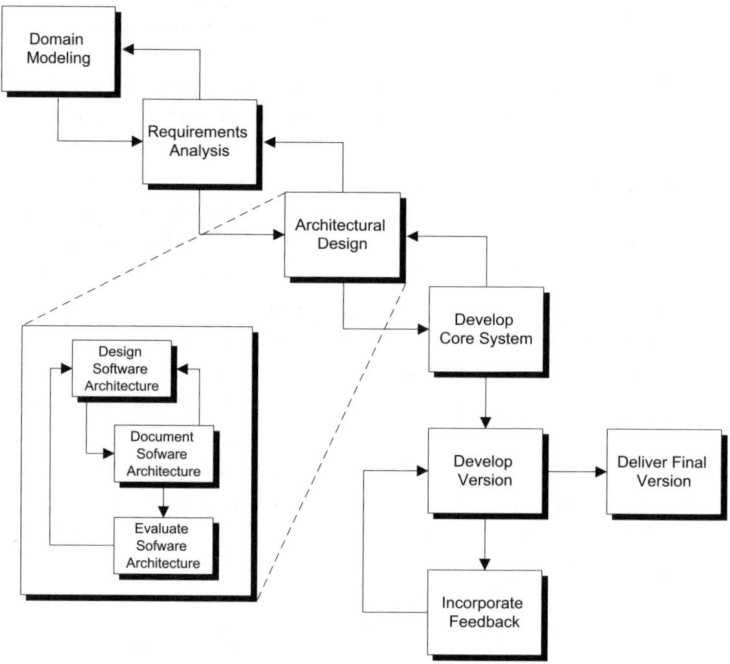

Fig. 1. Architectural design in the software development life cycle

In the first phase the core system is developed. This phase includes four activities: defining a domain model, performing a system requirements analysis, designing the software architecture, and developing the core system. Requirements analysis includes the formulation of functional requirements of the system as well as eliciting and prioritizing of the quality attributes requirements. Designing the software architecture includes the design and documentation of the software architecture, and an evaluation of the architecture. The development of the core system includes detailed design, implementation and testing. The software engineering process is an iterative process, the core system is developed incrementally, passing multiple times through the different stages of the development process. Fig. 1 shows how architectural design iterates with requirements analysis on the one hand, and with the development of the core system on the other hand. The output of the first phase is a domain model, a list of system requirements, a software architecture, and an implementation of the core of the software system.

In the second phase, subsequent versions of the system are developed until the final software product can be delivered. In principle there is no feedback loop from the second to the first phase although in practice specific architectural refinements may be necessary.

We now briefly look at architectural design and the activities it directly iterates with: requirements analysis and developing the core system.

Requirements Analysis. Gathering system requirements includes the elicitation of functional requirements as well as eliciting and prioritizing of the quality attributes requirements. Functional requirements of a system are typically expressed as use cases [25]. A use case lists the steps, necessary to accomplish a functional goal for an actor that uses the system. In our research, we also use scenarios that describe interactions among parts in the system—rather than interactions that are initiated by an external actor. An example is a scenario that describes the requirement of collision avoidance of automatic guided vehicles on crossroads. For the expression of quality requirements we use system-specific *quality attribute scenarios* [5]. A quality attribute scenario consists of three parts: (1) a stimulus: an internally or externally generated condition that affects (a part of) the system and that needs to be considered when it arrives at the system; (2) a context: the conditions under which the stimulus occurs; (3) a response: the activity that is undertaken—through the architecture—when the stimulus arrives. The response should be measurable so that the requirement can be tested. Here is an example of a quality attribute scenario:

> *An Automatic Guided Vehicle (AGV) gets broken and blocks a path under normal system operation. Other AGVs have to record this, choose an alternative route—if available—and continue their work.*

The stimulus in this example is "An Automatic Guided Vehicle gets broken and blocks a path", the context is "under normal system operation", and the response is "other AGVs have to record this, choose an alternative route—if available—and continue their work". Quality attribute scenarios provide a means to transform vaguely formulated qualities such as "the system shall be modifiable" or "the system shall exhibit acceptable flexibility" into concrete expressions. To elicit and prioritize quality attribute scenarios, we use *utility trees* [14]. An utility tree compels the architect and other stakeholders involved in a system to define the relevant quality requirements precisely. An utility tree consists of four levels. The root node of the tree is *utility* expressing the overall quality of the system. High-level quality attributes form the second level of the tree. Each quality attribute is further refined in the third level. Finally, the leaf nodes of the tree are the quality attribute scenarios. Eah scenario is assigned a ranking that expresses its priority relatively to the other scenarios. Criteria for prioritization include the importance of the scenario to the success of the system, and the difficulty to achieve the scenario. It is clear that the most important scenarios are those that have a high ranking on both criteria. [8] shows an example of a utility tree for the automatic transportation system we discuss in section 4.

Architectural Design. Architectural design includes the design and documentation of the software architecture, and an evaluation of the architecture (see Fig. 1).

Design. Designing a software architecture is moving from system requirements to architectural decisions. The various requirements are achieved by architectural decisions that are based on architectural approaches. One common architectural approach are architectural patters [31]. An architectural pattern is a description of architectural elements and their relationships that has proven to be useful for achieving particular qualities. Examples of architectural patterns are layers and blackboard. In our research, we have developed a reference architecture for MAS as a reusable architectural approach. This reference architecture integrates a set of architectural patterns that have proven their value in various MAS applications we have studied and built. A reference architecture provides an integrated set of architectural patterns the architect can draw from to select suitable architectural solutions.

Architectural design requires a systematic approach to develop a software architecture that meets the required functionality and satisfies the quality requirements. In our research, we use techniques from the Attribute Driven Design (ADD [10,6]) method to design the architecture for a software system with a reference architecture. ADD is a decomposition method that is based on understanding how to achieve quality goals through proven architectural approaches. Usually, the architect starts from the system as a whole and then iteratively refines the architectural elements, until the elements are sufficiently fine-grained to start detailed design and implementation. At that point, the software architecture becomes a prescriptive plan for construction of the system that enables effective satisfaction of the systems functional and quality requirements [21,13].

A reference architecture serves as a blueprint to *guide* the architect through the decomposition process. In particular, the ADD process can be used to iteratively refine the software architecture, and the reference architecture can serve as a guidance in this decomposition process. In addition, common architectural approaches have to be applied to refine and extend architectural elements when necessary according to the requirements of the system at hand.

Documentation. A software architecture is described by different *views*. Each view belongs to a *viewtype* [13]. A viewtype defines the elements and relationship used to describe the architecture of a software system from a particular perspective. We use three different viewtypes:

1. The module viewtype: views in this viewtype document a system's principal units of implementation.
2. The component-and-connector viewtype: views in this viewtype document the system's units of execution.

3. The deployment viewtype: views in this viewtype document the relationships between a system's software and its development and execution environment.

Documenting a software architecture comes down to documenting the relevant views of the software architecture for the application at hand. Each view is documented by means of a number of view packets [13]. A view packet is a small, relatively self-contained bundle of information of the reference architecture.

Evaluation. A software architecture is the foundation of a software system, it represents a system's earliest set of design decisions. Due to its large impact on the development of the system, it is important to verify the architecture as soon as possible. Modifications in early stages of the design are cheap and easy to carry out. Deferring evaluation might require expensive changes or even result in a system of inferior quality.

The evaluation of software architecture is an active research topic, see e.g. [4,27]. In our research, we use the Architectural Tradeoff Analysis Method [14] (ATAM). ATAM is a well-established method for software architecture evaluation developed at the Software Engineering Institute [2]. The ATAM incites the stakeholders to articulate specific quality goals and to prioritize conflicting goals; it forces the architect to provide a clear explanation and documentation of the software architecture; and especially it uncovers problems with the architecture that can be used to improve the quality of the software architecture in an early stage of the development cycle. An ATAM evaluation produces the following results:

- A prioritized list of quality attribute requirements in the form of a quality attribute utility tree.
- A mapping of architectural approaches to quality attributes. The analysis of the architecture exposes how the architecture achieves—or fails to achieve—the important quality attribute requirements.
- Risks and non-risks. Risks are potentially problematic architectural decisions, non-risks are good architectural decisions.
- Sensitivity points and tradeoff points. A sensitivity point is an architectural decision that is critical for achieving a particular quality attribute. A tradeoff point is an architectural decision that affects more than one attribute, it is a sensitivity point for more than one attribute.

[9] discusses our experiences with ATAM for the application discussed in Sect. 4.

Developing the Core System. The development of the core system includes detailed design, implementation and testing. The software architecture defines constraints on detailed design and implementation, it describes how the implementation must be divided into elements and how these elements must interact with one another to fulfil the system requirements. On the other hand, a software architecture does not *define* an implementation, many fine-grained design decisions are left open by the architecture and must be completed by designers and developers. For some tasks established techniques can be used such as design patterns or well-know algorithms. However, other—MAS specific—tasks require dedicated design guidelines, e.g. the detailed design of an agent communication language or a pheromone infrastructure.

3 Reference Architecture for Situated Multiagent Systems

In our research, we study the engineering of software systems with the following main characteristics and requirements:

- Stakeholders of the systems (users, project managers, architects, developers, maintenance engineers, etc.) have various—often conflicting—demands on the quality of the software. Important quality requirements are flexibility (adapt to variable operating conditions) and openness (cope with parts that come and go during execution).
- The software systems are subject to highly dynamic and changing operating conditions, such as dynamically changing workloads and variations in availability of resources and services. An important requirement of the software systems is to manage the dynamic and changing operating conditions autonomously.
- Global control is hard to achieve. Activity in the systems is inherently localized, i.e. global access to resources is difficult to achieve or even infeasible. The software systems are required to deal with the inherent locality of activity.

Example domains are mobile and ad-hoc networks, sensor networks, automated transportation and traffic control systems, and manufacturing control.

To deal with these requirement we apply the paradigm of situated MAS. During the last five years, we have developed several mechanisms of adaptivity for situated MAS, including selective perception [46], protocol-based communication [45], behavior-based decision making with roles and situated commitments [33], and laws that mediate the activities of agents in the environment [38]. We have applied these mechanisms in various applications, ranging from experimental simulations [37,39,36] and prototypical robot applications [44,33] up to an industrial transportation system for automatic guided vehicles [43,40,9].

Based on these experiences, we have developed a reference architecture for situated MAS. Motivations for the reference architecture are: (1) it integrates the different mechanisms for adaptivity. It defines how the functionalities of the various mechanisms are allocated to software elements of agents and the environment and how these elements interact with one another, (2) it provides a reusable design artifact, the reference architecture facilitates deriving new software architectures for systems that share the common base more reliably and cost effectively, and (3) the reference architecture embodies the knowledge and expertise we have acquired during our research. It conscientiously documents the know-how obtained from this research. As such, the reference architecture offers a vehicle to study and learn the advanced perspective on situated MAS we have developed.

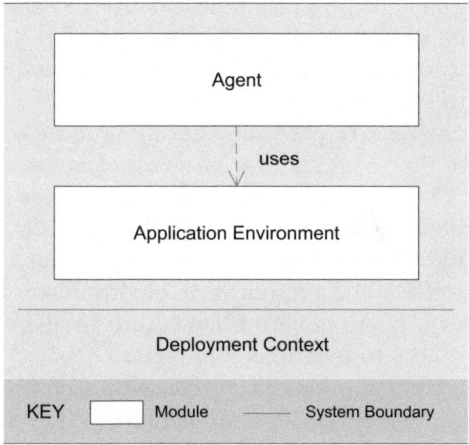

Fig. 2. Top-level module decomposition of a situated MAS

Fig. 2 shows the top-level module decomposition of the reference architecture of situated MAS that shows the main software units in the system.

A situated multiagent system is decomposed in two basic modules: **Agent** and **Application Environment**.

Agent is an autonomous problem solving entity in the system. An agent encapsulates its state and controls its behavior. The responsibility of an agent is to achieve its design objectives, i.e. to realize the application specific goals it is assigned. Agents are situated in an environment which they can perceive and in which they can act and interact with one another. Agents are able to adapt their behavior according to the changing circumstances in the environment. A situated agent is a cooperative entity. The overall application goals result from interaction among agents, rather than from sophisticated capabilities of individual agents.

A concrete MAS application typically consists of agents of different agent types. Agents of different agent types typically have different capabilities and are assigned different application goals.

The **Application Environment** is the part of the environment that has to be designed for a concrete MAS application. The application environment enables agents to share information and to coordinate their behavior. The core responsibilities of the application environment are:

- To provide access to external entities and resources.
- To enable agents to perceive and manipulate their neighborhood, and to interact with one another.
- To mediate the activities of agents. As a mediator, the environment not only enables perception, action and interaction, it also constrains them.

The application environment provides functionality to agents on top of the *deployment context*. The deployment context consists of the given hardware and software and external resources such as sensors and actuators, a printer, a network, a database, a web service, etc.

As an illustration, a peer-to-peer file sharing system is deployed on top of a deployment context that consists of a network of nodes with files and possibly other resources. The application environment enables agents to access the external resources, shielding low-level details. Additionally, the application environment may provide a coordination infrastructure that enables agents to coordinate their behavior. E.g., the application environment of a peer-to-peer file share system can offer a pheromone infrastructure to agents that they can use to dynamically form paths to locations of interest.

Thus, we consider the *environment* as consisting of two parts, the deployment context and the application environment [42]. The internal structure of the deployment context is not considered in the reference architecture. For a distributed application, the deployment context consists of multiple processors deployed on different nodes that are connected through a network. Each node provides an application environment to the agents located at that node. Depending on the specific application requirements, different application environment types may be provided. For some applications, the same type of application environment subsystem is instantiated on each node. For other applications, specific types are instantiated on different nodes, e.g., when different types of agents are deployed on different nodes.

In the next section, we zoom in on the collaborating components view of the reference architecture. For a description of other architectural views of the reference architecture and a formal specification of the various architectural elements we refer to [34].

3.1 Collaborating Components View Packets

The collaborating components view shows the MAS or parts of it as a set of interacting runtime components that use a set of shared data repositories to realize the required system functionalities. The elements of the collaborating components view are:

- *Runtime components.* Runtime components achieve a part of the system functionality. Runtime components are instances of modules described in the module decomposition view.
- *Data repositories.* Data repositories enable multiple runtime components to share data. Data repositories correspond to the shared data repositories described in the component and connector shared data view.
- *Component–repository connectors.* Component–repository connectors connect runtime components which data repositories. These connectors determine which runtime components are able to read and write data in the various data repositories of the system.
- *Component–component connectors.* Collaborating components require functionality from one another and provide functionality to one another.

Component–component connectors enable runtime components to request each other to perform a particular functionality.

The collaborating components view is an excellent vehicle to learn the runtime behavior of a situated MAS. The view shows the data flows between runtime components and the interaction with data stores, and it specifies the functionalities of the various components in terms of incoming and outgoing data flows.

We discuss two view packets of the collaborating components view. We start with the view packet that describes the collaborating components of agent. Next, we discuss the view packet that describes the collaborating components of the application environment.

A. Collaborating Components View Packet: Agent

Primary Presentation. The primary presentation is show in Fig. 3.

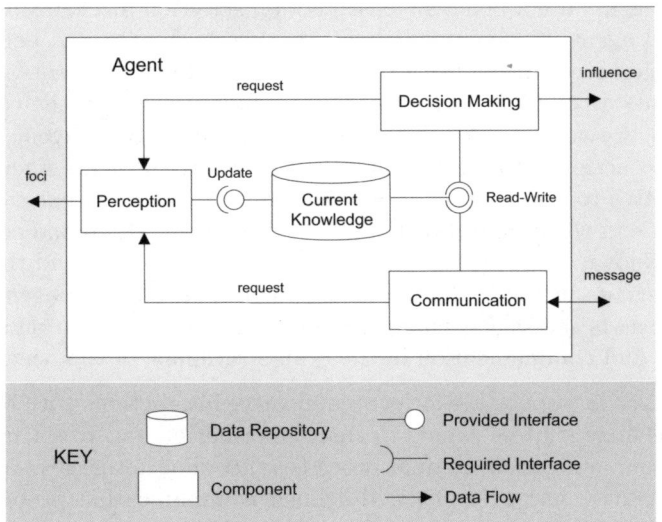

Fig. 3. Collaborating Components of Agent

Elements and their Properties. The Agent component (i.e. a runtime instance of the Agent module shown in Fig. 2) consists of three subcomponents: Perception, Decision Making, and Communication. These components share the Current Knowledge repository. We first give a brief explantion of the responsibilities of the components and then we explain the collaboration between the components and the shared data repository.

Perception is responsible for collecting runtime information from the environment (application environment and deployment context). The perception component supports selective perception [46]. Selective perception enables an agent to direct its perception according to its current tasks. To direct its perception an agent selects a set of foci and filters. Foci allow the agent to sense the environment only for specific types of information. Sensing results in a representation of the sensed environment. A representation is a data structure that represents elements or resources in the environment. The perception module maps this representation to a percept, i.e. a description of the sensed environment in a form of data elements that can be used to update the agent's current knowledge. The selected set of filters further reduces the percept according to the criteria specified by the filters.

Decision Making is responsible for action selection. The action model of the reference architecture is based on the influence–reaction model introduced in [15]. This action model distinguishes between influences that are produced by agents and are attempts to modify the course of events in the environment, and reactions, which result in state changes in the environment. The responsibility of the decision making module is to select influences to realize the agent's tasks, and to invoke the influences in the environment [38].

Situated agents use a behavior-based action selection mechanism [41]. To enable situated agents to set up collaborations, we have extended behavior-based action selection mechanisms with roles and situated commitments [44,33,45]. A role represents a coherent part of an agent's functionality in the context of an organization. A situated commitment is an engagement of an agent to give preference to the actions of a particular role in the commitment. Agents typically commit relative to one another in a collaboration, but an agent can also commit to itself, e.g. when a vital task must be completed. Roles and commitments have a well-known *name* that is part of the domain ontology and that is shared among the agents in the system. Sharing these names enable agents to set up collaborations via message exchange. We explain the coordination among decision making and communication in the design rationale of this view packet.

Communication is responsible for communicative interactions with other agents. Message exchange enables agents to share information and to set up collaborations. The communication module processes incoming messages, and produces outgoing messages according to well-defined communication protocols [45]. A communication protocol specifies a set of possible sequences of messages. We use the notion of a *conversation* to refer to an ongoing communicative interaction. A conversation is initiated by the initial message of a communication protocol. At each stage in the conversation there is a limited set of possible messages that can be exchanged. Terminal states determine when the conversation comes to an end.

The information exchanged via a message is encoded according to a shared communication language. The communication language defines the format of the messages, i.e. the subsequent fields the message is composed of. A message includes a field with a unique identifier of the ongoing conversation to which the message belong, fields with the identity of the sender and the identities of the addressees of the message, a field with the performative of the message, and a field with the content of the message. Communicative interactions among agents are based on an *ontology* that defines a shared vocabulary of words that agents use in messages. The ontology enables agents to refer unambiguously to concepts and relationships between concepts in the domain when exchanging messages.

Current Knowledge repository contains data that is shared among the data accessors. Data stored in the current knowledge repository refers to state perceived in the environment, to state related to the agent's roles and situated commitments, and possibly other internal state that is shared among the data accessors. Fig. 3 shows the interconnections between the current knowledge repository and the internal components of the agent. These interconnections are called assembly connectors [3]. An assembly connector ties one component's provided interface with one or more components' required interfaces, and is drawn as a lollipop and socket symbols next to each other. Provided and required interfaces per assembly connector share the same name.

The current knowledge repository exposes two interfaces. The provided interface Update enables the perception component to update the agents knowledge according to the information derived from sensing the environment. The Read-Write interface enables the communication and decision making component to access and modify the agent's current knowledge.

Collaborations. The overall behavior of the agent is the result of the coordination of two components: decision making and communication. Decision making is responsible for selecting suitable influences to act in the environment. Communication is responsible for the communicative interactions with other agents. When selecting actions and communicating messages with other agents, decision making and communication typically request perceptions to update the agent's knowledge about the environment. By selecting an appropriate set of foci and filters, the agent directs its attention to the current aspects of its interest, and adapts it attention when the operating conditions change.

To complete the agent's tasks, decision making and communication coordinate via the current knowledge repository. For example, agents can send each other messages with requests for information that enable them to act more purposefully. Decision making and communication also coordinate during the progress of a collaboration. Collaborations are typically established via message exchange. Once a collaboration is achieved, the communication module activates a situated commitment. This commitment will affect the agent's decision making towards actions in the agent's role in the collaboration. This continues until the commitment is deactivated and the collaboration ends.

The separation of functionality for coordination (via communication) from the functionality to perform actions to complete tasks has several advantages, including clear design, improved modifiability and reusability. Two particular advantages are: (1) it allows both functions to act in parallel, and (2) it allows both functions to act at a different pace. In many applications, sending messages and executing actions happen at different tempo; a typical example is robotics. Separation of communication from performing actions enables agents to reconsider the coordination of their behavior while they perform actions, improving adaptability and efficiency.

B. Collaborating Components View Packet: Application Environment

Primary Presentation. The primary presentation is show in Fig. 4.

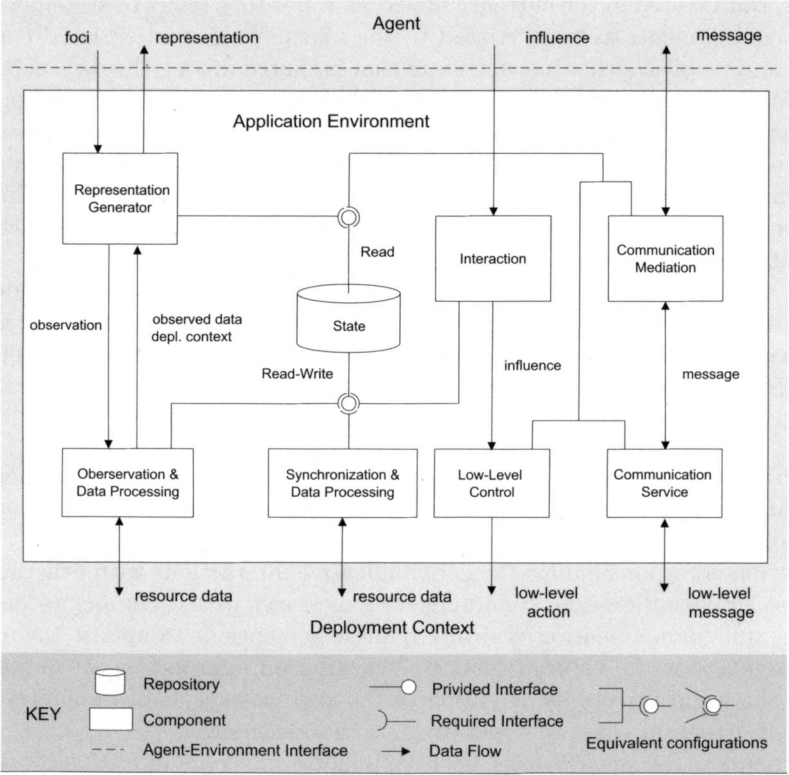

Fig. 4. Collaborating Components of Application Environment

Elements and their Properties The `Application Environment` component consists of seven subcomponents and the shared `State` repository. We discuss the responsibilities of each of the elements in turn. Then, we zoom on the collaboration between de components.

The **State** repository contains data that is shared between the components of the application environment. Data stored in the state repository typically includes an abstraction of the deployment context together with additional state related to the application environment. Examples of state related to the deployment context are a representation of the local topology of a network, and data derived from a set of sensors. Examples of additional state are the representation of digital pheromones that are deployed on top of a network, and virtual marks situated on the map of the physical environment. The state repository may also include agent-specific data, such as the agents' identities, the positions of the agents, and tags used for coordination purposes.

The **Representation Generator** provides the functionality to agents for perceiving the environment. When an agent senses the environment, the representation generator uses the current state of the application environment and possibly state collected from the deployment context to produce a representation for the agent. Agents' perception is subject to perception laws that provide a means to constrain perception [46]. For example, for reasons of efficiency a designer can introduce default limits for perception in order to restrain the amount of information that has to be processed, or to limit the occupied bandwidth.

Observation & Data Processing provides the functionality to observe the deployment context and collect date from other nodes in a distributed setting. The observation & data processing module translates observation requests into observation primitives that can be used to collect the requested data from the deployment context. Data may be collected from external resources in the deployment context or from the application environment instances on other nodes in a distributed application. The observation & data processing module can provide additional functions to pre-process data, examples are sorting and integration of observed data.

Interaction is responsible to deal with agents' influences in the environment. Agents' influences can be divided in two classes: influences that attempt to modify state of the application environment and influences that attempt to modify the state of resources of the deployment context. An example of the former is an agent that drops a digital pheromone in the environment. An example of the latter is an agent that writes data in an external data base. Agents' influences are subject to action laws [38]. Action laws put restrictions on the influences invoked by the agents, representing domain specific constraints on agents' actions. For example, when several agents aim to access an external resource simultaneously, an interaction law may impose a policy on the access of that resource. For influences that relate to the application environment, the interaction module calculates the reaction of the influences resulting in an update of the state of the application environment. Influences related to the deployment context are passed to the Low-Level Control module.

Low-Level Control bridges the gap between influences used by agents and the corresponding action primitives of the deployment context. Low-level control converts the influences invoked by the agents into low-level action primitives in the deployment context. This decouples the interaction module from the details of the deployment context.

The **Communication Mediation** mediates the communicative interactions among agents. It is responsible for collecting messages; it provides the necessary infrastructure to buffer messages, and it delivers messages to the appropriate agents. Communication mediation regulates the exchange of messages between agents according a set of applicable communication laws [45]. Communication laws impose constraints on the message stream or enforce domain–specific rules to the exchange of messages. Examples are a law that drops messages directed to agents outside the communication–range of the sender and a law that gives preferential treatment to high-priority messages. To actually transmit the messages, communication mediation makes use of the Communication Service module.

Communication Service provides that actual infrastructure to transmit messages. Communication service transfers message descriptions used by agents to communication primitives of the deployment context. For example, FIPA ACL message [16] enable a designer to express the communicative interactions between agents independently of the applied communication technology. However, to actually transmit such messages, they have to be translated into low-level primitives of a communication infrastructure provided by the deployment context. Depending on the specific application requirements, the communication service may provide specific communication services to enable the exchange of messages in a distributed setting, such as white and yellow page services. An example infrastructure for distributed communication is Jade [7]. Specific middleware may provide support for communicative interaction in mobile and ad-hoc network environments, an example is discussed in [30].

Synchronization & Data Processing synchronizes state of the application environment with state of resources in the deployment context as well as state of the application environment on different nodes. State updates may relate to dynamics in the deployment context and dynamics of state in the application environment that happens independently of agents or the deployment context. An example of the former is the topology of a dynamic network which changes are reflected in a network abstraction maintained in the state of the application environment. An example of the latter is the evaporation of digital pheromones. Middleware may provide support to collect data in a distributed setting. An example of middleware support for data collection in mobile and ad-hoc network environments is discussed in [29]. Synchronization & data processing converts the resource data observed from the deployment context into a format that can be used to update the state of the application environment. Such conversion typically includes a processing or integration of collected resource data.

Collaborations. Successively, we zoom in on the collaborating components for perception, interaction, communication, and the synchronization of state among nodes and with resources in the deployment context.

Perception. The representation generator collects perception requests from the agents and generates representations according to the given foci. Representation generator collects the required state from the state repository, and optionally it requests observation & data processing to collect additional data from the deployment context and possibly state of other nodes. State collection is subject to the perception laws. Observation & data processing returns the observed data to representation generator that generates a representation that is returned to the requesting agent.

Interaction. Interaction collects the concurrently invoked influences of agents and converts them into operations. The execution of operations is subject to the action laws of the system. Operations that attempt to modify state of the application environment are immediately executed by the interaction component. Operations that attempt to modify state of the deployment context are forwarded to low-level control that converts the operations into low–level interactions in the deployment context.

Communication. Communication mediation handles the communicative interactions among agents. The component collects the messages sent by agents, applies the communication laws, and subsequently passes the messages to the communication service. This latter component converts the messages directed to agents on other nodes into low–level interactions that are transmitted via the deployment context. Furthermore, communication service collects low–level messages from the deployment context, converts the messages into a format understandable for the agents, and forward the messages to communication mediation that delivers the messages to the appropriate agents. Messages directed to agents that are located at the same node are directly transferred to the appropriate agents.

State Synchronization. Synchronization & data processing performs its tasks independently of other components of the application environment. To synchronize the state of the application environment in a distributed setting, synchronization & data processing components on different nodes have to coordinate according to the requirements of the application at hand.

4 Excerpt of a Software Architecture for an AGV Transportation System

We now illustrate how we have used the reference architecture for the architectural design of an automated transportation system for warehouse logistics that has been developed in a joint R&D project between the DistriNet research group and Egemin, a manufacturer of automating logistics services in warehouses and manufactories [43,1]. The transportation system uses automatic guided vehicles (AGVs) to transport loads through a warehouse. Typical applications include distributing incoming goods to various branches, and distributing manufactured products to storage locations. AGVs are battery-powered vehicles that can navigate through a warehouse following predefined paths on the factory floor. The low-level control of the AGVs in terms of sensors and actuators such as staying on track on a path, turning, and determining the current position is handled by the AGV control software.

4.1 Multiagent System for the AGV Transportation System

In the project, we have applied a MAS approach for the development of the transportation system. The transportation system consists of two kinds of agents: transport agents and AGV agents. Transport agents represent tasks that need to be handled by an AGV and are located at a transport base, i.e. a stationary computer system. AGV agents are responsible for executing transports and are located in mobile vehicles. The communication infrastructure provides a wireless network that enables AGV agents at vehicles to communicate with each other and with transport agents on the transport base.

AGVs are situated in a physical environment, however this environment is very constrained: AGVs cannot manipulate the environment, except by picking and dropping loads. This restricts how AGV agents can exploit their environment. Therefore, a virtual environment was introduced for agents to inhabit. This virtual environment provides an interaction medium that agents can use to exchange information and coordinate their behavior. The virtual environment is necessarily distributed over the AGVs and the transport base, i.e. a local virtual environment is deployed on each AGV and the transport base. The local virtual environment corresponds to the application environment in the reference architecture. State on local virtual environments is merged opportunistically, as the need arises. The synchronization of the state of neighboring local virtual environments is supported by the ObjectPlaces middleware [29,30]. The AGV control system is developed on top of the .NET framework and programmed in C#.

As an illustration of the software architecture of the AGV transportation system, we zoom on the collaborating components view of the local virtual environment that is deployed on the AGVs.

4.2 Collaborating Components View of the Local Virtual Environment

Fig. 5 shows the collaborating components view of the local virtual environment.

The general structure of the local virtual environment is related to the structure of the application environment in the reference architecture as follows. The state repository corresponds to the state repository in the reference architecture, see Fig. 4 in section 3.1. The state elements are specific to the local virtual environment of an AGV control system. The perception manager provides the functionality for selective perception of the environment, similar to the representation generator in the reference architecture. Contrary to the representation generator, the perception manager interacts only with the state repository; the functionality of the observation & data processing component in the reference architecture is absent in the local virtual environment. The action manager corresponds to the interaction component of the application environment. Low-level control corresponds with E'nsor, i.e. the control software to interact with the sensors and actuators of the AGV. We fully reused E'nsor in the project. The communication manager integrates the responsibilities of communication mediation and the communication service of the application environment. The communication service handles the bidirectional translation of messages and manages message transmission via .Net remoting. Finally, the laws for perception, action, and communication, are integrated in the applicable components.

Elements and Their Properties

State. Since the virtual environment is necessarily distributed over the AGVs and the transport base, each local virtual environment is responsible to keep its

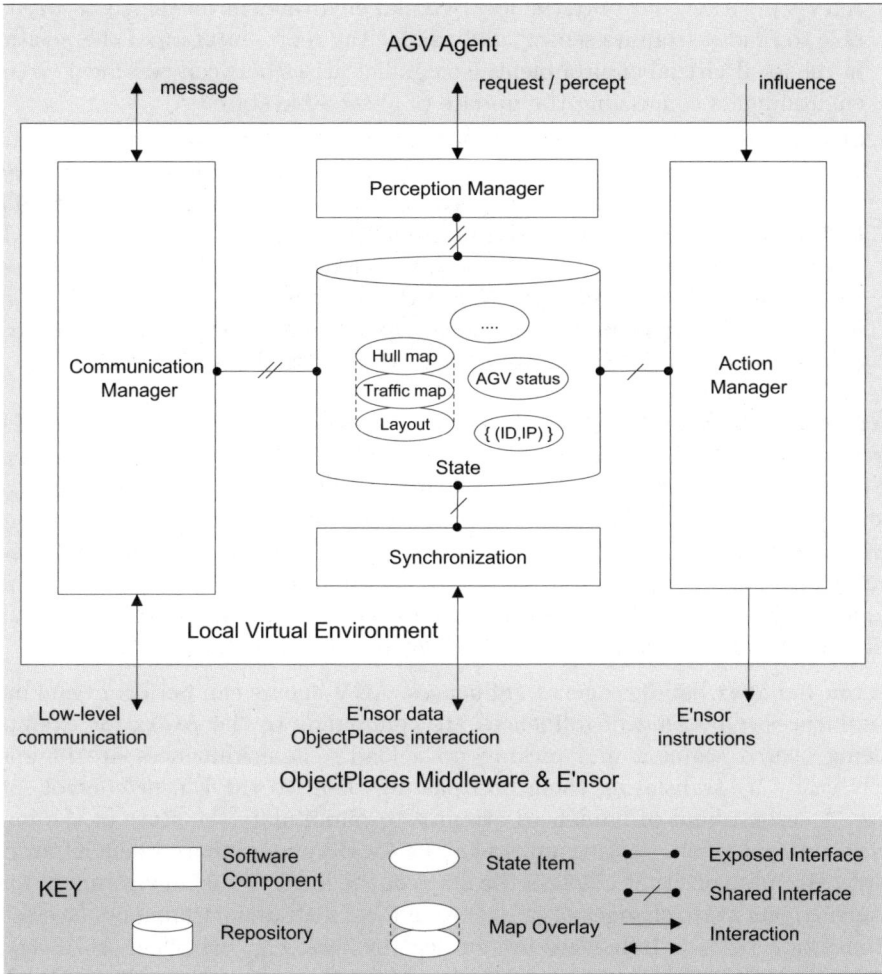

Fig. 5. Collaborating components view of the local virtual environment of AGVs

state synchronized with other local virtual environments. The state of the local virtual environment is divided into three categories:

1. Static state: this is state that does not change over time. Examples are the layout of the factory floor, which is needed for the AGV agent to navigate, and (AGV id, IP number) tuples used for communication. Static state must never be exchanged between local virtual environments since it is common knowledge and never changes.
2. Observable state: this is state that can be changed in one local virtual environment, while other local virtual environments can only observe the state. An AGV obtains this kind of state from its sensors directly. An example is an AGV's position. Local virtual environments are able to observe another

AGV's position, but only the local virtual environment on the AGV itself is able to read it from its sensor, and change the representation of the position in the local virtual environment. No conflict arises between two local virtual environments concerning the update of observable state.

3. Shared state: this is state that can be modified in two local virtual environments concurrently. An example is a hull map. AGV agents mark the path they are going to drive in their local virtual environment using hulls. The hull of an AGV is the physical area the AGV occupies. A series of hulls describe the physical area an AGV occupies along a certain path. AGV agents use hull for collision avoidance. When the local virtual environments on different machines synchronize, the local virtual environments must generate a consistent and up-to-date state in both local virtual environments.

Perception Manager handles perception in the local virtual environment. The perception manager's task is straightforward: when the agent requests a percept, for example the current positions of neighboring AGVs, the perception manager queries the necessary information from the state repository of the local virtual environment and returns the percept to the agent. Perception is subject to laws that restrict agents perception of the virtual environment. For example, when an agent senses the hulls for collision avoidance of neighboring AGVs, only the hulls within collision range are returned to the AGV agent.

Action Manager handles agents' influences. AGV agents can perform two kinds of influences. One kind of influences are commands to the AGV, for example moving over a segment and picking up a load. These influences are handled fairly easily by translating them and passing them to the E'nsor control software. A second kind of influences attempt to manipulate the state of the local virtual environment. Putting marks in the local virtual environment is an example. An influence that changes the state of the local virtual environment may in turn trigger state changes of neighboring local virtual environments (see Synchronization below). Influences are subject to laws, e.g., when an AGV agent projects a hull in the local virtual environment, this latter determines when an AGV acquires the right to move on. In particular, if the area is not marked by other hulls (the AGV's own hulls do not intersect with others), the AGV can move along and actually drive over the reserved path. In case of a conflict, the involved local virtual environments use the priorities of the transported loads and the vehicles to determine which AGV can move on. AGV agents monitor the local virtual environment and only instruct the AGV to move on when they are allowed. Afterwards, the AGV agents remove the markings in the environment. This example shows that the local virtual environment serves as a flexible coordination medium: agents coordinate by putting marks in the environment, and observing marks from other agents.

Communication Manager is responsible for exchanging messages between agents. Agents can communicate with other agents through the virtual environment. A typical example is an AGV agent that communicates with a transport agent to assign a transport. Another example is an AGV agent that requests the AGV agent of a waiting AGV to move out of the way. The communication manager translates the high-level messages to low-level communication instructions that can be sent through the network and vise versa (resolving agent names to IP numbers, etc.). Communication is subject to laws, an example is the restriction of communication range for messages used for transport assignment [35].

Synchronization has a dual responsibility. It periodically polls E'nsor and updates the state of the local virtual environment accordingly. An example is the maintenance of the actual position of the AGV in the local virtual environment. Furthermore, synchronization is responsible for synchronizing state between local virtual environments of neighboring machines. An example is the synchronization of hulls on neighboring AGVs.

Design Rationale

Changes in the system (e.g., AGVs that enter/leave the system) are reflected in the state of the local virtual environment, releasing agents from the burden of such dynamics. As such, the local virtual environment—supported by the ObjectPlaces middleware—supports openness.

Since an AGV agent continuously needs up-to-date data about the system (position of the vehicles, status of the battery, etc.), we decided to keep the representation of the relevant state of the deployment context in the local virtual environment synchronized with the actual state. Therefore, E'nsor and the ObjectPlaces middleware are periodically polled to update the status of the system. As such, the state repository maintains an accurate representation of the state of the system to the AGV agent.

5 Related Work

Current practice in agent-oriented software engineering considers MAS as a radically new way of engineering software. For example, in [10], Wooldridge et al. state "There is a fundamental mismatch between the concepts used by object-oriented developers and other mainstream software engineering paradigms, and the agent-oriented view. [...] Existing software development techniques are unsuitable to realize the potential of agents as a software engineering paradigm." As a result, numerous MAS methodologies have been developed [19]. Although some of the methodologies adopt techniques and practices from mainstream software engineering, such as object-oriented techniques and the Unified Modeling Language, nearly all methodologies take an independent position, little or not related to mainstream software engineering practice. The position of being a radically new paradigm for software development isolates agent-oriented software engineering from mainstream software engineering. In contrast, the architecture-centric perspective on MAS we follow in our research aims to integrate MAS in mainstream software engineering.

Related work that explicitly connects MAS with software architecture is rather limited. We briefly discuss a number of representative examples. In [32], Shehory presents an initial study on the role of MAS as a software architecture style. We share the author's observation that the largest part of research in the design of MAS addresses the question: given a computational problem, can one build a MAS to solve it? However, a more fundamental question is left unanswered: given a computational problem, is a MAS an appropriate solution? An answer to this question should precede the previous one, lest MAS may be developed where much simpler, more efficient solutions apply. Almost a decade later, the majority of researchers in agent-oriented software engineering still pass over the analysis whether a MAS is an appropriate solution for a given problem.

As part of the Tropos methodology [18], a set of architectural styles were proposed which adopt concepts from organization management theory [24,12]. The styles are modelled using the $i^\star$ framework [48] which offers modelling concepts such as actor, goal, and actor dependency. Styles are evaluated with respect to various software quality attributes. The specification of quality attributes is based on te notion of softgoal. [24] states that softgoals do not have a formal definition, and are amenable to a more qualitative kind of analysis. Whereas we use a utility tree to prioritize quality requirements and to determine the drivers for architectural design, Tropos does not consider a systematic prioritization of quality goals. In Tropos, a designer visualizes the design process and simultaneously attempts to satisfy the collection of softgoals for a system.

PROSA is an acronym for Product–Resource–Order–Staff Architecture and defines a reference architecture for a family of coordination and control application, with manufacturing systems as the main domain [47]. These systems are characterized by frequent changes and disturbances. PROSA aims to provide the required flexibility to cope with these dynamics. [20] presents an interesting extension of PROSA in which the environment is exploited to obtain BDI (Believe, Desire, Intention [28]) functionality for the various PROSA agents. The PROSA reference architecture embodies architectural knowledge of a particular *problem* domain. On the contrary, the reference architecture for situated MAS embodies architectural knowledge in terms of a particular *solution* approach.

In [17], Garcia et al. observe that several concerns such as autonomy, learning, and mobility crosscut each other and the basic functionality of agents. The authors state that existing approaches that apply well-known patterns to structure agent architectures—an example is the layered architecture of Kendall [22]—fail to cleanly separate the various concerns. This results in architectures that are difficult to understand, reuse, and maintain. To cope with the problem of crosscutting concerns, the authors propose an aspect-oriented approach to structure agent architectures. An aspect-oriented agent architecture consists of a "kernel" that encapsulates the core functionality of the agent (essentially the agent's internal state), and a set of aspects [23]. Each aspect modularizes a particular concern of the agent. Yet, it is unclear whether the interaction of the different concerns in the kernel (feature interaction [11]) will not lead to similar problems the approach initially aimed to resolve. Anyway, crosscutting concerns in MAS are hardly explored and provide an interesting venue for future research.

6 Conclusions

There is a close connection between MAS and software architecture, yet, this connection is often neglected or remains implicit. In our research, we have derived a reference architecture for situated MAS from various applications we have studied and built. This reference architecture provides a blueprint to develop new software architectures for systems that have similar characteristics and requirements as the systems from which it was derived.

The reference architecture shows how knowledge and experiences with MAS can systematically be documented and matured in a form that has proven its value in mainstream software engineering. Rather than considering MAS as a radical new way of engineering software, we believe that the integration of MAS in mainstream software engineering is a key to industrial adoption of MAS.

References

1. EMC[2]: Egemin Modular Controls Concept, Project Supported by the Institute for the Promotion of Innovation Through Science and Technology in Flanders (IWTVlaanderen), (8/2006), http://emc2.egemin.com/
2. Software Engineering Institute: Carnegie Mellon University, (8/2006), http://www.sei.cmu.edu/
3. The Unified Modeling Language: (8/2006), http://www.uml.org/
4. Al-Naeem, T., Gorton, I., Babar, M.A., Rabhi, F., Benatallah, B.: A Quality-Driven Systematic Approach for Architecting Distributed Software Applications. In: 27th International Conference on Software Engineering, Orlando, Florida (2005)
5. Barbacci, M., Klein, M., Longstaff, T., Weinstock, C.: uality Attribute Workshops. Technical Report CMU/SEI-95-TR-21, Software Engineering Institute, Carnegie Mellon University, PA, USA (1995)
6. Bass, L., Clements, P., Kazman, R.: Software Architecture in Practice. Addison Wesley, London, UK (2003)
7. Bellifemine, F., Poggi, A., Rimassa, G.: Jade, A FIPA-compliant Agent Framework. In: 4th International Conference on Practical Application of Intelligent Agents and Multi-Agent Technology, London, UK (1999)
8. Boucké, N., Holvoet, T., Lefever, T., Sempels, R., Schelfthout, K., Weyns, D., Wielemans, J.: Applying the Architecture Tradeoff Analysis Method to an Industrial Multiagent System Application. In: Technical Report CW 431. Department of Computer Science, Katholieke Universiteit Leuven, Belgium (2005)
9. Boucké, N., Weyns, D., Schelfthout, K., Holvoet, T.: Applying the ATAM to an Architecture for Decentralized Contol of a AGV Transportation System. In: Hofmeister, C., Crnkovic, I., Reussner, R. (eds.) QoSA 2006. LNCS, vol. 4214, pp. 180–198. Springer, Heidelberg (2006)
10. Buchmann, F., Bass, L.: Introduction to the Attribute Driven Design Method. In: 23rd International Conference on Software Engineering, Toronto, Ontario, Canada. IEEE Computer Society Press, Los Alamitos (2001)
11. Calder, M., Kolberg, M., Magill, E., Reiff-Marganiec, S.: Feature Interaction: A Critical Review and Considered Forecast. Comp. Netw. 41(1), 115–141 (2003)
12. Castro, J., Kolp, M., Mylopoulos, J.: Towards Requirements-Driven Information Systems Engineering: The Tropos Project. Inf. Syst. 27(6), 365–389 (2002)

13. Clements, P., Bachmann, F., Bass, L., Garlan, D., Ivers, J., Little, R., Nord, R., Stafford, J.: Documenting Software Architectures: Views and Beyond. Addison-Wesley, London, UK (2002)
14. Clements, P., Kazman, R., Klein, M.: Evaluating Software Architectures: Methods and Case Studies. Addison Wesley, London, UK (2002)
15. Ferber, J., Muller, J.: Influences and Reaction: a Model of Situated Multiagent Systems. In: 2nd International Conference on Multi-agent Systems, Japan. AAAI Press, Stanford, California, USA (1996)
16. FIPA: Foundation for Intelligent Physical Agents, FIPA Abstract Architecture Specification (8/2006), http://www.fipa.org/repository/bysubject.html
17. Garcia, A., Kulesza, U., Lucena, C.: Aspectizing Multi-Agent Systems: From Architecture to Implementation. In: Choren, R., Garcia, A., Lucena, C., Romanovsky, A. (eds.) Software Engineering for Multi-Agent Systems III. LNCS, vol. 3390. Springer, Heidelberg (2005)
18. Giunchiglia, F., Mylopoulos, J., Perini, A.: The TROPOS Software Development Methodology: Processes, Models and Diagrams. In: 1st International Joint Conference on Autonomous Agents and Multi-Agent Systems. ACM Press, New York (2002)
19. Henderson-Sellers, B., Giorgini, P.: Agent-Oriented Methodologies. Idea Group Publishing, USA (2005)
20. Holvoet, T., Valckenaers, P.: Exploiting the Environment for Coordinating Agent Intentions. In: E4MAS, Hakodate, Japan. LNCS, vol. 4389, pp. 51–66. Springer, Heidelberg (2006)
21. Jazayeri, M., Ran, A., van der Linden, F.: Software Architecture for Product Families. Addison Wesley Longman Inc. Redwood City,CA, USA (2000)
22. Kendall, E., Jiang, C.: Multiagent System Design Based on Object Oriented Patterns. Journal of Object Oriented Programming 10(3), 41–47 (1997)
23. Kiczales, G., Lamping, J., Menhdhekar, A., Maeda, C., Lopes, C., Loingtier, J., Irwin, J.: Aspect-Oriented Programming. In: Aksit, M., Matsuoka, S. (eds.) ECOOP 1997. LNCS, vol. 1241. Springer, Heidelberg (1997)
24. Kolp, M., Giorgini, P., Mylopoulos, J.: A Goal-Based Organizational Perspective on Multi-agent Architectures. In: Meyer, J.-J.C., Tambe, M. (eds.) ATAL 2001. LNCS (LNAI), vol. 2333, pp. 128–140. Springer, Heidelberg (2002)
25. Larman, C.: Applying UML and Patterns: An Introduction to Object-Oriented Analysis and Design. Prentice-Hall, Englewood Cliffs (2002)
26. McConell, S.: Rapid Development: Taming Wild Software Schedules. Microsoft Press, Redmond, Washington (1996)
27. Olumofin, F., Misic, V.: Extending the ATAM Architecture Evaluation to Product Line Architectures. In: 5th IEEE-IFIP Conference on Software Architecture, Pittsburgh, Pennsylvania, USA (2005)
28. Rao, A., Georgeff, M.: BDI Agents: From Theory to Practice. In: 1st International Conference on Multiagent Systems, San Francisco, California, USA. The MIT Press, Cambridge (1995)
29. Schelfthout, K., Holvoet, T.: Views: Customizable abstractions for context-aware applications in MANETs. In: Software Engineering for Large-Scale Multi-Agent Systems, St. Louis, USA (2005)
30. Schelfthout, K., Weyns, D., Holvoet, T.: Middleware that Enables Protocol-Based Coordination Applied in Automatic Guided Vehicle Control. IEEE Distributed Systems Online 7(8) (2006)
31. Shaw, M., Garlan, D.: Software Architecture: Perspectives on an Emerging Discipline. Prentice-Hall, Englewood Cliffs (1996)

32. O. Shehory. Architectural Properties of MultiAgent Systems. Technical Report CMU-RI-TR-98-28, Robotics Institute, Carnegie Mellon University, Pittsburgh, PA, 1998.
33. Steegmans, E., Weyns, D., Holvoet, T., Berbers, Y.: A Design Process for Adaptive Behavior of Situated Agents. In: Odell, J.J., Giorgini, P., Müller, J.P. (eds.) AOSE 2004. LNCS, vol. 3382. Springer, Heidelberg (2005)
34. Weyns, D.: An Architecture-Centric Approach for Software Engineering with Situated Multiagent Systems. Ph.D, Katholieke Universiteit Leuven (2006)
35. Weyns, D., Boucké, N., Holvoet, T.: Gradient Field Based Transport Assignment in AGV Systems. In: 5th International Joint Conference on Autonomous Agents and Multi-Agent Systems, AAMAS, Hakodate, Japan (2006)
36. Weyns, D., Helleboogh, A., Holvoet, T.: The Packet-World: a Test Bed for Investigating Situated Multi-Agent Systems. In: Agent-based applications, platforms, and development kits. Whitestein Series in Software Agent Technology (2005)
37. Weyns, D., Holvoet, T.: Model for Simultaneous Actions in Situated Multiagent Systems. In: Schillo, M., Klusch, M., Müller, J., Tianfield, H. (eds.) Multiagent System Technologies. LNCS (LNAI), vol. 2831. Springer, Heidelberg (2003)
38. Weyns, D., Holvoet, T.: Formal Model for Situated Multi-Agent Systems. Fundamenta Informaticae 63(1-2), 125–158 (2004)
39. Weyns, D., Holvoet, T.: Regional Synchronization for Situated Multi-agent Systems. In: Mařík, V., Müller, J.P., Pěchouček, M. (eds.) CEEMAS 2003. LNCS (LNAI), vol. 2691. Springer, Heidelberg (2003)
40. Weyns, D., Holvoet, T.: Architectural Design of an Industrial AGV Transportation System with a Multiagent System Approach. In: Software Architecture Technology User Network Workshop, SATURN, Pittsburg, USA, Software Engineering Institute, Carnegie Mellon University (2006)
41. Weyns, D., Holvoet, T.: From Reactive Robotics to Situated Multiagent Systems: A Historical Perspective on the Role of Environment in Multiagent Systems. In: Dikenelli, O., Gleizes, M.-P., Ricci, A. (eds.) ESAW 2005. LNCS (LNAI), vol. 3963. Springer, Heidelberg (2006)
42. Weyns, D., Omicini, A., Odell, J.: Environment as a First-Class Abstraction in Multiagent Systems. Autonomous Agents and Multi-Agent Systems 14(1) (2007)
43. Weyns, D., Schelfthout, K., Holvoet, T., Lefever, T.: Decentralized control of E'GV transportation systems. In: 4th Joint Conference on Autonomous Agents and Multiagent Systems, Industry Track, Utrecht, The Netherlands. ACM Press, New York (2005)
44. Weyns, D., Steegmans, E., Holvoet, T.: Integrating Free-Flow Architectures with Role Models Based on Statecharts. In: Choren, R., Garcia, A., Lucena, C., Romanovsky, A. (eds.) Software Engineering for Multi-Agent Systems III. LNCS, vol. 3390. Springer, Heidelberg (2005)
45. Weyns, D., Steegmans, E., Holvoet, T.: Protocol Based Communication for Situated Multi-Agent Systems. In: 3th Joint Conference on Autonomous Agents and Multi-Agent Systems, New York, USA. IEEE Computer Society Press, Los Alamitos (2004)
46. Weyns, D., Steegmans, E., Holvoet, T.: Towards Active Perception in Situated Multi-Agent Systems. Applied Artificial Intelligence 18(9-10), 867–883 (2004)
47. Wyns, J., Van Brussel, H., Valckenaers, P., Bongaerts, L.: Workstation Architecture in Holonic Manufacturing Systems. In: 28th CIRP International Seminar on Manufacturing Systems, Johannesburg, South Africa (1996)
48. Yu, E.: Modelling Strategic Relationships for Process Reengineering, PhD Dissertation: University of Toronto, Canada (1995)

Organization Oriented Programming: From Closed to Open Organizations

Olivier Boissier[1,*], Jomi Fred Hübner[2,**], and Jaime Simão Sichman[3,***]

[1] SMA/G2I/ENSM.SE, 158 Cours Fauriel
42023 Saint-Etienne Cedex, France
Olivier.Boissier@emse.fr
[2] GIA/DSC/FURB, Braz Wanka, 238
89035-160, Blumenau, Brazil
jomi@inf.furb.br
[3] LTI/EP/USP, Av. Prof. Luciano Gualberto, 158, trav. 3
05508-900 São Paulo, SP, Brazil
jaime.sichman@poli.usp.br

Abstract. In the last years, social and organizational aspects of agency have become a major issue in multi-agent systems' research. Recent applications of MAS enforce the need of using such aspects in order to ensure some social order within these systems. However, there is still a lack of comprehensive views of the diverse concepts, models and approaches related to agents' organizations. Moreover, most designers have doubts about how to put these concepts in practice, i.e., how to program them. In this paper we focus on and discuss about the literature on formal, top-down and pre-existent organizations by stressing the different aspects that may be considered to program them. Finally, we present some challenges for future research considering particularly the openness feature of those agents' organizations.

Keywords: Multi-agent Systems, MAS organizations, Open systems.

1 Introduction

Nowadays, current IT applications show the large scale interweaving of human and technological communities (e.g. Web Intelligence, Ambient Intelligence). The use of Multi-Agent System (MAS) technology introduces software entities that act on behalf of users and cooperate with those info-inhabitants. The complex systems' engineering needed to build such applications highlights and stresses requirements on *openness* in terms of the ability that takes into account several kinds of changes and the adaptation of the system configuration while it keeps running (29).

---

[*] Partially supported by USP-COFECUB, grant UC-98/04.
[**] Partially supported by FURB, Brazil and CNPq, Brazil, grant 200695/01-0.
[***] Partially supported by USP-COFECUB, grant UC-98/04 and by CNPq, Brazil, grants 304605/2004-2, 482019/2004-2 and 506881/2004-0.

In this paper, we are interested in the social and organizational aspects of agency. They have been a major topic of study in the multiagent domain since the seminal work of (14; 6). Moreover, they have become recently one of the major focus of interest in the MAS community (e.g. (28; 4)). Most designers, however, have doubts about how to put these concepts in practice, i.e., how to program them, while addressing the openness issue. Indeed, considering openness at the organization level introduces new challenges for dealing with the management of dynamic entry/exit of agents into/from organizations, online self-adaptation, control and regulation of the agents autonomy, etc.

Before going further in the presentation of these challenges, we will first sketch a comprehensive view on organizations in MAS (section 2). We refine this view by focusing on an organization oriented approach, i.e. formal, top-down, pre-existent organizations. We analyze some of the existants organizational models for programming organized multiagent systems (section 3) and the underlying programming approaches that are used (section 4). From this comprehensive understanding of organizations in MAS, we present (section 5) some research directions and challenges that need to be addressed to shift from closed to open agents' organizations.

2 Comprehensive View of Organizations in MAS

It is still missing a clear and unique definition of what is called "organization" in MAS. Its meaning (24) often varies between two basic views: (i) a collective entity with an identity that is represented by (but not identical to) a group of agents exhibiting relatively highly formalized social structures (32), (ii) a stable pattern/structure of joint activity that may constrain or affect the actions and interactions of agents towards some purpose (5). As we can see, organization refers, in a general sense, to a *cooperation pattern* that can be more or less formalized. As in Sociology (3), it may concern the expression of a division of tasks, a distribution of roles, an authority system, a communication system, or also a contribution-retribution system. According to (15), this range of topics may also be extended to knowledge, culture, memory or history.

Both views are generally not mutually exclusive and have led to different approaches in the domain. Let's focus on a few features in order to build a comprehensive view of them. First, we will take into account the "definition process" of the agents' organization (Sections 2.1) and then consider its "representation" within the agents´ minds (Section 2.2). As what happens with every classification attempt, the one proposed here has its limits and must be considered as an analysis grid of the different works and not as a normative view on agents' organizations in MAS.

2.1 Agent Centered View vs Organization Centered View

The first axis of the grid is an extension of the *agent centered* and *organization centered* points of view initially proposed in (25).

The *agent-centered* point of view takes the agents as the "engine" for the organization. Organizations only exist as observable emergent phenomena which state a unified bottom-up and objective global view of the pattern of cooperation between agents (see first row in Fig. 1-a-b). For instance (case (a)), in an ant colony (8), no organizational behavior constraints are explicitly and directly defined inside the ants. The organization is the result of the collective emergent behavior due to how agents act their individual behaviors and interact in a common shared and dynamic environment. Similar point of view may be considered in the different reactive self-organization approaches that exist in the litterature (30). In a more cognitive way (case (c)), the studies on coalition formation define mechanisms (within agents, e.g. social reasoning (33)), to build pattern of cooperation in a bottom-up process. In this view, the pattern of cooperation both structures and helps the agents in their collaborative activities.

The *organization centered* point of view sees the opposite direction: the organization exists as an explicit entity of the system (see second row in Fig. 1-c-d). It stresses the importance of a supra-individual dimension (15) and the use of primitives that are different from the agents' ones. The pattern of cooperation is settled by designers (or by agents themselves in self-organized systems) and is installed in a top-down manner in order to constrain or define the agent's behaviors. Let's note that, as in the first case, the observer of the system can obtain a description of the organization. For instance, in a school we have documents that state how it is organized. Of course, besides the explicit description of the organization, the beholder can also observe the real school's organization which is, possibly, different from the formal one.

2.2 Agents Know vs Agents Don't Know the Organization

¿From an agent architecture perspective, we can further refine these two points of view by considering an orthogonal axis regarding the agents' capabilities to represent and reason about its organization.

In the first column of Fig. 1, the agents *don't know anything about the organization*. In case (a) the agents don't represent the organization, although the *observer* can see an emergent organization. In some sense, they are not aware that they are part of an organization. In case (c), the organization exists as a specified and formalized schema, made by a designer but agents don't know anything about it and do not reason about it. They simply obey it as if the organizational constraints were hard coded inside them (e.g. the MAS resulting from some AOSE methodologies where the agent's code is generated from an organizational specification (23; 2)).

In the second column, we consider the cases where agents have some explicit representation of the organization in which they are executing. In case (b), each agent has an internal and local representation of cooperation patterns which it follows when deciding what to do (e.g. social networks for coalition formations (33)). This local representation is obtained either by perception, communication or explicit reasoning (e.g. social reasoning as in (33)) since, in an agent-centered view, there isn't, *a-priori*, any explicit global representation

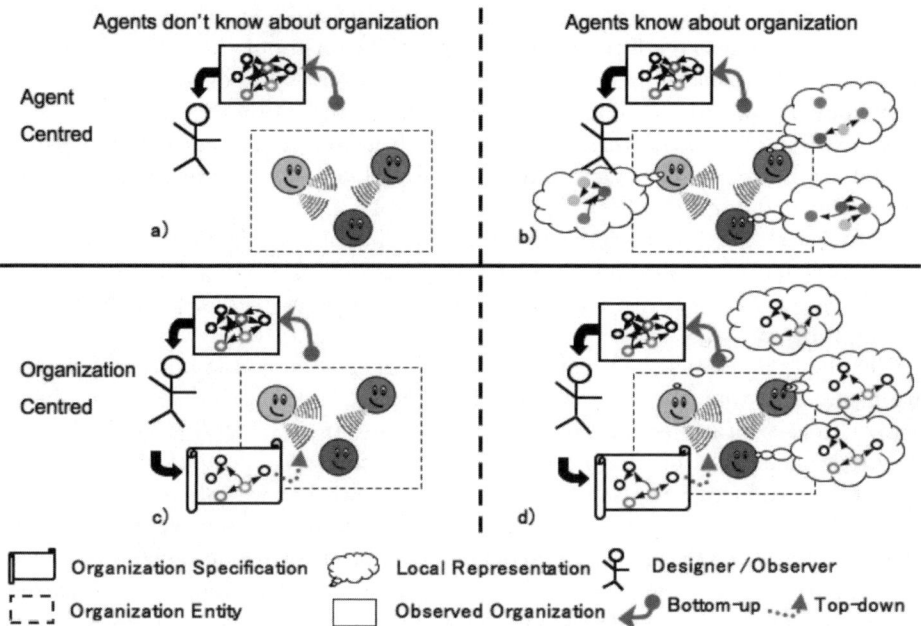

Fig. 1. Comprehensive view on organizations in MAS: (a) Emergence-based Designed MAS; (b) Coalition Oriented MAS; (c) Organization-based Designed MAS; (d) Organization Oriented MAS. Let's notice that the Designer/Observer may be the Developper/User (exogenous case) or a set of agents (endogenous case).

of the organization which is available to the agents. In case (d), agents have an explicit representation of the organization which has been defined (organization-centered view). The agents are able to reason about it and to use it in order to initiate cooperation with other agents in the system.

In the literature, some agents' organization approaches fit to a specific case shown in Fig. 1, others are lying on multiple cases. For instance, proposals concerning reorganization approaches for formal organizations may combine cases (b) and (d) in the sense that agents are using their internal mechanisms to adapt the organization that was imposed to the system. The bottom-up or top-down manipulation of the organization may be realized either *endogenously* (i.e. realized by the agents belonging to the organization themselves) or *exogenously* (i.e. by an external designer, a human or agents outside of the organization).

2.3 Organization Oriented Programming

In order to tackle the requirements for organizations as exposed in the introduction, we focus on the Organization Oriented MAS (Fig. 1-d). Indeed, such an approach provides us the possibility of expressing and making explicit one or more patterns of cooperation. This is made using a top-down approach to constrain and guide the agents behavior towards some purpose. Let's note that in this kind of systems, agents can practise organizational autonomy, in the sense that they are able to read, to represent, and to reason about the organization and may decide whether to follow the constraints stated by the organization or not. They may also decide to adapt and change the organization in a bottom-up process, installing a new pattern/structure. Such a functioning corresponds to the combination of Coalition Oriented MAS (Fig. 1-b) and Organization Oriented MAS (Fig. 1-d)) approaches.

In this context, if we consider the programming of organizations –called *Organization Oriented Programming (OOP)* in the sequel–, we consider the existence of an *organization modeling language* (OML) that is used to *specify* the organization(s) of an MAS.

This language is used to collect and express specific constraints and cooperation patterns that the designer (or the agents) have in mind, resulting in an explicit representation that we call an *Organization Specification* (OS). Finally the OS is executed and interpreted in an *Organization Implementation Architecture* (OIA), to install a collective entity in the MAS that we call an *Organization Entity* (OE): a set of agents that build the organization specified with an OS. Once created, the OE's history starts and runs, event by event. These events can be related to agents entering and/or leaving the organization, group creation, role adoption, goal commitment, etc. The OIA may be further divided (as we will see in section 4) into an agent part and an organization infrastructure part. In the following sections, we consider first OMLs for OOP and then OIAs that execute these OMLs.

3 Organizational Modeling Languages for MAS

Existing OML in the MAS domain may be compared along the way they constrain the agents' possible behaviors (B) shown in a behavior space (Fig. 2). At time t, a MAS has the purpose of maintaining its behavior in the set P, where P represents all behaviors that draw the MAS's global purposes at that time. The set E represents all the agents' possible behaviors given the current state of the environment at t. The organization (i.e. an entity composed of roles, groups, and links) constrains the agents' behaviors to those of the set O.

Depending on its definition, the set of possible behaviors according to the environment and to the organization ($E \cap O$), can become closer to P or not. Only agents are responsible for conducting their behaviors from a point in $((E \cap O) - P)$ to a point in P. All the difficulties of the definition of an OS is in the definition of the set O:

- A small $P \cap E \cap O$ certainly increases performance of the system by constraining agents' autonomy but may prevent the system self-adaptation;
- A great $P \cap E \cap O$ certainly increases the capability of the system adaptation by keeping agents' autonomy but may prevent the system high performance.

The first situation would led to a kind of agents' *rigor mortis*, i.e., they behave without any autonomy, like robots, following fully-specified policies. On the other

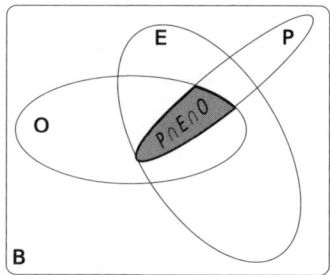

B: agents' possible behaviors.
P: agents' behaviors that lead to global purpose.
E: agents' possible behaviors constrained by the environment.
O: agents' possible/permitted/obliged behaviors constrained by the organization.

Fig. 2. Behavior space of MAS

hand, if the second situation is extreme, it makes the organization constraints useless, since a small $P \cap E \cap O$ tends to a small $P \cap E$. Let's illustrate now, how the set O may be defined using the OML $\mathcal{M}$OISE$^+$.

3.1 $\mathcal{M}$oise$^+$ Organization Modeling Language

$\mathcal{M}$OISE$^+$ (Model of Organization for multI-agent SystEms) (20) is an OML that explicitly decomposes the set O of Fig. 2 into *structural* (O_S) and *functional* (O_F) dimensions. The O_S defines the MAS structure with the notions of *roles*, *groups* and *links*. The O_F describes how the *global collective goals* should be achieved, i.e., how these goals are decomposed (in *plans*), grouped in coherent sets (by *missions*) to be distributed to the agents. A third *deontic* dimension is added in order to binds the structural dimension with the functional one by the specification of the roles' *permissions* and *obligations* for missions. Instead of being related to the agents' behavior space, the deontic dimension is related to the agents' autonomy. In order to give some concrete insights of this OML we depict part of the resulting OS for a simulated soccer game (see Fig. 3). A formal definition of $\mathcal{M}$OISE$^+$ is found in (20).

In the *structural dimension*, a *role* consists of a label, attached to agents, that constrains their behavior and their links.

A *group* is an instantiation in an OE of a *group specification* of the OS. A *group specification* consists of a set of links and roles. Well formed attributes may be ascribed to it. Their concerns are intra/extra group compatibility of roles, minimum and maximum number of role players inside a group. The *links* have direct effect on the agents behavior. They can be: *acquaintance* links (agents playing the source role are allowed to have a representation of the agents playing the destination role), *communication* links (agents are allowed to communicate with the target agents), *authority* links (source agents are allowed to control target agents). Fig. 3 shows the structural dimension of the soccer team's OS. It is formed by roles of the players (e.g. goalkeeper, leader, etc) that are distributed in two group specifications (defense and attack) which form the team group specification. In the defense group specification, three roles are allowed and any defense group will be well formed if there is one, and only one, agent playing the role goalkeeper, exactly three agents playing backs, and, optionally, one agent

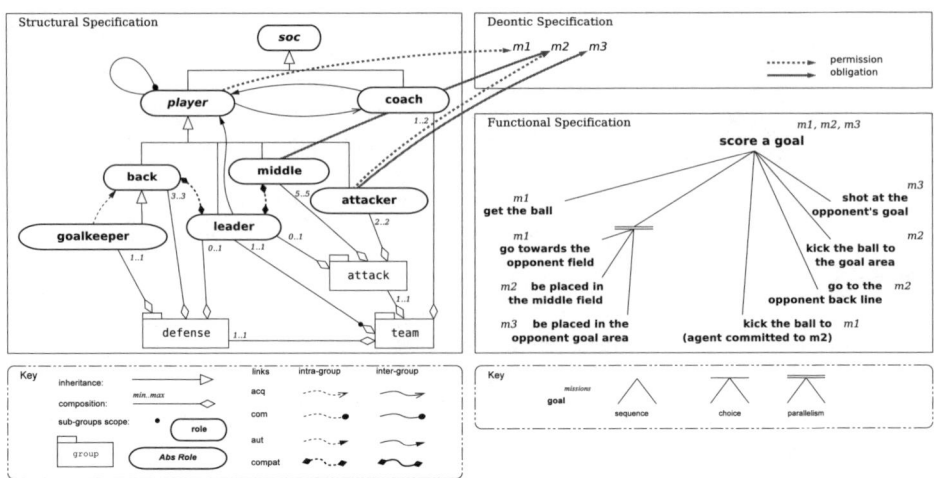

Fig. 3. Organization Specification of a Soccer Game with the $\mathcal{M}\text{OISE}^+$ Organization Modeling language

playing the leader role (see the composition relation in Fig. 3). The goalkeeper has authority on the backs. The leader player is also allowed to be a back since these roles are compatible. Due to the role specialization (see the back-goalkeeper inheritance relation in Fig. 3), the leader also can play the goalkeeper role. In the same example, a team is well formed if it has one defense sub-group, one attack sub-group, one or two agents playing the coach role, one agent playing the leader role, and the two sub-groups are also well formed. In this structure, the coach has authority over all players by an authority link. The players, in any group, can communicate with each other and are allowed to represent the coach (since they have an acquaintance link). There must be a leader either in the defense or attack group. The leader has authority over all players on all groups, since there is an authority link to the player role. For every authority link there is an implicit communication link and for every communication link there is an implicit acquaintance link.

In the *functional dimension*, the goals that are to be achieved by the organization are structured according to different *social schemes*. A social scheme is a goal decomposition tree where the root is the Scheme's goal. The operators that may appear in these plans express the execution in sequence/parallel and the possibility of choice. All the goals of a social scheme (root goal and subgoals) are divideded into *missions* (i.e. sets of coherent goals that are to be assigned to roles and that an agent can commit to). More precisely, if an agent accepts a mission m_i, it commits to all goals of m_i ($g_j \in m_i$) and the agent will try to achieve goal g_j only when the goal preconditions for g_j are satisfied. In the soccer example, suppose the team has a rehearsed play as the one specified in Fig. 3. This scheme has three missions (m_1, m_2, and m_3). When an agent commits to a mission, it is responsible for all this mission's goals. For example, an agent committed to the mission m_3 has the goals "be placed in the opponent goal area", "shot at the opponent's goal", and, a common goal, "score a goal". Each mission also has a cardinality constraint in the scheme that states how many agents should commit to it. In the soccer example, all three missions must be committed by exactly one agent. Further specifications are possible on this dimension to precisely constrain the execution of the social schema, but are outside of the scope of this paper (please refer to (20) for more details).

As shown in Fig. 4-e, both specifications (structural and functional) are two parts of the OS that constrains the agents' behaviors. They can be developed independently. To make explicit the semantic of the binding between them, a *deontic dimension* has been introduced. It explicitly states the *obligations* and *permissions* for agents to execute missions while playing roles (a temporal dimension is also specified within these operators, see (20)). For instance, in Fig. 3, permission $per(player, m_1)$ states that an agent playing the role *player* is allowed to commit to the mission m_1. Furthermore, the obligation $obl(middle, m_2)$ states that an agent playing the role *middle* ought to commit to m_2.

3.2 Organization Modeling Dimensions

Given this description of the modeling dimensions of the $\mathcal{M}$OISE$^+$ OML, let's consider other existing OMLs in the literature. We choose some of them to highlight the different organization modeling dimensions that they embed and to compare them with respect to the behavior space of Fig. 2.

TÆMS (26) addresses only the functional specification of the agents' organization. It proposes a domain independent language for defining hierarchical task structures (see O_F in Fig. 4-a). The task structures express the way to achieve different top-level goals, i.e. objective/abstract task including the deadline for its completion and the earliest time it can begin to be executed. Leaves are basic action instantiations (methods). Different constraints enrich those hierarchies: *quality-accumulation-functions*, logical constraints among tasks and their subtasks, expressing a precise, quantitative degree of achievement of tasks ; *task relationships* expressing inter-dependencies among tasks (enables/disables, hinders/facilitates) and basic actions or abstract task achievement affect task characteristics (e.g., quality and time) elsewhere in the task structure, *resource consumption* characteristics of tasks (e.g. consumes, limits).

AGR (Agent, Group, Role) (11) is the evolution of the AALAADIN model (13). Its minimalist structure-based model of organization is specified as a *role-group structure* imposed on the agents (see O_s in Fig. 4-b). A *group* is a set of agents sharing some common characteristics: it is the context for a pattern of activities and is used for partitioning organizations. A *role* is an abstract representation of an agent functional position in a particular group. An agent may request and play (multiple) roles within (several) groups. No constraints are placed upon the architecture of an agent or about its mental capabilities. The only constraint which is imposed by the role-group structure is concerned to communication: agents can communicate with each other if and only if they belong to the same group. AGR also says that agents can have their joint behavior orchestrated by *interaction protocols*, but the nature and the primitives to describe such protocols are left open.

TEAMCORE is part of the Shell for TEAM work (STEAM) (31). It uses the Joint-Intentions theory (27) and the Shared Plans theory developed by (17) to insure team synchronization. The first aspect of the organization modeling in TEAM-CORE consists of the definition of the structure of a team (see O_s in Fig. 4-c) as a hierarchy of roles. The leaves correspond to a role for an individual agent whereas the internal nodes correspond to (sub)teams. The second aspect (see O_F in Fig. 4-c) consists in specifying a hierarchy of reactive team plans in which joint activities are explicitly expressed with: (i) initiation conditions under which the plan is to be proposed, (ii) termination conditions under which the plan is to be ended, specifically, the conditions when the reactive team plan is achieved, irrelevant or unachievable and (iii) team-level actions to execute as part of the plan. The precise detail of how to execute a leaf-level plan is left unspecified. Domain-specific plan-sequencing constraints on the execution of team plans may also be specified. The third aspect of this OML consists in assigning agents to plans. This is done by first assigning the roles in the organization hierarchy to plans and then assigning agents to roles. A role inherits the requirements from each plan that it is assigned to. Agents may simultaneously take part in several different tasks, and corresponding roles. We will see during the description of the related OIA that the reorganization process is triggered in case of critical role failure, and it consists in reassigning critical roles.

In comparison to the previous models, ISLANDER (9) introduces a terminological shift from organizations to institutions. This declarative language for specifying *electronic institutions* is composed of four basic elements: (1) a *dialogic framework*, (2) *scenes*, (3) *performative structure*, and (4) *norms*. The three first elements express a reduced part of the structure in terms of roles and introduce a dialogical dimension of agents (see O_S and O_D in Fig. 4-d) as follows. The *dialogic framework* defines the valid illocutions that agents can exchange and which are the participant roles and their relationships. The valid illocutions are those that respect a common ontology, a common communication language and knowledge representation language. Each role defines a pattern of behavior within the institution and any agent within an institution is required to adopt some of them. A *scene* is a collection of agents playing different roles and interacting with each other in order to realize a given activity. Every scene follows a well-defined communication protocol. The *performative structure* establishes relationships among scenes. The idea is to specify a network of scenes that characterizes more complex activities. The connections among the scenes define which agents can move from one scene to another, depending on the role they are playing and on given constraints. Given these three specifications, the considered interactions of ISLANDER are defined purely by means of direct communication between agents. The normative element binds dialogical and structural dimensions with individual agent's actions. It is expressed in O_N (see Fig. 4-d). The *norms* component of an electronic institution define the *obligations* agents have while participating in some scene.

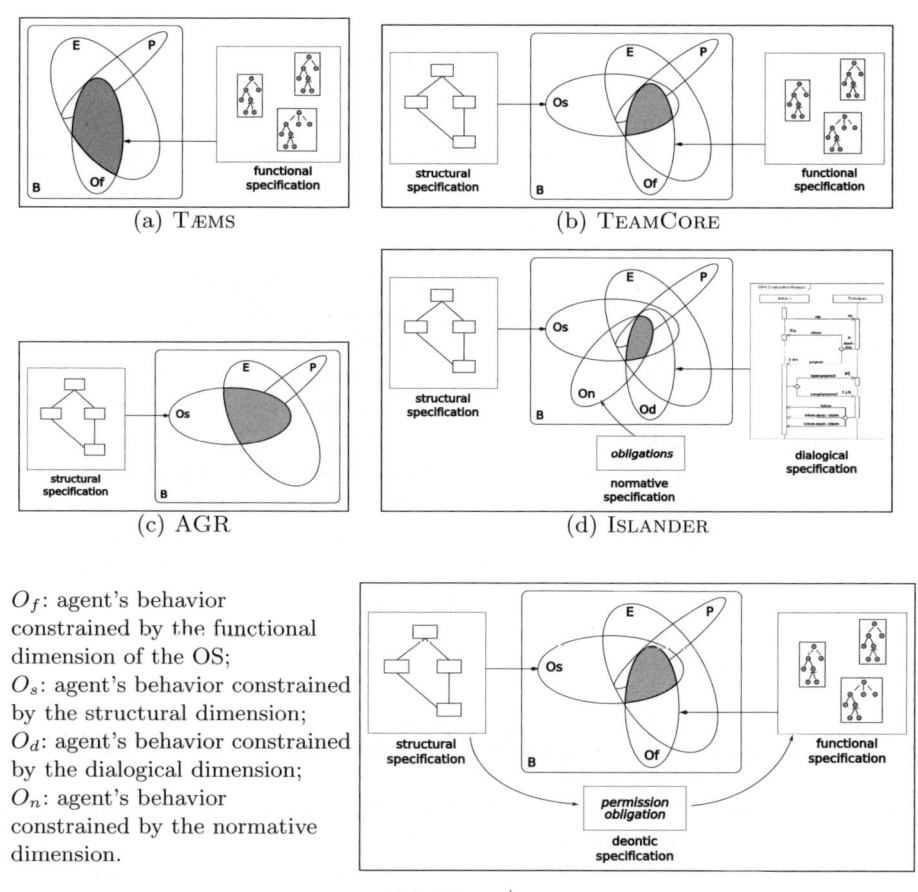

Fig. 4. Influences on the behavior space of an MAS of the - TÆMS (a), AGR (b), TEAMCORE (c), ISLANDER (d) MOISE+ (e) - Organization Modeling Languages

As shown above, existing OML exhibit a *structural* dimension that is related to structure of the collective level of an MAS, generally in terms of roles/groups/links; a *functional* dimension related to the global functioning of the system or a *dialogical* dimension whhicg specifies the interaction in terms of communications between the agents. Even if these dimensions are more commonly found, they are not the only ones. For instance, some models introduce an environmental dimension allowing to constrain the organization situation in an environment such as in AGRE (12), or a temporal dimension such as in $\mathcal{M}\text{OISE}^{Inst}$ (16). Besides those dimensions, the *deontic* and *normative* dimensions used respectively in $\mathcal{M}\text{OISE}^+$ and ISLANDER address the issue of agents´ autonomy. While in other OMLs the agents are supposed to be benevolent and compliant (de-facto) to the OS, these two models add the possibility for agents to develop explicit reasoning on their autonomy with respect to the organizational constraints.

We won't compare here all these OMLs, in terms of their primitives or modeling power that each one can offer (refer, for instance, to (24) for a systematic comparison of these models). Considering their influence on the behavior space, Fig. 4 briefly shows how an organization specified with each model could constrain the agents' behavior. In models such as TÆMS (case (a)) where only the functional dimension (O_F) is specified, the organization has nothing to "tell" to the agents when no plan or task can be performed (for instance, $E \cap O_F = \emptyset$). Otherwise, if only the structural dimension is specified (O_S) as in AGR (case (b)), the agents have to reason for a global plan every time they want to work together. Even with a smaller search space of possible plans, since the structure constrains the agents options, this may be a hard problem. Furthermore, the plans developed for a problem are lost, since there is no organizational memory to store these plans. Thus, in the context of open systems, we hypothesize that if the organization model specifies both dimensions as in $\mathcal{M}$OISE$^+$ (case (e)) or TEAMCORE (case (c)) or a third one as in ISLANDER (case (d)) then the MAS that follows such a model can be more effective in leading the group behavior to its purpose. On the agents' side, they can develop richer reasoning abilities about the others and their organization. Agents may gain more information about the possible cooperation (in terms of roles, groups, but also on the possible goals under achievement or about the performative structures that can be used) that may be conducted with other agents.

4 Organization Implementation Architectures for MAS

While the previous section was concerned with the several OMLs used to specify MAS' organizations, this one deals with the issues related to how these OMLs area easily programmed, i.e., their corresponding Organization Implementation Architectures (OIAs). As it happens with organizational models, implementations can also take an agent centered or a system centered point of view;[1] in (34), these points of view are called agent and institutional perspectives. In the former point of view, the focus is on how to develop agent reasoning mechanisms to interpret and reason about the OS and OE. We will call these mechanisms generally as *Agent's Organizational Reasoning* (AOR) methods. In the latter, the main concern is how to develop an Organization InfrastructureLayer (OIL) that ensures the satisfaction of the organizational constraints (e.g. agents playing the right roles, committing to the allowed missions). This point of view is important in heterogeneous and open systems where agents that enter into the system can have unknown architectures. This will be the focus of this section. Of course, to develop the overall MAS, the former point of view is necessary since the agents probably need to have access to an organizational representation that enable them to reason about it. However, the agents should follow the OS despite their organizational reasoning abilities.

---

[1] We prefer here to use the term system-centered instead of organization-centered in order to avoid confusion between OMLs and OIAs, even if, as we have seen, the organization is reified in an OE.

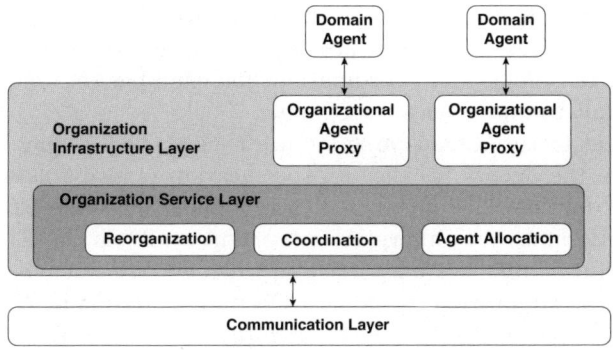

Fig. 5. Common Organization Implementation Architecture for open MAS

Many implementations of the OIL follow the general two-layer architecture depicted in Fig. 5. Domain (or application) agents are responsible for achieving organizational goals and using an *Organizational Agent Proxy* (OAP) component to interact with the organization. The *Organizational Services Layer* (OSL) is responsible to bind all agents in a coherent system and provides some services for them. The communication layer is responsible for connecting all components of the infrastructure in a distributed and heterogeneous applications.

Next sub-sections briefly present the OIL supporting the OML presented in section 3. To compare them, the following features are considered: (i) the OSL's available organizational services, (ii) how easily can heterogeneous agents enter, leave and execute in the organization, (iii) how the OE compliance to the OS is controled and ensured by the OIL services, (iv) if the OS description is available to the agents, (v) the support for reorganization and, finally, (vi) the kind of communication layer supporting the OIL, since FIPA-ACL and KQML are more suitable for open systems than TCP/IP, CORBA and RMI, for example.

4.1 $\mathcal{S}$-$\mathcal{M}$oise$^+$

$\mathcal{S}$-$\mathcal{M}$oise$^+$ is an open source implementation of an OIL that supports the $\mathcal{M}$oise$^+$ OML (22). The OAP is called *OrgBox* and it consists of an API that agents use to access the OSL services, provided by a special systen agent called *OrgManager*. The OrgManager receives and manages all the messages from the agents' OrgBox asking for changes in the OE state (e.g. role adoption, group creation, mission commitment). Those changes bring about the OrgManager to modify the OE only if they do not violate an organizational constraint. For instance, when some agent asks for a role adoption in a group g, the OrgManager ensures that: (1) the role belongs to a specified group g; (2) the number of players in g is lesser or equals than the maximum number of players defined in the group's compositional specification; (3) each role ρ_i that the agent already plays is compatible with the new role in g. Although most of the organizational constraints coming from the three dimensions are ensured, the OrgManager doesn't verify the achievement of committed goals by agents. A sanction system is under development in a new version of $\mathcal{M}$OISE$^+$ (see (16)). Besides the organizational compliance, the OrgManager also provides useful information for the agents' organizational reasoning and coordination, for example the missions they are forced to commit to and goals it can pursue.

An important feature of this architecture is that the OS may be interpreted at run-time by the agents because its specification is available to them. Thus agents can be developed as hardwired programmed for some particular OS or they can be programmed to interpret the current available OS. This last feature is not only useful in open systems, but also when one considers a reorganization process, since the adaptation of the new OS may be dynamic. $\mathcal{S}$-$\mathcal{M}$OISE$^+$ does not require any specific type of agent architecture. Although agents normally use the OrgBox to interact with the system, an agent could even interact with the OrgManager directly using KQML. Organizational adaptation by means of an endogenous reorganization process is presented in (21).

4.2 Other Organization Infrastructure Architectures Approaches

T$\mathcal{E}$MS has many related tools and those that better fit our general architecture of Fig. 5 are *GPGP* (7) for the OSL and *JAF* (35) for the OAP. While GPGP main concern is coordination, JAF is more than an organizational proxy. The aim at JAF is to develop a framework allowing the rapid development of different types of agents, and to facilitate the adoption of the new technology. To this aim, a component-based design is adopted, where developers can plug new components for domain dependent requirements, e.g. a default implementation of a local scheduler is provided but it can be replaced whenever it is needed, by new scheduler. Besides this extensibility feature, the agent architecture can be applied to several domains. It is mainly due to its use of the T$\mathcal{E}$MS domain-independent OML which specifies how agents can achieve their goals. Agents are not constrained in their actions to ensure that they are obeying the organization. Nevertheless, once agents are allocated and coordinated by GPGP, it is supposed that they will perform the allocated tasks. If they do not, this fact is considered as a failure, and not an autonomous act. A diagnosis systems is proposed to deal with this situation in (19), whose proposed solution is to change the organization specification.

MadKit (18) is not simply an OIA: it is a complete agent development platform that supports AGR. AGR minimalistic approach was also adopted in Mad-Kit: it follows a micro-kernel approach. Thus, it provides the minimal set of features for the maintenance of the OE state and ensures some constraints. Although the kernel maintains the OE, this information is not available for agents. Coordination services are not available because the model is focused on the structural dimension. The communication layer is provided by MadKit itself, although FIPA-ACL compliance is planned for future releases.

KARMA supports the TEAMCORE OML and thus considers both the structural and the functional dimensions. Its main focus is the functional dimension, specifically aiming the coordination of heterogeneous agents in open systems that should perform global plans. As described in (31), KARMA was successfully used to build an evacuation rehearsal application composed by agents developed by different research groups, languages, and operating systems. The architectural approach to achieve this openness follows the general schema depicted in Fig. 5, where the OSL is provided by KARMA itself (with coordination and agent allocation services), the OAP is called *teamcore proxies* (these proxies do not requite any change in domain agents). The communication layer is based on KQML. Monitoring tools are available to support some adaptation level. For instance, when some agent fails, other agents' proxies may decide to adopt its role to maintain the system running, a kind of exogenous reorganization at the entity level. This monitoring tools could also be used to enforce compliance to the specification, but the original paper does not focus on this issue, as this feature was not necessary for the applications described. Regarding the agents' autonomy, domain agents can decide whether to accept the team task or not. The OAP even learns the suitable level of autonomy, so that it does not ask the domain agent for every simple decision, but just for the most important. However, domain agents are organizationally unaware – as in Fig. 1-c. They do not have access to the OS nor OE and therefore can not reason about or change it. Every organizational operation is performed by the teamcore proxy. For instance, since the agents do not know their roles in some global plan when accepting or refusing a task, the decision of accepting or refusing must be done based only on non-organizational information. Due to this organization unawareness, agents can not enter in the system and perform a reorganization. KARMA has some level of reorganization, but it is the proxy responsibility. Domain agents are thus task-autonomous but not organization-autonomous as we defined it in section 2.

AMELI supports the ISLANDER (10) OML. It aims at facilitating the participation of heterogeneous agents while ensuring that the organizational constraints are enforced. The constraints are based on the dialogical dimension. AMELI also follows the general architecture of Fig. 5. A component called *governor* is the OAP and three special system agents implement the OSL and the coordination for a scene execution: Institution Manager, Transition Manager and Scene Manager. The communication layer is provided by JADE (1), a FIPA-ACL compliant implementation. The specification of the organization is available for agents in XML format generated by ISLANDER Editor.

We summarize in table 4.2 how each organization infrastructure considers the criteria given at the begining of this section. Since all the described OIL were developed for heterogeneous and open systems, they follow in some extent the architecture depicted in Fig. 5. They naturally differ in their underlying OML and thus the set of available organization modeling dimensions that they manage. In the existing OILs, where an OS is made available to the agents, the requirements for an incoming agent usually are that it knows the OML, and it is capable of reasoning about it. As we can see, organizational compliance and reorganization are not satisfactory treated: in some proposals, compliance is not a matter because the agents' autonomy regarding the organization is not a matter. Moreover, in other models this compliance was not fully developed.

Table 1. Comparison of organizational infrastructures of the Organization Implementation Architecture

	Services	Hetero. Agents	Org. Compliance	Specification Available	Reorganization	Communication layer
$\mathcal{S}$-$\mathcal{M}$OISE$^+$	Coordination	yes	yes	yes	endogenous specification	KQML
GPGP/JAF	Coordination	yes	no	yes	exogenous specification	TCP/IP
MadKit		yes	yes	no	no	—
KARMA	Coordination	yes	no	no	exogenous entity	KQML
AMELI	Agent allocation Coordination	yes	yes	yes	no	FIPA-ACL

5 From Closed Organization to Open Organization

As seen in the two previous sections, building a MAS with an OOP approach installs an explicit and formal OS on the agents that may help or prevent the openness properties. In the following, we briefly focus on three aspects related to this issue: organization permeability, reorganization and ability of the organization to control agents' autonomy.

5.1 Organization Permeability

Open agents' organization should be permeable in the sense of allowing the dynamic arrival/exit of agents into/from it. As shown in section 4 an explicit OS is specially useful for the MAS to integrate non centrally designed agents: even not knowing what kind of agents are trying to enter the system, the organization could enforce some restrictions on their behaviors. No dedicated integration management mechanism is available in the existing organization approaches (integration is still done in an adhoc manner, e.g. (16)). However, as shown in Fig. 5, some OILs make the OS available. It is of crucial importance since arriving agents should be able to read it, to reason about it in some cases and thus improve their collaboration in the system.

Besides such a mechanism, different features of the OML have to be considered. Assigning abstract roles rather than actual agents to plans or interactions provides a useful level of abstraction for the quick (re)assignment of new agents in the organization when needed. Existing OMLs offer multiple dimensions (see sec. 3 and Fig. 4). Combining the multiple OML modeling dimensions in a declarative OS offers the opportunity to tackle with the heterogeneity of the agents architecture and reasoning mechanisms. In the same way, the degree of abstraction with respect to the agents' architecture may help their integration: functional specification in $\mathcal{M}$OISE$^+$ or TEAMCORE, for instance allows the use of agents with different skills to execute the unspecified leaf-level plan.

Even if some progress was made in the infrastructure/middleware level for puting these information available to the agents, most of the current agent programming is purely ad-hoc. Organizational standards may be a good direction towards openness at this level, as it has been done at the communication level by the FIPA-ACL standard. Another approach could be developping an AOR mechanism based on those models in order to enable an agent to read an organization modeling, written in some known OML, and to decide what to do.

5.2 Reorganization Ability

Openness, in the sense of the system configuration change, maps directly to the problematic of changing the current state of the OS or OE into a new one. As in previous section, delivering the explicit organization to the agents opens new possibilities for the adaptation of the agents' organization while the system keeps running. As a consequence, an endogenous process could occur *on-line*, in the sense that the designers should be replaced by the *agents* themselves to control and to make on-line changings in the organization.

Reorganization may be at the OE level (e.g. allocation of agents to roles are changed) or at the OS level (where some dimension of the OS changes). Depending on the multiplicity of existing modeling dimensions and also on the richness of each of these dimensions, there may be a wide spectrum of change types. However the mutual independence of each dimension has to be considered. For instance, while ISLANDER and TEAMCORE closely bind the different modeling dimensions, $\mathcal{M}$OISE$^+$ enforces the mutual independence of its structural and functional dimensions. It binds explicitly and declaratively both dimensions with a deontic specification that maintains a suitable independence for the reorganization process. This binding allows the MAS to change its structure without changing its functioning, and vice versa: the system only needs to adjust its deontic relations.

5.3 Ability of the Organization to Control Autonomy

As said in (10) "Openness without control may lead to chaotic behavior". Installing openness requires the ability to install observation and control of the system. Both of these processes can be exogenous or endogenous. It is worth to notice that the control should be performed outside the agent (since its "inside" is unknown), normally as an OSL service (see Fig. 5) of the OIL.

However, as shown in the previous section, agents' organizations should control the agents' behaviors while permitting their self-adaptation in response to evolving environmental constraints or changes in the systems' goals. Balancing openness and control is a difficult task that depends on the richness and the degree of the constraints that the OS may express. As stated in section 3, identifying a good size for the set O must conciliate collective constraints with the agent autonomy, i.e. with the possibility for the agents on deciding what to do.

Delivering OS enables organizational autonomy as defined in section 2. However, as we have seen in previous section, this is not sufficient. The more the OML is rich in terms of dimensions, the more it is able to constrain the agents' behavior space. The drawback is that when the agents need to adapt their behaviour due to changing conditions, they may violate some organizational constraints. In some sense, the richness of the OML may hinder adaptation of the agents to changes. As for permeability, the degree of abstraction with respect to the agent's architecture (available in the dimensions of the OML) is an important point to consider. For instance, in TEAMCORE and $\mathcal{M}$OISE$^+$, it is possible to express leaf-level goals in their functional dimension instead of plans of actions, and this fact allows the agents to decide locally which action to commit to for executing the committed plans. Another important feature deals with the use of primitives dedicated to the explicit management of agents autonomy in the OML. $\mathcal{M}$OISE$^+$ and ISLANDER address such a problematic with their deontic and normative dimensions, stating flexibility in the expression of the constraints bearing on the roles played by the agents.

Reflexivity of the control expression and of the reorganization process is an important and crucial issue that has also to be considered. It has direct repercussion on the permeability of the organization: the use of OML to specify the reorganization (cf (21)), and the supervision infrastructure itself (cf. (16)). Agents entering the organization are nonetheless able to understand the organization in which they enter but also the organization structuring the reorganization or the supervision process.

6 Conclusion

As we have shown in this paper, organization is a complex and rich dimension in MAS. We have focused our study on the Organization Oriented Programming approach that answers the best, to our point of view, to the openness requirements of current and future applications of MAS. This approach aims at combining agent centered and organization centered points of view to derive a mixed of top-down and bottom-up definition/adaptation of the organization. Different modeling dimensions are mobilised to program rich organizational patterns to control or to help the cooperation among agents in the system: structural, functional or dialogical. As noticed, these dimensions are not exclusive and some dimensions are still being proposed (e.g. environment, context). The agents' autonomy concern has recently been added to these OMLs to build normative organizations. Dealing with the support to program these OMLs, different OIAs have been presented: OS may be interpreted within the agent architecture, in an OIL existing outside of the agent in multi-agent platform, or in both kinds of components. As noticed, the agents' autonomy, adaptation and reorganization abilities are also important topics in these approaches.

In order to create real open organizations with an OOP approach, combination of agent centered and organization centered approaches are necessary. To this aim, AOR mechanisms and OIL services should be used to define the OIA of such systems. Some challenges still need to be considered and solved. Among them, we can cite: (i) decentralization of the organization implementation architecture (e.g. the centralized architecture of $\mathcal{S}$-$\mathcal{M}$OISE$^+$ and AMELI prevent them to address the scaling problem) is one important topic, (ii) development of reasoning abilities in order to integrate top-down predefined organizations -organization-centered- with bottom-up emergent organizations -agent-centered-, (iii) a better undertsanding of each of the dimensions of the existing OMLs, leading to organization ontology standards to enable interoperation, (iv) reorganization issues in general (how to evaluate? how to change?), and (v) building of mixed models for human and artificial agents working together.

References

[1] Bellifemine, F., Bergenti, F., Caire, G., Poggi, A.: JADE – a java agent development framework. In: Bordini, R.H., Dastani, M., Dix, J., El Fallah Seghrouchni, A. (eds.) Multi-Agent Programming: Languages, Platforms, and Applications, number 15 in Multiagent Systems, Artificial Societies, and Simulated Organizations, ch. 5. Springer, Heidelberg (2005)

[2] Bergenti, F., Gleizes, M.P., Zambonelli, F.: Methodologies and Software Engineering for Agent Systems. Kluwer, Dordrecht (2004)

[3] Bernoux, P.: La sociologie des organisations. Seuil, 3ème edn. (October 1985)

[4] Boissier, O., Padget, J., Dignum, V., Lindemann, G., Matson, E., Ossowski, S., Sichman, J.S., Vázquez-Salceda, J. (eds.): Coordination, Organizations, Institutions, and Norms in Multi-Agent Systems. LNCS (LNAI), vol. 3913, pp. 25–26. Springer, Heidelberg (2006)

[5] Castelfranchi, C.: Modeling social action for AI agents. Artificial Intelligence 103, 157–182 (1998)

[6] Corkill, D.D.: A Framework for Organizational Self-Design in Distributed Problem Solving Networks. PhD thesis, University of Massachusetts, Amherst (1983)

[7] Decker, K.S., Lesser, V.: Designing a family of coordination algorithms. Umass computer science technical report 1994-14, Department of Computer Science, University of Massachusetts, 1995. UMAss Computer Science Technical Report 1994-14

[8] Drogoul, A., Corbara, B., Lalande, S.: MANTA: New experimental results on the emergence of (artificial) ant societies. In: Gilbert, N., Conte, R. (eds.) Artificial Societies: the Computer Simulation of Social Life, pp. 119–221. UCL Press, London (1995)

[9] Esteva, M., Padget, J., Sierra, C.: Formalizing a language for institutions and norms. In: Meyer, J.-J.C., Tambe, M. (eds.) ATAL 2001. LNCS (LNAI), vol. 2333, pp. 348–366. Springer, Heidelberg (2002)

[10] Esteva, M., Rodríguez-Aguilar, J.A., Rosel, B., Joseph, L.: AMELI: An agent-based middleware for electronic institutions. In: Jennings, N.R., Sierra, C., Sonenberg, L., Tambe, M. (eds.) Proceedings of the Third International Joint Conference on Autonomous Agents and Multi-Agent Systems (AAMAS'2004), pp. 236–243. ACM, New York (2004)

[11] Ferber, J., Gutknecht, O., Michel, F.: From agents to organizations: an organizational view of multi-agent systems. In: Giorgini, P., Müller, J.P., Odell, J.J. (eds.) Agent-Oriented Software Engineering IV. LNCS, vol. 2935. Springer, Heidelberg (2004)

[12] Ferber, J., Michel, F., Baez, J.: AGRE: Integrating environments with organizations. In: Weyns, D., Parunak, H.V.D., Michel, F. (eds.) E4MAS 2004. LNCS (LNAI), vol. 3374, pp. 48–56. Springer, Heidelberg (2005)

[13] Ferber, J., Gutknecht, O.: A meta-model for the analysis and design of organizations in multi-agents systems. In: Demazeau, Y. (ed.) Proceedings of the 3rd International Conference on Multi-Agent Systems (ICMAS'98), pp. 128–135. IEEE Press, Los Alamitos (1998)

[14] Fox, M.S.: An organizational view of distributed systems. IEEE Transactions on Systems, Man, and Cybernetics 11(1), 70–80 (1981)
[15] Gasser, L.: Organizations in multi-agent systems. In: Pre-Proceeding of the 10th European Worshop on Modeling Autonomous Agents in a Multi-Agent World (MAAMAW'2001), Annecy (2001)
[16] Gâteau, B., Boissier, O., Khadraoui, D., Dubois, E.: Moiseinst: An organizational model for specifying rights and duties of autonomous agents. In: Third European Workshop on Multi-Agent Systems (EUMAS 2005), Brussels Belgium, December 7-8, 2005, pp. 484–485 (2005)
[17] Grosz, B.J., Kraus, S.: Collaborative plans for complex group action. Artificial Intelligence 86, 269–357 (1996)
[18] Gutknecht, O., Ferber, J.: The MadKit agent platform architecture. In: Agents Workshop on Infrastructure for Multi-Agent Systems, pp. 48–55 (2000)
[19] Horling, B., Benyo, B., Lesser, V.: Using self-diagnosis to adapt organizational structures. In: Proceedings of the Fourth International Conference on MultiAgent Systems (ICMAS'2000), pp. 397–398. IEEE, Los Alamitos, CA (2000)
[20] Hübner, J.F., Sichman, J.S., Boissier, O.: A model for the structural, functional, and deontic specification of organizations in multiagent systems. In: Bittencourt, G., Ramalho, G.L. (eds.) SBIA 2002. LNCS (LNAI), vol. 2507, pp. 118–128. Springer, Heidelberg (2002)
[21] Hübner, J.F., Sichman, J.S., Boissier, O.: Using the $\mathcal{M}$OISE$^+$ for a cooperative framework of MAS reorganisation. In: Bazzan, A.L.C., Labidi, S. (eds.) SBIA 2004. LNCS (LNAI), vol. 3171, pp. 506–515. Springer, Heidelberg (2004)
[22] Hübner, J.F., Sichman, J.S., Boissier, O.: $\mathcal{S}$-$\mathcal{M}$OISE$^+$: A middleware for developing organised multi-agent systems. In: Boissier, O., Padget, J., Dignum, V., Lindemann, G., Matson, E., Ossowski, S., Sichman, J.S., Vázquez-Salceda, J. (eds.) Coordination, Organizations, Institutions, and Norms in Multi-Agent Systems. LNCS (LNAI), vol. 3913. Springer, Heidelberg (2006)
[23] Iglesias, C., Garrijo, M., Gonzalez, J.: A survey of agent-oriented methodologies. In: Proceedings of the 5th International Workshop on Intelligent Agents V: Agent Theories, pp. 317–330. Springer, Heidelberg (1999)
[24] Boissier, O., Coutinho, L., Sichman, J.S.: Modeling dimensions for multi-agent systems organizations. In: Dignum, V., Dignum, F., Edmonds, B., Matson, E. (eds.) Agent Organizations: Models and Simulations (AOMS), Workshop held at IJCAI 07 (2007)
[25] Lemaître, C., Excelente, C.B.: Multi-agent organization approach. In: Garijo, F.J., Lemaître, C. (eds.) Proceedings of II Iberoamerican Workshop on DAI and MAS (1998)
[26] Lesser, V., Decker, K., Wagner, T., Carver, N., Garvey, A., Horling, B., Neiman, D., Podorozhny, R., NagendraPrasad, M., Raja, A., Vincent, R., Xuan, P., Zhang, X.Q.: Evolution of the gpgp/taems domain-independent coordination framework. Autonomous Agents and Multi-Agent Systems 9(1), 87–143 (2004)
[27] Levesque, H.J., Cohen, P.R., Nunes, J.H.T.: On acting together. In: Dietterich, T., Swartout, W. (eds.) Proceeding of the Eight National Conference on Artificial Intelligence (AAAI-90), Menlo Park, pp. 94–99. AAAI Press / MIT Press (1990)
[28] Modeling Autonomous Agents in a Multi-Agent World (MAAMAW'2001). In: Pre-Proceeding of the 10th European Workshop on Modeling Autonomous Agents in a Multi-Agent World (MAAMAW'2001) (2001)
[29] Omicini, A., Ricci, A., Goldin, D.: Introduction to the workshop. In: Second InternationalWorkshop on Theory and Practice of Open Computational Systems (TAPOCS 2004) (2004)

[30] Picard, G., Glize, P.: Model and Analysis of Local Decision Based on Cooperative Self-Organization for Problem Solving . Multiagent and Grid Systems 2(3), 253–265 (2006)
[31] Pynadath, D.V., Tambe, M.: An automated teamwork infrastructure for heterogeneous software agents and humans. Autonomous Agents and Multi-Agent Systems 7(1–2), 71–100 (2003)
[32] Scott, W.R.: Organizations: rational, natural and open systems, 4th edn. Prentice-Hall, Englewood Cliffs (1998)
[33] Sichman, J.S., Conte, R., Demazeau, Y., Castelfranchi, C.: A social reasoning mechanism based on dependence networks. In: Cohn, T. (ed.) Proceedings of the 11th European Conference on Artificial Intelligence, pp. 188–192 (1994)
[34] Vázquez-Salceda, J., Aldewereld, H., Dignum, F.: Norms in multiagent systems: some implementation guidelines. In: Proceedings of the Second European Workshop on Multi-Agent Systems (EUMAS 2004) (2004)
[35] Vincent, R., Horling, B., Lesser, V.: An Agent Infrastructure to Build and Evaluate Multi-Agent Systems: The Java Agent Framework and Multi-Agent System Simulator. In: Wagner, T.A., Rana, O.F. (eds.) Infrastructure for Agents, Multi-Agent Systems, and Scalable Multi-Agent Systems. LNCS (LNAI), vol. 1887, pp. 102–127. Springer, Heidelberg (2001)

Modelling and Executing Complex and Dynamic Business Processes by Reification of Agent Interactions

Marco Stuit and Nick B. Szirbik

Department of Business & ICT, Faculty of Management & Organization,
University of Groningen, Landleven 5, P.O. Box 800, 9700 AV Groningen, The Netherlands,
Tel.: +31 50 363 7083
{M.Stuit,N.B.Szirbik}@rug.nl

Abstract. *Interaction* refers to an abstract and intangible concept. In modelling, intangible concepts can be embodied and made explicit. This allows to manipulate the abstractions and to build predictable designs. Business processes in organisations are in fact reducible to interactions, especially when *agent-oriented* modelling methods are employed. Business processes represented as interaction structures can appear at different levels of abstraction. There is a compositional coupling between these levels, and this necessitates a method that allows dynamic de/re-composition of hierarchically organised interactions. We introduce the novel concepts that allow interaction-based diagramming and explain the syntax and semantics of these constructs. Finally, we argue that a business process composition with interactions allows more organisational flexibility and agent autonomy, providing a better approach in complex and dynamic situations than current solutions.

1 Introduction

Each modelling approach has a specific focus. It could be the "entity" concept, or the "activity", or "location". In our approach, the most important is the concept of *interaction* that allows for an easier modelling of complex and distributed business processes (BPs). Classic BP modelling takes a centralistic standpoint; an external observer sees the entire process model as a monolithic whole. In our approach, due to the agent-orientation, each participant has its own local view of the process. These local views, known as *interaction beliefs* (IBs), are linked to explicit depictions of named interactions. These beliefs are used by agents that play the *roles* that are attached to an interaction. Interactions are compositional; they can be grouped and divided. A method for decomposition of interactions is used to identify 'smaller' interactions in a top-down analysis, or the 'smaller' interactions can be grouped in more complex ones in a bottom-up approach (or a mix of these two).

This paper is organised as follows. Section 2 introduces and justifies the use of the interaction concept by analysing it from both a philosophical and pragmatic perspective. The relation with similar approaches is discussed. In section 3, we explain the syntax and semantics of our interaction diagrams. Section 4 presents the de/re-composition of interactions. Patterns to model the de/re-composition are discussed, a de/re-composition symbol is introduced and compared to UML, and the usage and execution of the de/re-composition is explained, together with an example. In the last section we discuss the applicability of our approach, we indicate some directions for future work, and we conclude the paper.

2 Agents, Roles, and Reification of the Interaction Concept

The application of our research in multi-agent systems is in the area of BPs. We analyse and model organisations that we view as multi-agent systems, and use the conceptual models to build designs for agent systems that support the BPs within these organisations. We are developing a framework (AGE – i.e. **A**gent **G**rowing **E**nvironment, see [18]), which uses a language (TALL – i.e. **T**he **A**gent **L**ab **L**anguage, see [21, 22]) for multiple functions: analysis, specification, conceptual modelling, interactive simulation/gaming and dynamic visualisation, design, implementation, and testing.

2.1 Using Agents and Roles for Business Process Modelling

There are three main concepts in the TALL language: agent, role, and interaction. Many approaches tend to associate the role played by an agent in an organisation with the agent itself (as a property) or define specialised types of agents. However, we adhere to the ontological distinction made by Steimann [25], who considers that roles should be defined separately from agents. He states that in software development roles should be a sort of interfaces; in organisational modelling they should be dynamic placeholders for the actor/agent participants. The main reason is that a role itself has no meaning outside its context and has to be filled with persons, things or places to have a meaning.

These persons, things or places are named *natural types*. In general, an agent in TALL belongs to a natural type (i.e. is an instance of the natural type), under all circumstances and all times. An agent can never drop its natural type without losing its identity. For example in TALL, all agents belong to the natural type *Agent* that cannot be dropped without loosing their identity as an agent. Roles are filled with agents and these agents can leave their roles without losing their identity. Another difference between agents and roles according to Steimann [25] is that roles, unlike classes, have "something dynamic" about them. Unlike entity types, playing several roles simultaneously is not unusual for the instances of certain natural types. For example, a person can be a customer, supplier, and stakeholder at the same time.

In an organizational context, agents belong to agent groups (i.e. they are members of a certain group) that are generic sets of agents with similar skills – not necessarily based on the function assigned formally to the agent in the organisation. Agents can belong to more than one group, thus, these sets can overlap; also specialisation (subsets) can be used. The agent groups can be used to define an authorisation scheme in the organisation.

In TALL, roles are types with a dynamic nature, which cannot be instantiated. They can add properties to the agents that play the role, and may regulate the behaviour of these agents. We denote that an agent plays a role like in Fig. 1a. This diagramming method can be at a generic level, denoting what agent groups can play the role (authorisation), but also at an instance level, denoting what agent members are

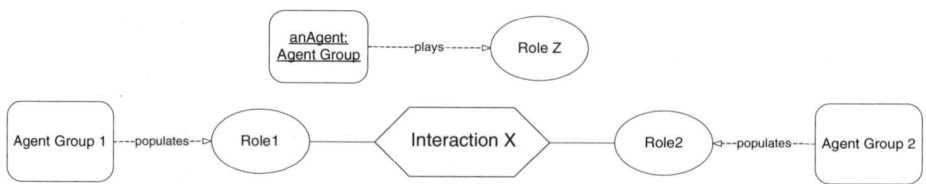

Fig. 1a & 1b. Example of an agent member playing a role (top) and an interaction representing the dynamic connection between two roles (bottom)

currently assigned to that role. Moreover, the interpretation of an instance level diagram is temporally dependent. It can illustrate a plan ("these are the agents who will play the role"), or a snapshot ("currently, these are the agents playing the role"), or a trace ("these have been the agents that played the role").

Within any organisation, by playing the roles, agents execute certain activities and always communicate with other agents – they cannot exist in isolation. That introduces the necessity to relate the roles. Agents that play "connected" roles interact. We depict the fact that agents interact as in Fig. 1b (here at a generic level). The dynamic connection is illustrated between the roles, and not the agents. To explain this choice in our language, we need to discuss the interaction concept in more depth.

2.2 Reification of the Interaction Concept for Agent Modelling

Interaction is an observable action that happens when two or more entities have an effect upon one another. From an organisational and social perspective, the entities are usually humans (or groups of humans). When viewed via the agent paradigm, the interaction can occur between two artefacts (like software agents, or simulated humans/organisations). The concept of interaction is an intangible one. By reification, a symbol can be introduced to capture the presence of an interaction in a model.

Sowa (see [24] pp. 60-62) discusses Peirce's *basic categories* that comprise the triad of *Firstness*, *Secondness*, and *Thirdness*. We argue that an interaction should be represented by a modelling concept that belongs to the Thirdness category, that is, interactions are conceptualised as the mediating Thirdness. We see the agents in a multi-agent system (artefactual or human/organizational agents) as natural types who correspond to the Firstness category. An agent exists independent of any external relationship. Roles played by the agents are Secondness because roles depend on an external relationship to agents. Interactions "bring" the roles and the agents into mediation.

The three main concepts that we use in our universe of discourse: agents, roles, and interactions together form a triad. In a model in TALL, or during an agent simulation in AGE, interactions are explicit and "exist". The modeller or the experimenter can create and manipulate the interactions, as reflections of observed interactions in a real BP. One can relate it to "reification", meaning to make concrete and perceptible. In a formal context, reification of an abstraction raises the need for identity. Thus, each interaction has to get a name (label). Naming can be achieved, according to the object-oriented approach, in a two level scheme: generic name (or interaction class identity) and specific name (or interaction instance identity). The first level is useful for the beliefs of the agents, who should "know" what interactions are possible within the organisation, and the second is useful for simulation and BP monitoring purposes. "Name/identity" is the first property of an interaction.

An interaction can be seen as a "piece" of a BP. An interaction is performed to achieve something, and this leads to other properties. Some agent-oriented methods propose to define "external goals" [20]. Another simple solution is to define preconditions and post-conditions. There is a generic precondition for all interactions. There should be at least two roles involved. The roles can be linked to an interaction statically or dynamically. In a static framework, an interaction will have always the same pre-defined roles. For more flexibility, it is better to have the possibility to dynamically change the number of roles involved in an interaction. For example to link new roles that did not exist (in the specific universe of discourse) before the interaction started. Roles linked to interactions can be considered as implicit properties. Of course, roles are more than that; they help describe the organisation's structure and nature.

Current BP support approaches put the description of the process outside the actors/agents that execute that process. In a typical agent-oriented approach, the description appears as beliefs of the agents, and it is inherently distributed. That gives no "power" to the organisation to enforce behaviour. For a modeller of BPs, a process description that resides solely in the beliefs of the agents is not really making sense, unless an agent has the "authority" to enforce behaviour over other agents. We believe that a simpler way to enforce behaviour is to attach an interaction description to the interaction entity. In the organisational paradigm, this description is called a protocol. These protocols can also be considered as properties of the interaction. We discuss protocols in more detail in section 3. Being "pieces" of BPs, interactions serve the purpose to build a description of a BP. This induces also a temporal dependency. The overall process description can be a plan, a runtime snapshot or a log of a past process. The agents in the organisation should have beliefs about potential interactions, in order to initiate them. We explain later how various triggers induce the agents to initiate interactions.

2.3 Related Work

2.3.1 Implicit and Explicit Interaction Representation

The triad of Firstness, Secondness, and Thirdness can be compared to the *Relation-Ship Service* in CORBA [13] in which entities and relationships between entities can be explicitly represented. Two or more CORBA objects that participate in a relationship are called related objects and these are connected with the relationship via a role. A role is an object, which characterizes a related object's participation in a relationship type. These relationships can appear as a *relationship type* (relationships sharing the same semantics) or a *relationship instance* with an identity.

In UML, the same triad can be represented by using the constructs *class/object*, *association class* and *rolename*. Agents will be represented by classes/objects and interactions will be represented by an association class that is attached to a binary or n-ary association/link between two or more (agent) classes/objects. By using association classes for representing interactions the interactions can be assigned class-like properties such as names, attributes and operations. Finally, the roles appear as role names at the association ends to indicate the role played by the (agent) class/object. This means that the TALL diagrams, at the generic level as well as at the instance level, can always be reduced to UML class/object diagrams. However, representing roles explicitly allows one to model attributes specific to a role and also allows one to model dependencies between roles that represent a hierarchical scheme or organizational structure (describing also reporting, authorization and appraisal relations between roles).

Other works [9, 11] pointed out the importance of agent interaction and devised ways to capture the patterns and execution procedures for interaction. In [8] it is explained that it is not sufficient to concentrate on the local behaviour of agents as the interaction execution is often the most difficult. In UML for example, sequence, collaboration and activity diagrams are showing how messages flow between participants. AUML (agent UML, see [15]) uses the same diagramming techniques. Wagner [27] introduces the interaction pattern diagrams and interaction sequence diagrams. In the former, the nature of the interaction is explained (showing what kind of information flows can occur), and in the latter, the particular sequence of messages is indicated, like in UML behaviour diagrams. Still, these techniques all take a centralistic point of view. Except for the interaction pattern diagram of Wagner, all indicate a priori the execution procedure for the interaction, as it is enforced on the participants.

MESSAGE (see [5]) introduces the explicit interaction view. A special symbol is used to denote interaction. Roles, represented by special symbols, are linked to the interaction. The technique also shows the flows of information using UML notation. The execution description is separated, by using an Interaction Protocol description, corresponding to a UML sequence diagram. The collection of Interaction Protocols defines how agents are performing their overall activities. It adopts still a centralistic approach. The main step forward in MESSAGE is the decoupling between generic interactions and the way they are performed. Flexibility can be achieved by having more – alternative - interaction protocols for each kind of interaction. It is not explicitly stated, but also in Wagner's approach, it is possible to have more than one Interaction Sequence Diagram for one Interaction Pattern.

2.3.2 Similarities with Other Agent-Oriented Modelling Approaches

AML [3, 4] is a visual modelling language for specifying, modelling and documenting MAS systems. It provides the ability to model the "social behaviour" of agents using extensions of UML behavioural models. Although AML regards the social ability of agents as one of the most important concerns in their language, it is focused on modelling all aspects of software development with multi-agent systems. Hence, it is advertised as a useful tool for software architects. Clearly, it is not an explicit BP modelling language. In contrast, TALL is specifically focused on modelling BPs and eventually building a software system based on multi-agent technology that supports the BPs of an organization.

Horling and Lesser [6] made the observation that under real conditions, the complete view of a unique, central, omniscient manager in a multi-agent organisation is practically impossible - the resources of central managers are always limited, communication takes time and the environment gives an arbitrarily large number of inputs. They emphasise that dynamically changing the manner in which the agents interact can change the local and global behaviour. In their approach, based on the

ODML organizational design language, they allow dynamic agent-role binding and behavioural learning. Roles bring responsibility and tasks to the agents bond to them.

However, the decision to formalize the representation of reality via general nodes, and also the view on roles from a quantitative perspective, hinders the applicability of this approach for the modelling and simulation of discrete business processes that are executed mainly by humans. They demonstrate the successful simulation of a sensor fusion system based on an agent organisational design, which is rather far from a MAS supporting BPs.

The explicit modelling of interaction and the subsequent mechanism we intend to use to enact agent-to-agent coordination can be also seen as another form of dynamic team formation, as in [26]. Our approach differs from team formation in the sense that the interaction composition information (locally stored as models) shows patterns of pieces of BPs and it is not goal or task oriented as it is viewed in team formation. Team formation with explicit goals is suitable in BPs where the tasks are common (more agents are working effectively together, like a football team). Except for meetings and conversations, BPs have individually defined tasks and high communication. We appreciate that to reason locally about a model that specifies interactions instead of goals is simpler to implement. Interaction compositions capture more than just goals because it also shows how interactions could be enacted. Goals allow to specify what should be done, whereas interactions specify how it should be done [5].

2.3.3 Similarities with Business Process Modelling Approaches

Besides the approaches discussed above, a multitude of BP modelling languages exist that are not advertised as agent- and/or interaction-oriented modelling languages. Still, these languages could be used to model agent's behaviour and the interactions between them. BPs are often supported by WfM (workflow management) systems and/or BPM (business process management) systems. Petri nets and various extensions are used here to model the BPs. The traditional workflow community is moving to a more decentralized approach that is more capable of handling dynamic change. The adaptive, ad-hoc, dynamic workflow approach can handle deviations from the process model that allows them to model, and enact more than just rigidly structured BPs [see 7, 17]. These approaches are closer to our agent-oriented modelling approach because they can also handle processes that are not completely defined in advance and they can cope with process changes. Still, these models present a global model perspective in which the modeller acts as a sort of 'overseer'.

In TALL and in AGE the behaviour of agents is explicitly modelled with an extension of Petri nets: Behaviour Nets [10]. In our approach, the interaction is performed after the local behaviours (known as Interaction Beliefs) are aligned. There is no centralistic view and a lack of consistency between beliefs is allowed.

The Business Process Modeling Notation (BPMN, see [16, 28]) is one of the new standardized graphical notations for drawing BPs in a workflow. Just as our Behaviour Nets, BPMN is suited for modelling (complex) processes that span multiple organizations by using special notations to depict message-based events and message passing between two participants (business entities or business roles). BPMN consists of one Business Process Diagram (BPD) to model a BP as a set of graphical elements. BPMN is capable of modelling BPs from a single point of view. These private (internal) BPs are considered self-contained processes and message flows can be used to show interactions with other participants that are considered black boxes. It is also possible to show (from a global point of view) message passing between multiple private BPs. Moreover, public processes that show only the 'touch-points' between the participants that are visible to the public can be modelled (more combinations are possible). The focus on interactions implies that each participant has its own local process (view).

2.3.4 Discussion

Although modelling from a local viewpoint (with the activities of other participants being a black box or also detailed) is possible in BPMN, the local viewpoints are not allowed to conflict a priori. Furthermore, none of the above approaches considers that an agent can have explicit beliefs about the interaction. Moreover, they consider that all agents behave according to the pre-defined sequences (protocols). This implies that the agent design is going through the "external-to-internal route" or, as Wagner calls it, the internalisation process [27]. First, the modeller will have to understand from a global point of view what is happening, and design the internal behaviour of the agent accordingly. We argue that it is easier to understand first what an agent believes about the interaction and build an internal model, and only after try to achieve coherence, not necessarily by aligning a-priori all these "local behaviours" of the agents. Complex BPs are difficult to understand from a global perspective. In these cases, the analysis can start from a local point of view, and global understanding of the process is not even necessary a priori. The agent approach enables the modelling and support of a distributed, heterogeneous, constantly evolving and not "well known" process; the humans who are carrying out the process do not have a precise overall view either. In addition, the agent paradigm allows for conflicting process views. Most human organisations function because the participants know partially what to do, and nobody has necessarily a global view. Even the high-level managers have a limited view, because they see the process from the "highest point", but they do not have to understand the details. They delegate this to lower level managers. Especially in flat organisations, delegation and decisional autonomy restrict the view of one participant. Complex organisations tend to be anyway incomprehensible for single humans, and efforts to model "everything" are usually futile attempts [19].

3 Modelling Interactions in TALL

3.1 Organisational Modelling with TALL

The modeller has the choice either to identify first what kinds of interactions are observable, either to enact a role scheme (usually a hierarchy) and "embody" the interactions that are observable between roles. In the first approach, roles are to be defined (or identified) in the second step. This choice between "roles-first" or "interaction labels first" depends on the kind of organisation that it is being modelled. Bureaucracies for example allow the modeller to easily identify first the roles and an eventual hierarchical tree of these. On the other side of the organisational spectrum, in the so-called "ad-hocracies" [12], which are flat organisations, the analysis should start first with the identification of interactions and only after roles can be "discovered". Real organisations are somewhat in between these two extremes, and the modeller generally identifies interactions and role names in parallel.

The agents that are forming the organisation can be represented as a *pool of agent members*. This pool can be structured in groups (as discussed in paragraph 2.1). Currently, these agent groups are rigidly associated to roles in the organisational hierarchy. For example, an agent with secretarial skills is assigned to the secretary role. Unfortunately, such a static association will assign an agent group to all the interactions where that role appears.

In TALL, we adopt a more flexible structure. It is possible to assign agent groups (depicted like generic agents) to the roles that are linked to an interaction (as in Fig. 1b). This denotes a constraint, showing that only agents of that group are authorised to play that role in that interaction. This enhances considerably the flexibility of the authorisation scheme, allowing the same role to have different authorisation schemes, depending on the interactions linked to this role. As we have stated already, even if the agent groups are not illustrated there, the "universal agent" (the set encompassing the whole agent pool) is always assigned to any role. Fig. 1b is an example of a generic level ARI diagram, showing that the agent members of those agent groups are authorised to play the depicted roles in that particular interaction called X.

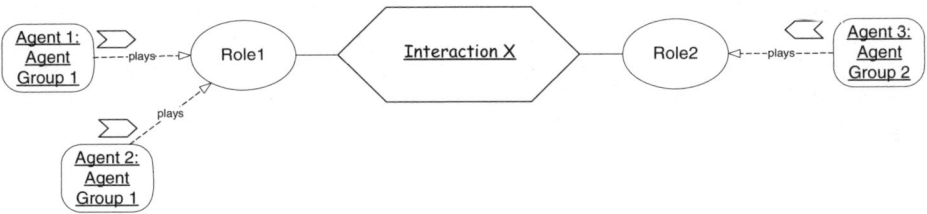

Fig. 2. Example of an ARI diagram (at instance level) with two roles attached that are being played by agents according to their own local view

ARI diagrams can also be expressed at instance level, as in Fig. 2. The ARI instance level diagram depicts an interaction instance and shows agent members that are dynamically assigned to play the roles. In the diagram, roles are never depicted as instances. The local views of the agents participating in the interactions are represented by the so-called Interaction Beliefs (IBs). This concept is not thoroughly presented in this paper, more about its meaning and content can be found in [23]. Basically, the execution of an interaction is the coordinated execution of (local) IBs provided by the participating agents, as is depicted in Fig. 4. The IBs are represented in ARI diagrams by a chevron symbol that appears at the agent side of the association end of the agent-role connector. The connector, an arrow with a hollow triangular head, with an interrupted line as shaft, defines an "`[agent] playing [role]`" – relationship, at the instance level. It is convenient to speak of an agent group as "populating" a role (by using the same arrow symbol) and of a member as "playing" a role [25]. The generic level ARI diagram can also show multiplicity between agents and roles. In an organisation, an agent that wants/needs to play a role must be authorized to play that role.

This authorization is determined at run time by checking (by consulting the generic level ARI diagram) if the agent group has the relation "populates" with that specific role.

3.2 Interaction Beliefs and Protocols

The TALL Interaction Belief (IB) defines the (intended) behaviour of an agent playing a role using a swimlane workflow-like description (based on Petri nets) of the activities, states, and messaging channels as seen by an agent member in a specific interaction (see figures 4, 7, and 9). At the modelling level, an IB is a representation of that "piece" of the BP that is performed as an interaction in the way it is seen from a local agent perspective. An agent has an intended behaviour (do not confuse this with the IB itself), but also a belief or acquaintance model about the behaviour of other agents participating in the interaction. This part of the IB is called expected behaviour. If all agents are linking their intended behaviours together, a process view about how the interaction should be performed is enacted. Of course, if these behaviours are not matching, the interaction will fail. One of the main features of the AGE framework is that it allows these behaviours to be "aligned". Alignment refers here to the soundness of the resulting Petri Net [1]. It is possible that processes deadlock, due to different reasons. In AGE, the Petri Net that results from the combination of IBs can be checked and modified in a way that will ensure the soundness and termination of the interaction seen as a sub-process description [10]. This alignment can be realised automatically or manually, depending on the level of severity of the deadlock situation. Some simple deadlocks can be identified and formalised, allowing for automatic alignment, but others will necessitate human intervention.

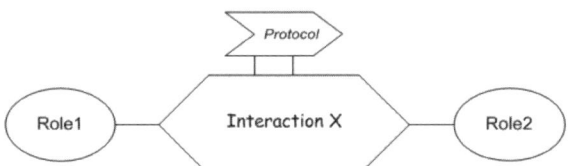

Fig. 3. Example of an interaction with a protocol attached to it

In TALL, the chevron symbol is used for two denotational purposes. The first has been described and depicted in Fig. 2. The other one is that it can also graphically represent a *protocol*. In this case, only the position of the symbol is different, as shown in Fig. 3. It shows a generic interaction X that is regulated by a protocol. The description of a protocol is also a Petri Net with swimlanes, where for each role, certain activities and routings are imposed. The meaning of this concept is centralistic, like the interaction protocols in MESSAGE [5] and interaction diagrams in AORml [27]. If agents perform an interaction that has a protocol (or more) attached to it, the AGE framework will first try to make the agents obey the protocol. They have to try to adopt the behaviour that is described in the associated swimlane to the role they are playing. However, in our framework we allow the agents to overrule the protocol, especially if it is not possible to apply it in an unexpected context. In addition, they can change the content of the protocol, if this is considered necessary at that moment.

In an extreme case, protocols can be attached to *all* the interactions. Actually, this is exactly what MESSAGE and AORml propose. This is also very similar to a workflow enactment machine with agents (but allowing less flexibility). In AGE, even when all interactions have associated protocols, the agents can still overrule them. On the other extreme, none of the interactions is regulated by protocols. Here, agents have to rely only on their own IBs. In real organisations, some protocols can be useful, by regulating those interactions that are routine and predictable, enabling "newcomer" agents to interact even if they do not yet posses the necessary "knowledge"/beliefs (i.e. IBs).

4 Interaction De/Re-composition

Interactions, occurring between agents playing roles, can be identified at different levels of abstraction. On a high level of abstraction, a BP can be conceptualized as a single, over-encompassing interaction (see Fig. 4, the top interaction). This should not be confused with an operational description of a BP, as a workflow. It just shows that a BP has a number of roles involved, and together with the ARI diagrams indicates what agents can be involved. From a modelling perspective, such a representation is very simple but at the same time, its simplicity reduces the expressiveness and richness of the model. From the perspective of the agents performing the roles, such a conceptual interaction is very complex and undesirable, because agents will have to maintain huge IBs that help them perform such a complex interaction.

4.1 Patterns to De/Re-compose Interactions

To be useful, it is necessary to decompose a high-level interaction into 'smaller' interactions in a top-down analysis fashion. Alternatively, it is possible to have a set of interactions where all interactions are "elementary" (i.e. these have no children). Such a set of interactions is not very useful either, because the triggering of an interaction may occur at a higher level of abstraction. Besides, we want to be able to show a coupling between the interactions. There is clearly a need to define interactions that are "in between" the overall BP-describing interaction and the elementary interactions, which are structured on different levels of abstraction.

One can define a priori a complete compositional scheme between the interactions that appears on successive levels of abstraction, looking as a tree structure, where elementary interactions are grouped in interactions that are more complex until the root interaction is reached. However, this will reduce drastically the flexibility of the execution of the BP as a set of on-the-fly executed interactions. Such a tree becomes just another way to illustrate the structure for a workflow execution, conditioning rigidly the way the BP is executed. Our approach aims at defining various "partial recipes" of compositionality, insuring the existence of different possible interaction compositions. In this way, we build a supplementary pool of diagrams that shows potential ways of how some of the interactions can be deconstructed or

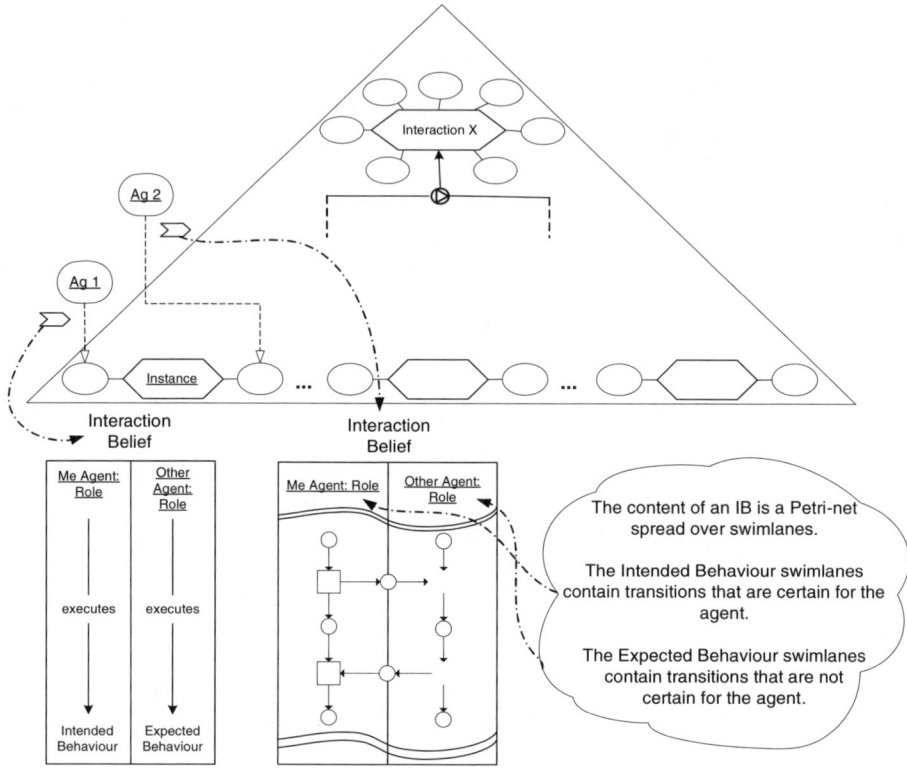

Fig. 4. The link between the interaction composition and the Interaction Beliefs of agents

(re)constructed on different levels. We call this pool the *patterns pool*. The modelling of a prescriptive diagram can be done either top-down or bottom-up. In the first case, the modeller identifies the root interaction (which can represent the whole BP) and then decomposes it into 'smaller' interactions to create the decomposition.

In the second case, the modeller starts identifying 'smaller' interactions that occur between people and/or computer systems at a low level. The lower-level interactions are then assembled into more generic interactions to create the (re)-composition. Of course, practical modelling exercises show that a mixed method (top-down and bottom-up combined) is the best.

4.2 The FlowSet Symbol

For the patterns pool, a new diagram type called the Interaction De/Re-composition Diagram has been introduced in TALL. This IDR diagram shows prescriptively how a higher-level interaction can be structured as a graph of lower level interactions. Fig. 5 shows an example of a *Sales* interaction that is "formed" by sub-interactions. These are linked to the higher-level interaction by a symbol that we call TALL FlowSet (FS). This has been created to depict that sub-interactions together form a set of 'flowing' interactions that are executed in a certain topological order. Parallel execution of the child interactions is allowed. The order of execution is determined during the coordinated execution of the IBs. The diagram is "weakly" prescriptive; it is not always necessary that all child interactions are executed or executed in that specific order, in order to complete a parent interaction. There can be more than one prescriptive IDR diagram for a (high-level) interaction, allowing more recipes for the execution of a complex interaction. The whole set of prescriptive IDR diagrams will form the patterns pool.

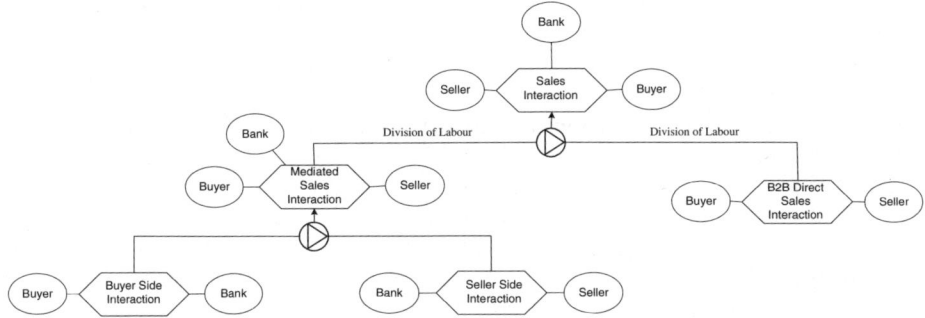

Fig. 5. Example of an interaction composition of a Sales interaction by a TALL IDR diagram

4.3 Comparison of the FS Symbol with "similar" Symbols in UML

The FS symbol is comparable with the UML *aggregation* symbol. In UML, there are two forms of the aggregation relationship. The first is the strong form of aggregation called composition. In [14, p.38] it is stated that: "Composite aggregation is a strong form of aggregation that requires a part instance be included in at most one composite at a time. If a composite is deleted, all of its parts are normally deleted with it". The second level of aggregation proposed in UML corresponds to a weak form of aggregation or shared aggregation. This is a special form of association that specifies a whole-part relationship between the aggregate and a component part. Thus, the part may be included in several aggregates and the owner of the part can change over time. In addition, it does not imply the deletion of the parts when an aggregate referencing is deleted [14].

Lower-level interactions should be, in our view, able to exist on their own. This is not possible when using the strong composite relationship. When the higher-level interaction is deleted, it should be possible to retain the interactions that have been part of this interaction. In addition, it is not desirable that a low-level interaction is only owned by a single high-level interaction. It should be possible that during run-time the interaction graph is reconstructed in another way, which means sub-interactions could become part of other interactions as well.

It can be argued that that the explicit *FlowSet* symbol in the interaction composition is similar to the weak form of UML aggregation and therefore redundant. The main difference with our approach is that both these forms of UML aggregation are bound to a static view. The TALL IDR diagrams show also an (implicit) order between the interactions in the interaction composition. According to [2] UML class diagramming is a non-temporal conceptual modelling technique. For the interaction composition, we want to express a dynamic phenomenon.

4.3 Different Ways to Use IDRs

From an organisational perspective where roles usually have precedence over interactions, depending on the kind of organisation that is investigated, the IDR diagram shows a division of labour. Other approaches [20] emphasise that a certain plan or organisational or individual behaviour exists because of a *goal*. The difference between the IDR diagram and a goal decomposition is that goals focus on "what" should be done, whereas interactions specify also the structure, possibly the order, and the potential participants. The main difference is that a goal describes motivation (internal behaviour) whereas an interaction describes a "collective activity" and its "external behaviour".

For another purpose, the IDR diagrams can be used to track a running process. The AGE framework allows simulations of BPs that are enacted as dynamic sets of interactions. The active (instantiated) interactions can be illustrated in a sort of "cockpit-view" diagram. Each new interaction that is triggered by an agent can appear dynamically in the picture, and the agents that are playing the roles can be shown. Interactions can be connected dynamically to a higher-level interaction, or they can be decomposed into "smaller" interactions. The way the de/re-composition is done can be based on a prescriptive IDR diagram from the patterns pool, or can be determined on the fly. When the process simulation has ended, the IDR diagrams can be used for a "post-mortem" of a BP simulation. This "trace" of IDRs shows what interactions have occurred and who played the roles in these interactions. We assume that each interaction instance could be different from another of the same type, thus allowing a high degree of flexibility. However, certain (partial) patterns occur, and these can be "mined" from a repository of traces and can be added to the patterns pool – which will become in time a repository of prescriptive de/re-compositions. However, the information about "how" these should be executed resides as beliefs of the agents – with the reserve that some protocols can be added to the ARI diagrams and can be emphasised in IDR diagrams from the patterns pool.

4.4 Executing the Process as a Dynamic Set of Interactions

At run time, the initiation of an interaction that may belong to an IDR from the patterns pool occurs ad-hoc. It starts when a specific trigger "activates" an agent (member). The agent "knows" what interaction(s) is(are) necessary at this moment and what role(s) it is required to play. The interaction is instantiated and other agents are "called" to play the remainder of the roles. This mechanism produces a *cascade* of interactions that could be described in a prescriptive IDR. Because of the run-time context, it is possible to have interactions present in the IDR that are not executed. In addition, it is possible to have multiple threads (of the same interaction) that are started independently by different agents. Fig. 4 shows the link between the interaction composition on the upper side and the diagrams that depict the execution of the IBs of agents on the bottom side. If Fig. 4 would depict a single prescriptive IDR (except for the bottom left side interaction that has started to execute) this would mean that the organisation has a single BP and all the process instances are similar. In this case, the interaction composition will be a single tree and the execution is similar to a traditional workflow-based system (except for the agents' flexibility that always allows them to deviate from the process definition).

At the bottom left side of Fig. 4, we show how a cockpit view IDR looks like. For example, the agent **Ag_1** is triggered to start an elementary interaction. It will instantiate an interaction and it will select the appropriate role for itself. The AGE environment will act as an "interaction mediator" and it will find an agent to play the other role (in this case agent **Ag_2**). Both agents will provide an IB that will perform the interaction by executing their intended behaviours.

An agent can trigger an interaction that is not elementary. In this case, this interaction has to be decomposed (either guided by a prescriptive IDR or completely ad-hoc) into component interactions until the elementary level is reached. Agents that trigger an interaction, at any level, have the necessary IBs for this (complex) interaction. It could be that in a specific organisation there is an agent that "understands" the whole BP. That means that this agent has a complete IB that can be applied to the top-level interaction in Fig. 4, and this interaction can be triggered directly by this "omniscient" agent. In this situation, the execution will be performed top-down, and it will be equivalent to a workflow execution. In flat organisations, which are our target domain, interactions are typically triggered at the middle and lower levels of the IDR diagrams.

4.5 Example of an Interaction

The generic ARI-diagram of a *Sales* interaction is shown in Fig. 6 and it defines three agent groups and it enacts an authorization scheme. The multiplicity in the diagram shows that for a successful interaction the Buyer and Seller roles must be played by at least one agent. In addition, the Bank role is not always required in a Sales interaction.

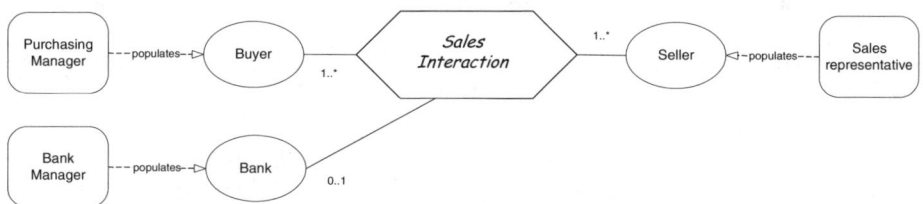

Fig. 6. The generic ARI diagram for the Sales Interaction

A prescriptive IDR-diagram for this Sales interaction has already been presented in Fig. 5. Because the involvement of a bank is not always required, two different Sales interactions can be distinguished. Fig. 5 shows that the Sales interaction is decomposed into two "smaller" interactions that represent the two different Sales interactions. *The B2B Direct Sales Interaction* does not require a bank and the *Mediated Sales Interaction* requires mediation of a bank in order to successfully complete the interaction.

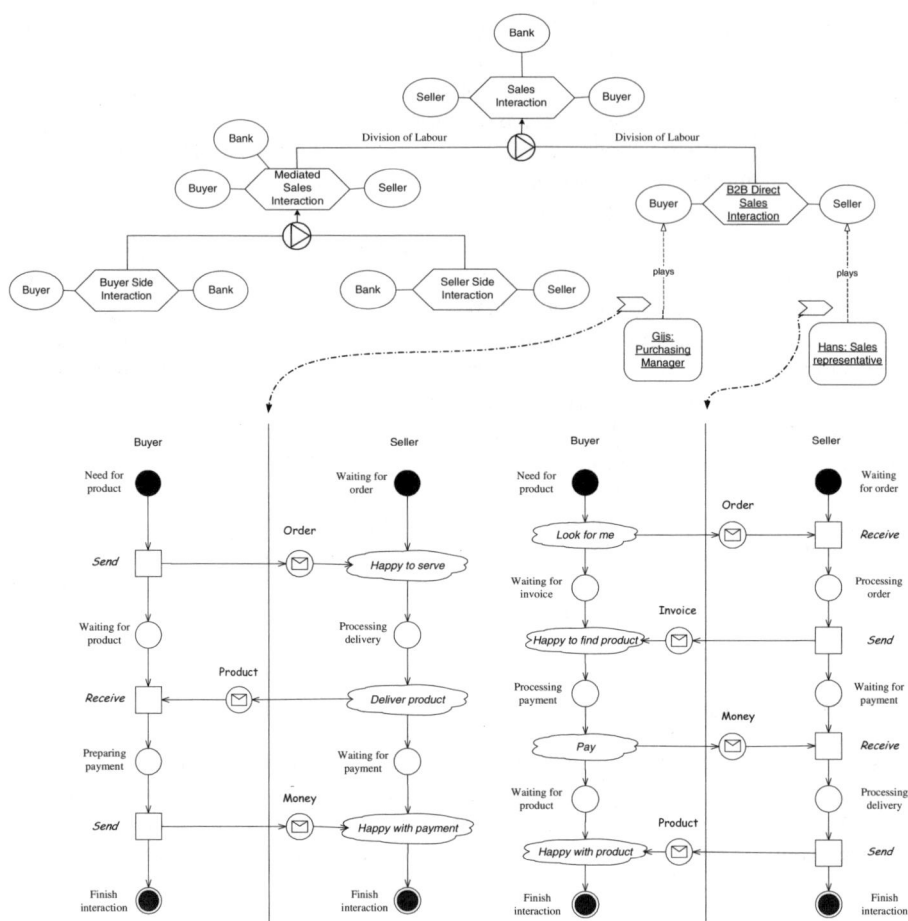

Fig. 7. A run-time snapshot IDR diagram with the Interaction Beliefs of the buyer and seller specified in an IB diagram through Behaviour Nets

Fig. 7 shows a run-time IDR diagram in which the *B2B Direct Sales interaction* has started to execute. Here, two agents are executing the interaction according to their own IBs that appear in the run-time diagram. Both IBs can be exploded from the chevron symbols into the IB diagrams that are also depicted here. These Interaction Beliefs consist of Behaviour Nets representing intended and expected behaviours. The seller is waiting for an order, and after the order is received, it will send an invoice to the buyer. It expects that the buyer will pay, and when the payment is done, only then it will send the product. However, the buyer expects to receive the product first and pay after. It is obvious that two "rigid" agents with these kinds of beliefs and behaviours will never succeed to complete the intended Sales interaction.

The buyer believes the interaction cannot successfully end, and it may want to involve a mediator assuming that he has a prescriptive mental model of the interaction composition (as depicted in Fig. 5). Therefore, it is possible that the buyer agent will trigger the *Mediated Sales interaction* during execution of the *B2B Direct Sales interaction*. Of course, it is also possible the *B2B Direct Sales interaction* is not executed at all because both parties believe (from their experience) that a Sales interaction can only successfully complete with a mediator involved. If the bank becomes involved as a mediator, then the cockpit view will look like in Fig. 8.

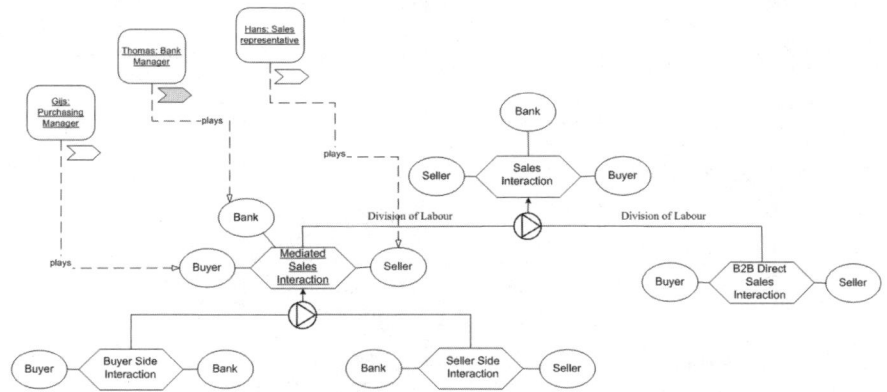

Fig. 8. A run-time snapshot IDR Diagram in which the *Mediated Sales interaction* has started to execute

In the *Mediated Sales interaction*, we assume that the bank has the experience (behaviour) to solve the problem of the buyer and seller without modifying their basic behaviours (only minor additions and/or revisions are necessary). *The Mediated Sales interaction* is triggered at run-time and the bank becomes part of the interaction. The bank will provide the seller with the money for the delivery of the product and the buyer will, after receiving the product, pay the bank instead of directly paying the seller. Both parties will have to pay a small fee for this service.

Fig. 9 depicts the Interaction Belief of the bank for the *Mediated Sales interaction*. This can be seen in a real-time IDR diagram as two separate interactions, one between the seller and the bank and one between the buyer and the bank (as depicted in the prescriptive IDR diagram of Fig. 5 – *Buyer Side Interaction* and *Seller Side Interaction*). Fig. 9 depicts a typical IB where the middle swimlane shows the intended interaction and the other two swimlanes show what the bank believes about the behaviour of the other two roles involved. Essential here is that what the bank believes matches the intended behaviours of the other two agents. It is also possible that these two agents align their behaviours (e.g. the buyer accepts to pay first) and they can fulfil the interaction without the intervention of the bank.

The knowledge about the prescriptive structure of the interaction composition could be in any of the agents involved (in our case, we said that the buyer is the one that "knows" an alternative where the bank can be involved). Alternatively, we can introduce central views "through the back door" and define the pool of interaction patterns as a central resource.

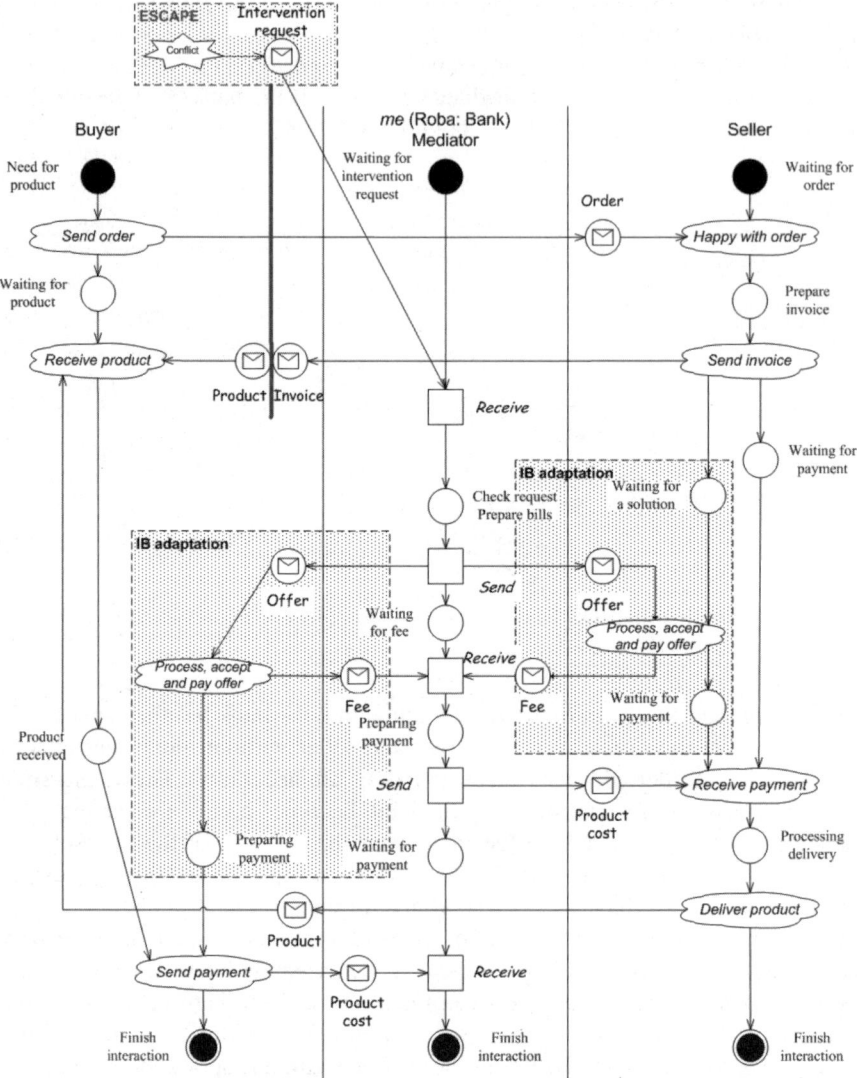

Fig. 9. The Interaction Belief of the Bank

5 Discussion, Future Work, and Conclusions

An organisation that adopts an agent-based software solution supporting its BPs, can best use it for executing the "low level" interactions. These tend to be more stable and simple, and/or can be regulated by protocols, making their automatic execution easier. Typically, higher-level interactions are triggered by humans, and at the lower levels of composition, software agents will be delegated to take over the routine interactions. This will allow for flexibility on the higher levels but will also ensure stability on the lower levels – especially if the organisational learning process has enacted various protocols at this level. However, it is possible in organisations that are characterized by formal procedures and standard ways of doing things the situation is reversed. In this case, generic protocols will dominate at higher levels whereas at lower levels the agents have full autonomy to enact the necessary interactions. Even in this case, interaction-centric, agent-based systems can be useful, especially if they are linked to (owned by) whole organisations and not by individuals. However, it can be argued that these agents will be just high-level workflow enactment systems.

The current target for our BP supporting multi-agent system (MAS) is not the bureaucratic and centralised organisation but the flexible, decentralised organisation that emphasises horizontal collaboration. In these organisations, BP participants do not perform work according to pre-defined tasks but they act in empowered roles part of a dynamic social context. This context is usually characterized by complex, dynamic and emergent BPs. Although such processes can easily be modelled on a high-level by using centralistic models they are not able to capture the "hidden" and "implicit" knowledge/behaviour of the BP participants. Modelling such BPs by using role-based interactions between agents allows organizations to better understand the intricacies of their BPs, visualized as a set of interactions. Each agent having its own IB mirrors the way business is done in dynamic social contexts where everyone decides on its own 'piece' of business. Still, the participants need to interact because no one has all data or skills to complete the BP alone.

If the social contexts are stable, experience from previous "runs" can lead to the requirements for a stable software package developed in a "classical" fashion. However, in the dynamic social contexts described above the requirements change all the time and it becomes difficult to develop a useful solution that can support these dynamic social contexts that make up organisations. By using TALL to rapidly identify the behaviours of the actors and by creating a weakly descriptive pool of interaction patterns, software developers that are using AGE as an interactive gaming/simulation tool can develop simulated agents that exhibit the identified (emergent) behaviour - "captured" from the playing actors. Moreover, they can use these agents as the base for the MAS that will support the human actors in the real process. The development process can be iterative and the simulation and the usage cycle can change and enrich the behaviour of the agents as well as the global behaviour of the organisation. Our strong belief is that the main advantage of the presented method is the speed of the requirements analysis, design, development, and integration in a dynamic organisation.

Future Work. The next step in our research will be to formalise the de/re-composition process, in order to have a precise operational semantics, allowing the agents in the AGE environment to make use of the patterns pool. This is also necessary for visualisation purposes in order to have clear semantics for what is exactly shown in the cockpit view during simulations. Another interesting future research direction is towards building IDR based representations for the post-mortem analysis. This will allow the analysts to investigate past simulations and infer new IDRs that can be added to the patterns pool.

Currently, we have no explicit representation of goals in our framework. We consider that there is a strong conceptual link between BPs, interactions and goals, and we intend to explore the nature and applicability of this link. A more elaborated mechanism for authorisations and explicit representations of responsibility is also necessary. Finally, a methodology to apply the models in a consistent way is needed.

Conclusion. We have argued that the interaction concept can be useful to analyse, model, and simulate organisations in BP-oriented and agent-oriented ways. In addition, semantics for the execution of these interactions can form the basics for implementing agent-based software support systems for BPs. Interaction, as an abstract and intangible concept, can be explicitly represented as an agent-role-interaction triad. We also argue that the modelled organisation should build a set of interactions, in the form of ARI diagrams. In addition, based on patterns that emerge with experience, a pool of prescriptive recipes that prescribe how to dynamically structure these interactions (IDR diagrams) can be built. The syntax and semantics of an interaction composition has been described, showing how interactions can be triggered and chained in top-down and bottom-up fashion.

BPs can be represented as static structures, like Petri Nets, and workflow representations. This ensures discipline, stability and repeatability in an organisation. However, it produces rigidity, centralistic views, and obsolescence. We argued that a novel way to enact complex, distributed and dynamic BPs, based on de/re-composition of interactions, ensures more organisational flexibility and more autonomy for the participants. We believe that the first step to achieve this is by adopting an agent-role-interaction based view.

References

1. Van Der Aalst, W.M.P.: Loosely coupled interorganizational workflows: modeling and analyzing workflows crossing organizational boundaries. Information & Management 37(2), 67–75 (2000)
2. Cabot, J., Olive, A., Teniente, E.: Representing Temporal Information in UML. In: Stevens, P., Whittle, J., Booch, G. (eds.) «UML» 2003 - The Unified Modeling Language. Modeling Languages and Applications. LNCS, vol. 2863, pp. 44–59. Springer, Heidelberg (2003)
3. Cervenka, R., Trencansky, I., Calisti, M., Greenwood, D.: AML: Agent Modeling Language Toward Industry-Grade Agent-Based Modeling. In: Odell, J.J., Giorgini, P., Müller, J.P. (eds.) AOSE 2004. LNCS, vol. 3382, pp. 31–46. Springer, Heidelberg (2005)
4. Cervenka, R., Trencansky, I., Calisti, M.: Modeling Social Aspects of Multi-Agent Systems: The AML Approach. In: Proceedings of AAMAS05, AOSE, Utrecht, The Netherlands, pp. 85–96 (2005)
5. Eurescom, Message: Methodology for Engineering Systems of Software Agents, Deliverable 1, Initial Methodology. Heidelberg, Germany (Project P907-GI) (2000)
6. Horling, B., Lesser, V.: Using ODML to Model Multi-Agent Organizations. In: Proceedings of IAT 2005, pp. 72–80 (2005)
7. Klein, M., Dellarocas, C., Bernstein, A.: Introduction to the special issue on adaptive workflow systems. Computer Supported Cooperative Work 9(3-4), 265–267 (2000)
8. Klügl, F., Oechslein, C., Puppe, F., Dornhaus, A.: Multi-agent Modelling in comparison to Standard Modelling. In: AIS 2002, pp. 105–110. SCS Publishing House (2002)
9. Koning, J., Romero-Hernandez, I.: Thoughts on an Agent Oriented Language Centered on Interaction Modeling. In: Proceedings ISADS 2002, Holland. Springer, Heidelberg (2002)

10. Meyer, G.G., Szirbik, N.B.: Behaviour Alignment as a Mechanism for Anticipatory Agent Interaction. In: Proceedings of ABiALS'06 (2006), Available from:http://www.rug.nl/staff/g.g.meyer/p4.pdf?as=pdf
11. Miles, S., Joy, M., Luck, M.: Designing Agent-Oriented systems by Analysing Agent Interactions. In: Ciancarini, P., Wooldridge, M.J. (eds.) AOSE 2000. LNCS, vol. 1957, pp. 171–184. Springer, Heidelberg (2001)
12. Mintzberg, H.: Structure in fives: designing effective organizations. Prentice Hall, Englewood Cliffs, New Jersey (1983)
13. Object Management Group (OMG): Relationship Service Specification (CORBA) (2000)
14. Object Management Group (OMG): UML 2.0 Superstructure Specification (2004)
15. Odell, J., van Dyke Parunak, H., Bauer, B.: Extending UML for Agents. In: Proceedings AOIS 2000, Stockholm, pp. 3–17 (2000)
16. Owen, M., Raj, J.: BPMN and Business Process Management: Introduction to the New Business Process Modeling Standard. Popkin Software (2003)
17. Reijers, H.A., Rigter, J.H.M., Van Der Aalst, W.M.P.: The Case Handling Case. International Journal of Cooperative Information Systems 12(3), 365–391 (2003)
18. Roest, G.B., Szirbik, N.B.: Intervention and Escape Mode: Involving Stakeholders in the Agent Development Process. In: Proceedings AAMAS06, AOSE, Hakodate, Japan, pp. 109–120 (2006), Available from: http://www.rug.nl/staff/g.b.roest/GijsBRoest&NickBSzirbik_Escape_Intervention.pdf?as=pdf
19. Simon, H.A.: Models of bounded rationality. MIT Press, Cambridge, MA (1982)
20. Simon, G., Mermet, B., Fournier, D.: Goal Decomposition Tree: An Agent Model to Generate a Validated Agent Behaviour. In: Baldoni, M., Endriss, U., Omicini, A., Torroni, P. (eds.) DALT 2005. LNCS (LNAI), vol. 3904, pp. 124–140. Springer, Heidelberg (2006)
21. Snoo De, C.: Modelling planning processes with TALMOD. University of Groningen(2005), Available from: http://tbk15.fwn.rug.nl/tal/downloadfile.php?id=28
22. Stuit, M.: Modelling organisational emergent behaviour using compositional role-based interactions in an agent-oriented language. University of Groningen (2006), Available from: http://www.rug.nl/staff/m.stuit/p1.pdf?as=pdf
23. Stuit, M., Szirbik, N.B., De Snoo, C.: Interaction Beliefs: a Way to Understand Emergent Organisational Behaviour. In: Proceedings ICEIS 2007, Madeira, Portugal (2007)
24. Sowa, J.F.: Knowledge Representation: logical, philosophical, and computational foundations. Brooks/Cole, Pacific Grove, CA (2000)
25. Steimann, F.: Role = Interface: A merger of concepts. Journal of Object-Oriented Programming 14(4), 23–32 (2001)
26. Tambe, M., Adibi, J., Al-Onaizan, Y., Erdem, A., Kaminka, G.A., Marsella, S.C., Muslea, I.: Building agent teams using an explicit teamwork model and learning. Artificial Intelligence 110, 215–239 (1999)
27. Wagner, G.: The Agent-Object-Relationship metamodel: towards a unified view of state and behavior. Information Systems 28(5), 475–504 (2003)
28. White, S.A.: Introduction to BPMN. IBM Cooperation (2004)
29. Yan, Y., Maamar, Z., Shen, W.: Integration of Workflow and Agent Technology for Business Process Management. In: Proceedings of CSCW in Design 2001, Ontario, Canada, pp. 420–426 (2001)

Model Driven Development of Multi-Agent Systems with Repositories of Social Patterns

Rubén Fuentes-Fernández, Jorge J. Gómez-Sanz, and Juan Pavón

Universidad Complutense Madrid, Dep. Ingeniería del Software e Inteligencia Artificial[*]
28040 Madrid, Spain
{ruben,jjgomez,jpavon}@fdi.ucm.es
http://grasia.fdi.ucm.es

Abstract. Design patterns are templates of general solutions to commonly-occurring problems in the analysis and design of software systems. In mature development processes, engineers use and combine these patterns to work out those parts of their systems that correspond to well-identified issues in their domains. The design of new structures is just concerned with those aspects that are specific for their projects and with the glue between different components. Model driven development approaches can benefit of design patterns to improve the building of models and their transformations; at the same time, design patterns can take advantage in this kind of approaches of a better integration in the overall development process. In the case of Agent-Oriented Software Engineering, design solutions for agents and multi-agent systems have been also described in the literature. However, their application and transformation to code largely relies on manual processes. This paper proposes a framework that includes repositories of patterns that can be reused in different projects and processes to generate models and code for multi-agent systems on different target platforms. Instead of focusing on low-level issues, our approach positions the abstraction level of these design patterns at the intentional and social features that characterize multi-agent systems. The paper illustrates this framework with a case study about the development of the models of an agent-based system for collaborative filtering of information.

1 Introduction

Design patterns [7] are proven solutions to usual problems in software design. At this stage of a development process, engineers have to consider matters that may not become visible until later in the implementation. Reusing design patterns helps to prevent subtle issues that can cause major problems and improves models and code readability for architects and coders familiar with the patterns. The application of design patterns has been usually a manual work. Available automated support for their use just covers the generation of some views of models or code that the development team must integrate with the remaining information about the system.

---

[*] This work has been funded by Spanish Ministry of Science and Technology under grant TIN2005-08501-C03-01.

The spread of Model Driven Development in Software Engineering improves the integration of design patterns in the whole development cycle. Model Driven Development (MDD) advocates for building systems from models through automated transformations. The Object Management Group (OMG) proposes a standardized approach to MDD, the Model Driven Architecture (MDA) [10]. MDA mainly considers three kinds of models: Computational Independent Models (CIMs), Platform Independent Models (PIMs) and Platform Specific Models (PSMs). CIMs are intended for domain practitioners. They state the requirements of the system without knowledge about the models or artefacts to realize them. PIMs further specify the system to be with details about its software architecture and design. Finally, PSMs give the details necessary to build the system for a specific platform. The main advantage of this approach is that it layers the development in levels of abstraction that includes the models. From this departure point, its processes are intended to encourage reuse of design solutions and to save effort in the migration of systems to different target platforms. In this setting, design patterns are not longer external knowledge that forges the specifications of the system, but they become an integral part of the models of the system and are transformed with the remaining information to running code.

Agent-Oriented Software Engineering (AOSE) has already applied both design patterns and MDD to build Multi-Agent Systems (MAS). Literature has documented research for its reuse in different designs. These patterns specify aspects like communication protocols, power relationships between individuals, or planning modules in individual agents. Examples of these patterns are the FIPA protocols for standardized auctions between agents (*http://www.fipa.org*). About MDD, the works described in [2, 4, 8, 16] are state of the art applications to build MAS integrating a MDD perspective. Nevertheless, many of these MAS developments with MDD are *ad-hoc* solutions [2, 4,], with focus on the generation of very specific portions of code, like the support for a protocol or a planning module. The generation of code for other aspects of MAS or domains usually needs a complete new development from scratch. Anyway, none of the studied approaches truly integrate design patterns, as developers must manually bring them into the specifications. That is, developers, departing from their own knowledge about the patterns or some abstract specification of them, must describe and customize those patterns for their specific methodology and project. Readers interested in more information about AOSE and MDD can read about several researches in [9].

From our point of view, the application of design patterns in MDD for MAS can be regarded at a higher level of abstraction than it is done in current practices. Instead of engineering isolated elements, we consider design patterns for intentional and social issues that traverse several aspects in the architecture of the whole MAS or individual agents. These patterns are intended to describe the ruling principles in the architecture of the MAS rather than the individual elements encompassed by that architecture. Given this target, we propose a partial framework for model driven development based on repositories of *social patterns* that act as predefined CIMs. Sets of automated transformations described declaratively will enable in the framework the generation of the PIMs for specific methodologies from these *social patterns*.

A *social pattern* is a proven solution for a design problem about the intentional and social architecture of MAS. They are the equivalent to design patterns at the

Knowledge Level [12] that characterizes the agent paradigm. Two are the sources of knowledge for these patterns in our framework. First, the practice of the agent community in the development of MAS, which already provides well-established solutions to some recurrent problems, like the organization of distributed problem solvers for planning or reasoning for auctions. Second, knowledge from Social Sciences, specifically from the Activity Theory. The Activity Theory (AT) [11] is a paradigm for interdisciplinary human research based on the socio-cultural approach initiated by Vygotsky [18] in early 1920s. Previous works [5, 6] have proven the advantages of AT as a source of expert knowledge about intentional and social aspects that can be applied in the development of MAS. Besides, AT is the foundation of the language UML-AT [6] that our framework uses to describe the diagrams of *social patterns*. This avoids the introduction of bias to specific methodologies and the need of rewriting patterns for every method. Section 0 provides a brief overview of UML-AT along a description of the structure used to describe *social patterns*. This structure is based on those usually applied in the software patterns community. That section also presents a pair of *social patterns* to be used in the case study.

The other component of our framework is the MDD process that generates new views or code from *social patterns*. This process and its support tools are described for models in Section 3. The process for code is slightly different [16] and not considered in this paper.

The application of the framework is shown in Section 4, where *social patterns* help to generate the specification of an application for collaborative filtering. This application defines agents that support users in order to know the level of trust deserved by the information from other users. According to this case study and additional experimentation, Section 5 presents a discussion about the use of *social patterns* in this MDD framework, including its potential benefits and limitations and a roadmap of future work.

2 Describing *Social Patterns*

A *social pattern* is depicted using a structure very similar to that of design patterns [7]. Modifications of this format are mainly intended to ease a more automated processing, both to decide its applicability and to integrate it in existing MAS specifications. Parts of the description of a design pattern related to their implementation in a concrete programming language are removed for *social patterns*. In this MDD framework, code is generated from *social patterns* using general mechanisms to transform modelling primitives. These mechanisms are explained in Section 3. The proposed structure for *social patterns* considers the elements in Fig. 1.

A *social pattern* includes a unique *pattern name* that identifies it. The *intent* is a textual description of the motivation to use the pattern. It says the kind of intentional or social setting that the pattern models. This section is intended to clarify the kind of information that the pattern can add to the specifications of a MAS and the mutual influences between a MAS that exhibits the features of this pattern and its human environment. The *foundation* relates the pattern with its theoretical basis in AT research. This knowledge is useful to guide further insights in its meaning.

Pattern name	
1. **Intent**	
2. **Foundation**	
3. **Use**	
• Applicability Participants + Structure + Collaboration	• Solution Participants + Structure + Collaboration
(More use pairs with Applicability + Solution)	
4. **Consequences**	
5. **Examples**	
6. **Related patterns**	

Fig. 1. Structure to describe *social patterns*

The *use* section of the structure is concerned with the situations where the pattern can be applied and how it is considered. It is composed by *use pairs*; each of ones represents a different interpretation of the pattern. This allows giving a precise description of the *pattern* for different target domains. As *social patterns* are applied in a Software Engineering context and with the intent of having automated support tools, the *use* pairs have a twofold representation: textual and with UML-AT diagrams (see Section 2.1 for more details about this language). Both of these descriptions are used to explain the pre and post-conditions of the pattern to developers. However, the UML-AT diagrams have the additional purpose of enabling the automated processing of the *social patterns*. UML-AT diagrams in *use pairs* are templates with variables. To check applicability, these diagrams must match against the MAS specifications; for the application of patterns, variables are instantiated with information from the specifications. This will be described with more details in Section 3.

A *use pair* comprehends *applicability* and *solution* sections. The *applicability* component determines the pre-conditions that must hold in the specifications for the pair being applicable. The *solution* component describes the information to add to the specifications in order to include the pattern. If the *applicability* restrictions are satisfied, the *solution* can be automatically instantiated. However, a user can choose to apply the pair ignoring the prerequisites and instantiating by hand the *solution* Each component of a pair can have sections for *participants*, *structure*, and *collaboration*, following usual decompositions in software modelling. The *participants* describe the elements that appear in the pattern, the *structure* describes static relations between these elements, and the *collaboration* section reports the scenarios where these elements interact, that is, the dynamics of the pattern.

The *consequences* of the pattern are its results, drawbacks, side effects, and trade offs. The *examples* section includes real usages of the pattern described with *use pairs*. Finally, there is also a list of *related patterns*. These are patterns that can apply instead of this one in certain settings or be used along with it. This section also accounts for the differences between these *related patterns* and the one currently reported. *Consequences* and *related patterns* are described with natural language.

After this introduction to the structure to represent *social patterns*, the following sub-sections elaborate on some of its elements. Section 2.1 gives an introduction to

UML-AT and the underlying concepts of AT and Section 2.2 describes parts of two examples of *social patterns*.

2.1 UML-AT

Activity Theory (AT) [11] is a framework for the study of different forms of human practices and their evolution in a social and historical context. Its analysis focuses on the interactions and conflicts between individuals and their environment, which includes their societies.

The intention of employing AT concepts in AOSE demands that they are expressed in a language understandable and applicable by development teams. With this purpose, AT concepts have been described using the UML profile extension mechanism [13]. The resulting language is called UML-AT. UML-AT includes the core concepts of AT (which are represented in Fig. 2) and additional ones to ease reasoning about specifications (e.g. contribution relationships or the *artifact* concept). The full specification of UML-AT can be found at *http://grasia.fdi.ucm.es/at/uml-at*. Examples of its use appear in the figures of this paper.

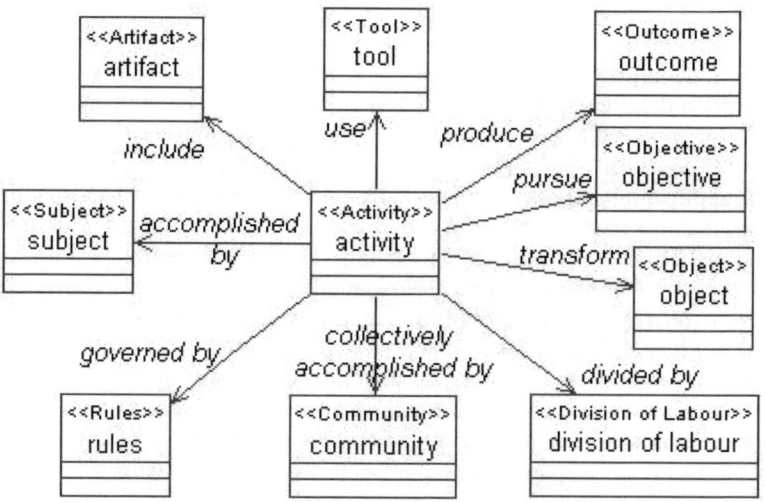

Fig. 2. UML-AT representation of AT core concepts

AT builds around the concept of *activity* [1]. The *activity* reflects a process, with individual and social levels interleaved. At the individual level, the *activity* focuses on the *subject* that carries out the *activity* to obtain the *outcome* that satisfies the needs represented by his *objectives*. The *outcome* is the result of the transformation of an *object* using *tools*. *Tools* always mediate *subject*'s interaction with the environment and can be both material and mental. The social level has as its key concept that of *community*. The *community* represents those *subjects* who share the same *object*. An *activity* may concern many *subjects* and each *subject* may play one or more roles and have multiple motives. *Rules* influence the behaviour of *subjects* in their *communities*.

They include both explicit and implicit norms and social relationships within a *community*. The *division of labour* describes how the *community* is organised as related to the specific *activity*. All of these concepts are interconnected with *mediation relationships*. Examples of these relations are the already mentioned mediations of *tools* between *subjects* and *objects* or *rules* governing *subjects* in their *communities*.

2.2 Examples: Selfish Actors and Community of Experts

This section presents two examples of *social patterns* that will be applied in the case study of this paper. These are the descriptions of the kind of mental attitudes in *selfish actors* (this pattern was first stated in AT studies in [3]) and the organization known as *community of experts* (as described in [17]). In both cases, the description is focused on the UML-AT representation of the *solution* component in a *use pair*, specifically on the *structure* and *participants* components (see Section 2). In the patterns in Fig. 3 and Fig. 4, the names of the entities are variables and the remaining properties (i.e. the stereotypes and names of relationships) are constants. *Applicability* sections are not considered for these examples because they will not be used in the use case of Section 4, where the user will choose the instantiation of variables in *solution* components. More about how a pair can be added to the specifications appears in Section 3.

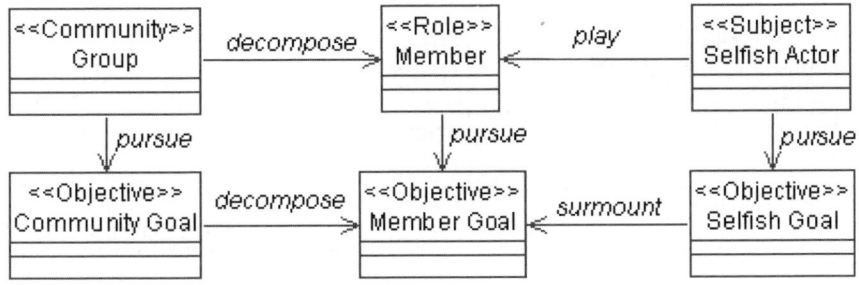

Fig. 3. Mental attitudes of selfish actors described with UML-AT

Selfish actors [3] are those that put their own interests in front of those of the community to which they belong. That is to say that these actors consider their individual and particular goals more important than those goals that they pursue for their group. Fig. 3 models this situation. The *Selfish Actor* has an individual *objective Selfish Goal* that does not emerge from the *community Group*. Besides, the actor plays the *role Member*. This *role* represents the fact that the actor belongs to the *community Group*. The *role* has goals that are intended to satisfy the overall objectives of the *community*. For instance, the *Group* pursues the *Community Goal* that is satisfied through the *Member Goal* pursued by *Members*. As a consequence of playing the *role*, the actor also pursues the *role*'s *objectives*. The selfishness of the actor appears in the relation *surmount* from the *Selfish Goal* to the *Member Goal*. Despite of the circumstances, the *Selfish Actor* always considers his own goals as more relevant than those emerging from the *community*.

The second *social pattern* in this section describes a kind of organization called the *community of experts* [17]. A *community of experts* is a flat organization where different actors coordinate through mutual adjustment of their solutions so that overall coherence can be achieved. The agents interact by pre-established rules of order and behavior. This organization appears in Fig. 4. The *Community of Experts* is composed by actors who play the role *Expert*. The *object* of the *community*, and also its *outcome*, is the *Knowledge* that its interaction builds. Related with this *Knowledge*, an *Expert* carries out two different *activities*: he modifies the common knowledge with the *activity Contribute* and he uses it with the *activity Consume*. The *activity Contribute* pursues the *Expert Goal* to improve the shared *Knowledge*. The *activity Consume* pursues the *Individual Goal* that covers the utility that the *Expert* obtains from the *Knowledge*. Pay attention to the fact that this last *objective* is not necessarily positive for the *community*. The *Expert* could use the knowledge for his own reasons, without profit for the *community*. The positive contribution of the *Expert* to the group is represented just by the *Expert Goal*.

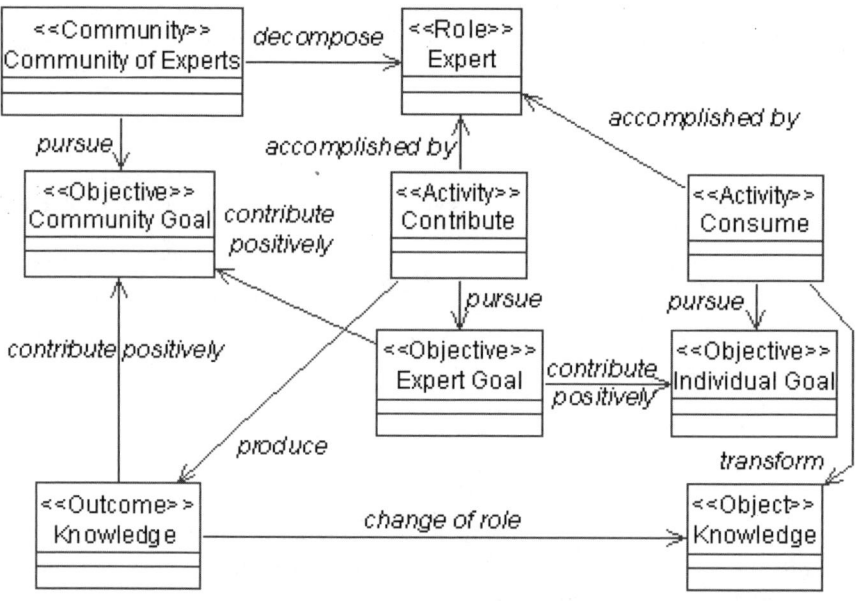

Fig. 4. An organization of *community of experts* described with UML-AT

The pattern in Fig. 4 does not consider the rules to build and update the *Knowledge*, which largely depend on the domain of the community. In this sense, Fig. 4 represents a domain independent version of the pattern that dismisses part of its related information. Domain specific instances that consider those rules should be described as different *use* pairs for this *social pattern*. From the AT point of view, these norms would be *division of labour* if directly emerging from this *activity* context or *rules* if coming from the surrounding environment. Anyway, they are not included here for the sake of brevity.

3 MDD with *Social Patterns*

The *social patterns* of the previous section provide the knowledge used in our MDD framework to add new features to MAS specifications in an automated way. The process is depicted in the activity diagram of Fig. 5. It is conceived with the goal of enriching the support for MAS development provided by available agent-oriented methodologies. So, it is always applied as a complement of a target development process. Some of the tasks are justified by the intended neutrality of the process about methodologies, as it does not impose restrictions about languages, models, or tools. The remaining of this section explains the tasks in Fig. 5.

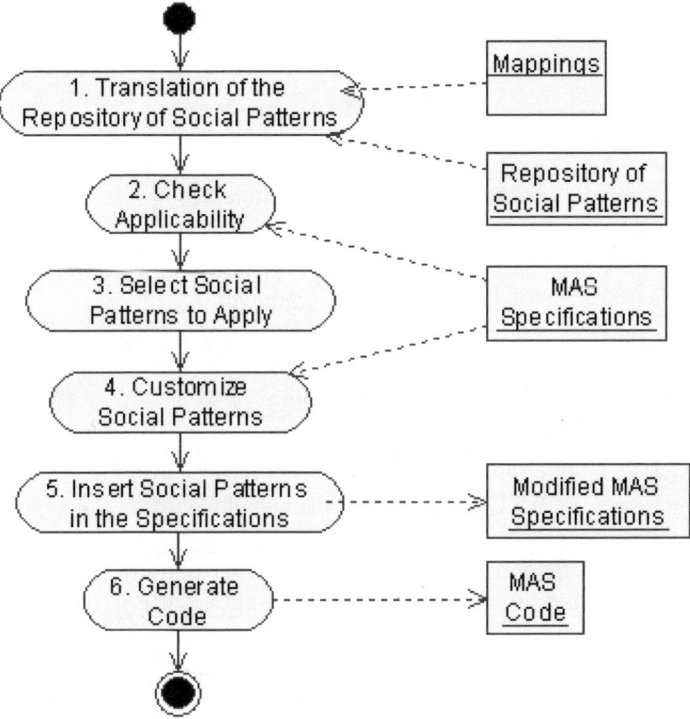

Fig. 5. UML activity diagram for the MDD process with *social patterns*

Task 1 is the *Translation of the Repository of Social Patterns*. As section 2 explains, *social patterns* are described with text and UML-AT diagrams. The representation with UML-AT provides the methodology neutrality and avoids the need of rewriting the patterns for every agent-oriented methodology. However, patterns described with UML-AT cannot be directly applied in the target development process. Using the integration framework for specifications coming from different methodologies described in [6], the repository of *social patterns* is translated to the language of the target process. This translation process is summarized in Fig. 6.

The translation in Fig. 6 is based on mappings between UML-AT and other languages. Mappings describe correspondences between structures (i.e. *source* and *target patterns*) that share variables. Structures in mappings are composed by entities, and relationships. All of these elements can have related properties that are key-value pairs. The properties contain fixed values or variables. Some examples of simplified mappings appear in Table 1. The translation departs in state 1 from the source specifications and the mappings to perform the translation according to the involved languages. For every mapping, it looks for instances of the *source pattern* in the available models (task 3 in Fig. 6). When a match is found, there is a correspondence between variables in the source *pattern* and the values from the specifications that appear in the found instance. This correspondence from variables to values is the result of task 4 and it allows grounding in task 5 those variables in the *target pattern* that are shared with the *source pattern*. The resulting structure is added to the translation in task 6.

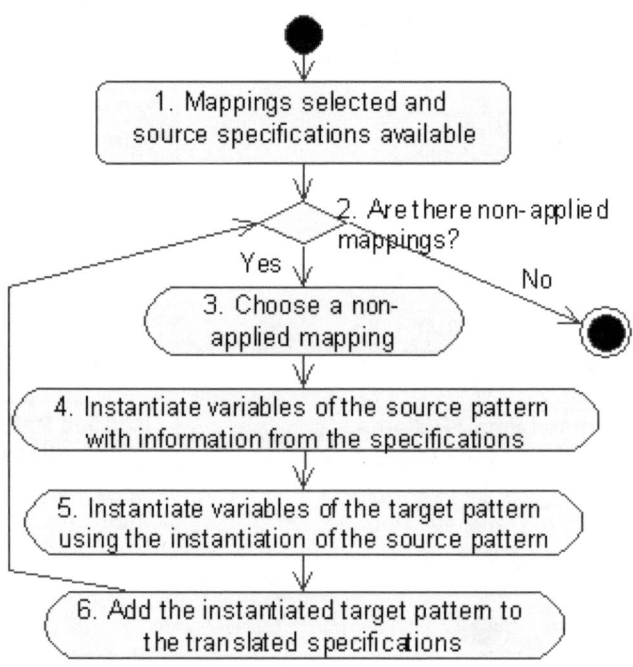

Fig. 6. UML activity diagram for the translation process

After the study of the translation process, this discussion returns to Fig. 5. With the repository available in the target modelling language, the process can carry out task 2 (i.e. *Check Applicability*). If a *social pattern* has UML-AT descriptions in the *applicability* section, those diagrams can be checked against the specifications. This is a process of pattern matching between graphs described in the language of the target methodology: the *social pattern* is applicable to a MAS if a group of elements in its specifications corresponds to the diagrams in the *applicability* section of some of the *use pairs* of that *social pattern*. This pattern matching process was already described in [5]. Briefly, two graphs correspond if they have entities of the same types connected by the same relationships, and their properties are compatible. Two properties are compatible if they have the same values or at least one of them is non-ground variable. If a non-ground variable must be compatible with a value, it becomes ground to that value. In the case that the *applicability* section of the *social pattern* only contains a textual description, it is considered as applicable by default and the decision about it is left to the user. Besides, the usercan decide to apply the pattern ignoring the *applicability* sections.

In task 3, the user selects the *social patterns* to insert in the specifications, among those that are applicable or he manually selects. As it was stated in Section 2, *social patterns* contain variables that users must instantiate to customize them. Some of the variables in the *solution* sections can be ground from their *applicability* sections in task 2; nevertheless, it is possible that after task 2 there are already non-instantiated variables and the user must choose their values. This is done in task 4. Turning back to the examples of *solutions* in Section 0, users should decide who the selfish actors are or what piece of knowledge the community of experts collaboratively builds. Besides, if several patterns are added at the same time, users could decide to link different variables to the same value. For instance, they could decide that the *roles* of *Member* and *Expert* in the previous examples will be played by the same agent. The result of task 4 is therefore a set of *solutions* of *social patterns* where all of their variables have been ground, either with information from the specifications or with values directly provided by the user.

Task 5 inserts the customized *solutions* in the specification of the MAS. New architectural principles about the intentional and social aspects of the system are added in this way to the original specification.

At this point, the users of the process can follow their own development process or can use some of the other support tools available in the integration framework based on mappings [6]. Its ability to represent the specifications of a system in the language of choice at every moment allows, among others, detecting inconsistencies in the specifications [5] and generating code for the MAS [16] (this is task 6 in Fig. 5). This MDD framework is supported by a tool called the *Activity Theory Assistant* (ATA) (*http://ingenias.sourceforge.net*). The general purpose of the ATA is to help MAS engineers and developers with AT-based techniques.

4 Case Study: Building a Network of Trust

The application of the MDD framework is illustrated with a case study about a recommender system, which relies on collaborative filtering techniques. Collaborative filtering assumes that if a user finds interesting a piece of information, then other users with similar opinions and preferences may also find interesting the same piece of information. Therefore, this system implements workflows to distribute and evaluate information among users, who are grouped in communities. There are agents to represent users and communities in the system. The full specification of this system, which has been developed with the INGENIAS methodology [15], can be found at *http://grasia.fdi.ucm.es/ingenias*.

An important issue for the collaboration of agents in the recommender system is the level of trust that the community has in its members. Trusted members propose relevant information; non-trusted ones usually give information of bad quality and can even be expelled from the community. With the original organization of the system, community agents manage the updates of the score of members in a centralized way. This situation appears in the model from INGENIAS of Fig. 7. The *Personal Agent* plays the roles of *Advisor* and *Suggester*. As a *Suggester*, the agent gives *Suggestions* (i.e. information) to the community of users, what pursues the goal *Provide Interesting Documents*. On the other side, the agent pursues the goal *Preserve Document Quality* as an *Advisor*. The interest of this role is that the community deals with relevant information. The workflow shows that the *Community Agent* initiates *Share Documents* that selects *Advisors*, requests them to evaluate a *Suggestion*, and recollects the evaluations. In other workflow that does not appear in Fig. 7, the *Community Agent* also expels members with low scores.

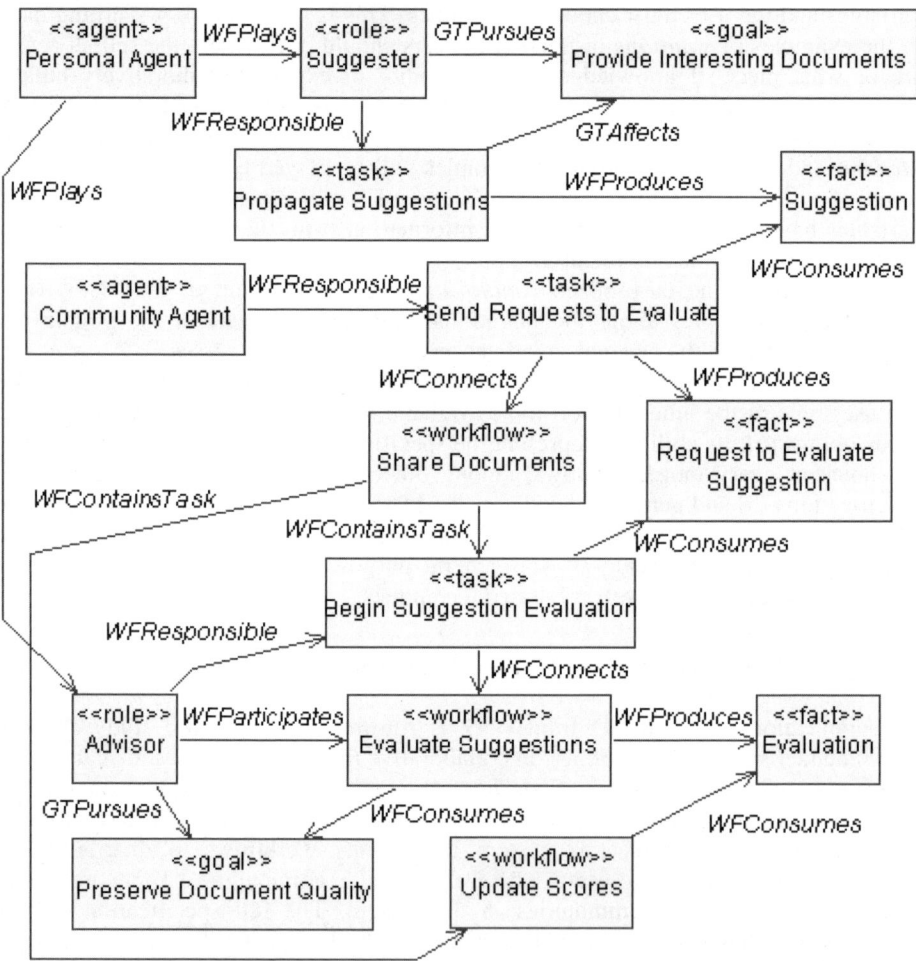

Fig. 7. A representation of the collaborative filtering community with INGENIAS concepts (which are represented as UML stereotypes)

The new organization proposed in this case study promotes that *Personal Agents* as *Advisors* directly share the information about other agents. The role of the *Community Agent* as intermediary in the evaluation process would disappear. This situation corresponds to the decentralized organization previously mentioned in this paper of the *community of experts* (see Section 2.2 and Fig. 4).

Table 1. Mappings used in the translation from UML-AT to INGENIAS

Source pattern (AT)	Target pattern (INGENIAS)
community → [decompose] → role 1	- Organization Model group(community) → [OHasMember] → role(role 1)
activity → [accomplished by] → role activity → [transform] → object activity → [pursue] → objective	- Agent Model role(role) → [WFResponsible] → task(activity) role(subject) → [GTPursues] → goal(objective) - Tasks and Goals Model task(activity) → [GTAffects] → fact(object) task(activity) → [GTSatisfies] → goal(objective)
activity → [accomplished by] → role activity → [produce] → outcome activity → [pursue] → objective	- Agent Model role(subject) → [WFResponsible] → task(activity) role(subject) → [GTPursues] → goal(objective) - Tasks and Goals Model task(activity) → [WFProduces] → fact(outcome) task(activity) → [GTSatisfies] → goal(objective)
artifact → [contribute positively] → objective	- Tasks and Goals Model *task 1*[1] → [GTSatisfies] → goal(objective) *task 1* → [WFProduces] → fact(artefact)
objective 1 → [contribute positively] → objective 2	- Tasks and Goals Model goal(objective1)→[ContributePositively]→goal(objective2)
community → [pursue] → objective	- Organization Model group(community) → [GTPursues] → goal(objective)

The first step to apply the *social pattern* according to the process in Fig. 5 is to translate it from UML-AT to INGENIAS (task 1) following the process in Fig. 6. This translation uses the mappings in Table 1. These mappings describe structures to identify over the original specifications (in this case over the *solution* of the *social pattern*) and the structures to insert in the translation. The result of the application of these mappings appears in Fig 8. For instance, let see the process with the first mapping in the table. It would be selected in task 3 of Fig. 6. This mapping states that a *source pattern* with a *decompose* relation between a *community* and a *role* (same name for the AT concepts and the variables) in UML-AT corresponds to a *target pattern* with a *OHasMember* relation between a *group* and a *role* in INGENIAS. In Fig. 4, this *source pattern* appears for the *community Group* and the *role Member*, what would be found by task 4 and determine the instantiation of variables *community* and *role*; thus, the instantiated *target pattern* with a *group Group* and a *role Member*

---

[1] Elements in italics are new ones introduced by the mappings.

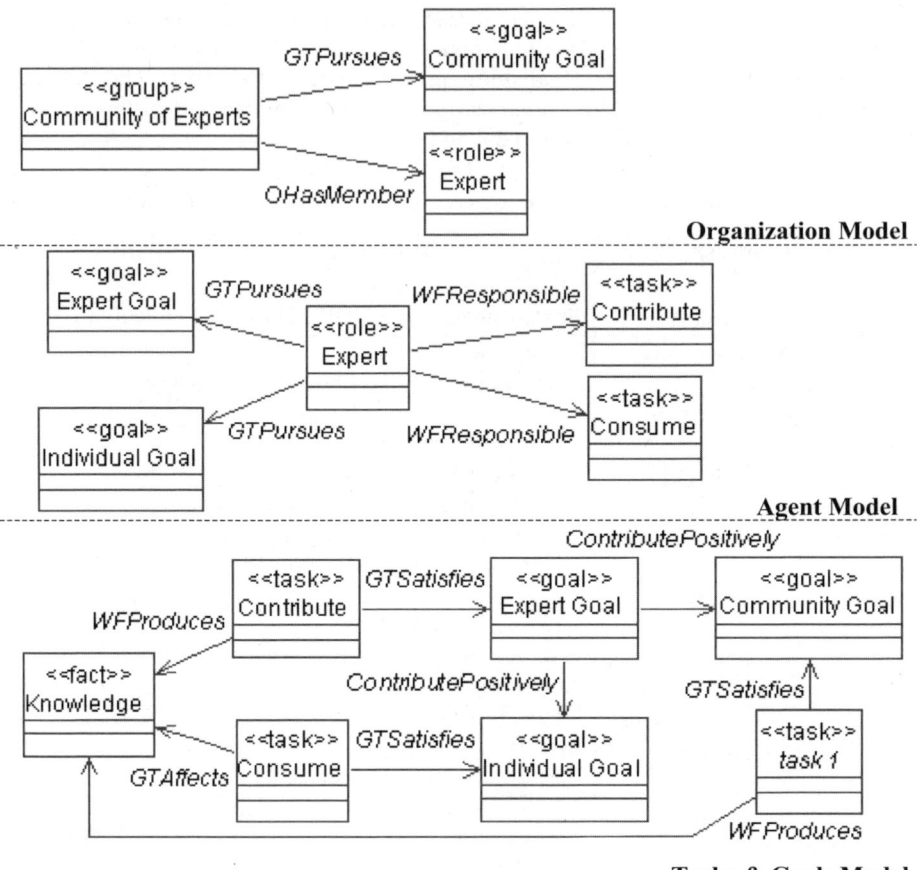

Fig. 8. Translation to INGENIAS of the *community of experts solution* component

would be created by task 5 and inserted in an Organization model from INGENIAS with task 6. More about the mappings and the translation process can be found in [6].

According to Fig. 5, the next step would be to check the applicability of the translated *social patterns* (task 2). In this case, the user needs to add new elements to the specifications, the new organization of the community of experts and their related elements. These concepts do not exist previously in the specifications and the pattern is not directly applicable. However, the user can ignore task 2 and decide that he wants to apply the pattern in any case in task 3.

The *solution* component in Fig 8 is a template of model that must be customized for this specific case study in task 4. The names of the entities are variables that the engineer has to instantiate with specific information for the current collaborative filtering problem. As the user decides in task 3 the application of the pattern from scratch, there are no previously ground variables because of a match between the *applicability* component of the *solution* and the specifications of the MAS. The choices in this case can be seen in Table 2.

The result of task 4 allows generating instances of the translated solution components (those in Fig 8) where all the variables are substituted according to the previous choices. The resulting ground diagrams are inserted in the specifications with task 5. As a consequence of this addition, the *social pattern* of the community of experts is incorporated to the specifications. The result of this last task can be seen in Fig 9.

Table 2. Instantiation of the *community of experts* to the recommender system with INGENIAS concepts

Community of Experts	Recommender System
group(Community of Experts)	group(*Community*)
role(Expert)	role(*Advisor*)
goal(Community Goal)	goal(Preserve Document Quality)
goal(Expert Goal)	goal(Preserve Document Quality)
goal(Individual Goal)	goal(*Know User's Reliability*)
fact(Knowledge)	*User's Score*
task(Contribute)	workflow(Evaluate Documents)
task(Consume)	task(*Evaluate User*)

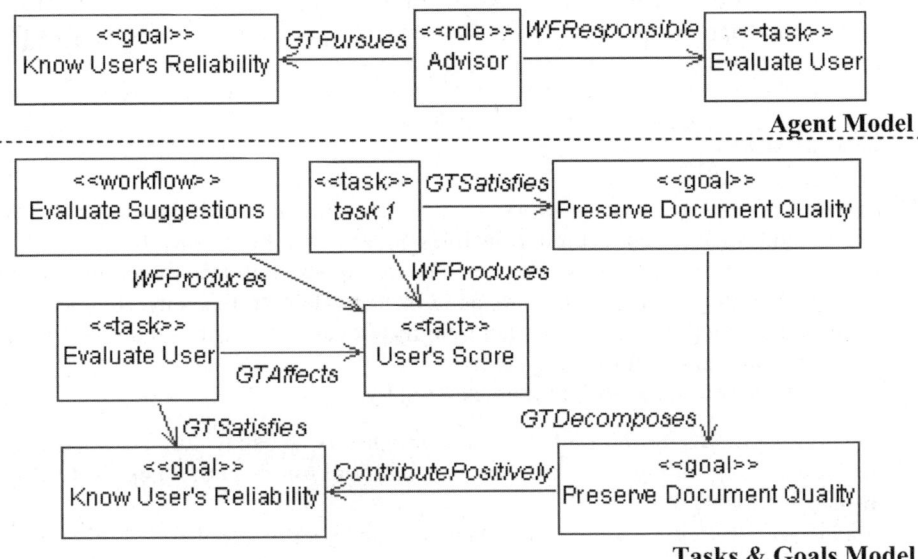

Fig. 9. Recommender system with a *community of experts*

Some remarks must be done about the result. The first one is that the *social pattern* is not only instantiated with information from the specifications (like in the case of the already existing role *Expert*), but also can generate completely new elements (like in the case of the *Community*). Elements in italics in Table 2 correspond to new concepts. The second one is that when adding the information, engineers must be conscious of possible changes in the semantics of some elements. For instance, the workflow *Evaluate Suggestions* was initially intended to receive a *Suggestion* from the *Community Agent* and generate an *Action* that evaluated it. With the new organization, this workflow would receive the *Suggestion* directly from an agent playing the role of *Advisor*, would generate the *Action*, and would also update the score of the suggester (i.e. the *fact User's Score*). The third is that the MDD process can take advantage of semantic equivalences between structures in modelling languages. In Fig 8 the *Expert Goal* contributes positively to the *Community Goal*; this relation is substituted by *GTDecomposes* in Fig 9, since a goal that is the result of the decomposition of some original goal is considered as contributing positively to that original goal. This kind of semantic equivalencies are considered through mappings of the translation process (see Fig. 6).The fourth and final remark is that inserting a *social pattern* in an existing specification has a potentially deep impact in those models. The insertion of the *community of experts* changes the overall organization of the agents in the recommender system and their mental attitudes, as new goals, tasks, and relations may come with the pattern.

In this system, besides the social pattern of the *community of experts*, agents want to keep their scores as high as possible. A high score would allow them, for instance, acquiring social power, becoming more difficult to be expelled, and biasing the community to their own interests. This corresponds to the *social pattern* of selfish agents that also appears in Section 2.2. It would be applied in the same way as the *community of experts*.

5 Conclusions

This paper has shown a framework to generate partial specifications that add intentional and social architectural principles to existing MAS specifications. The framework is built over previous infrastructure to integrate information coming from different methodologies and to generate code from models. In this way, it supports a model driven development process where engineers can work with predefined high-level behavioural patterns for their systems.

This model driven framework is characterized by:

- *Social patterns* as meaningful reusable components of behaviour. These patterns are focused on ruling principles of organizations or mental states, instead of isolated elements of the MAS.
- *Social patterns* as customizable templates for specific problems. Variables in *social patterns* are intended to be substituted for information from the specifications in order to connect the pattern with the existing specifications.
- Integration of information through mappings between languages. *Social patterns* are described with a neutral language called UML-AT. Its application to a concrete methodology relies on correspondences between the structures of UML-AT and those of the modelling language of the methodology.
- Generation of code through mappings between languages. Correspondences between languages are also the key to define the transformations from models to

code in this approach. In this process they are called *templates*. These *templates* are slightly different from the *mappings* used to integrate models in order to optimize the coding. More information can be found in [16].

The main advantage of this framework is the availability of a repository of *social patterns*. It gives engineers the possibility of building their specifications (and therefore the running MAS) from blocks that determine the main issues of their systems. The goal is not to specify the details of agents or organizations, but the main principles that the system has to satisfy. These patterns crystallize the knowledge of the AOSE community about the intentional and social aspects that are key in the development of MAS.

An additional advantage of the framework over other approaches with design patterns is the integration of these patterns in a whole development process supported by automated tools. In this way, the development team just selects and customizes *social patterns*, leaving to tools the integration of the new information.

The current research has raised some limitations that will be the object of future work. The first one is about deciding the applicability of *social patterns*. In the current repository, the *applicability* component of the *use* section of a *social pattern* can contain textual or UML-AT information. Although further enrichment of UML-AT will improve the specification of these restrictions, it is not enough to express, for instance, statements about quantification or properties of the elements in UML-AT. OCL [14] is being studied as a possible complement for this purpose. Besides, only complete matches of the *applicability* component of *social patterns* with the specifications are considered. Allowing partial matches will reduce the user's workload to customize the *solutions* of the patterns to insert. The second one is the need to change some aspects in the definition of *social patterns*. The current structure is inspired by that common for design patterns. However, in a model driven development process, more aspects need to be considered. For instance, issues like side effects or the connection with related patterns are described with plain text, what makes difficult adding automated support. The foreseen solution is to enrich the definition of these aspects with UML-AT too.

References

1. Bednyi, G.Z., Meister, D.: The Russian Theory of Activity: Current Application to Design and Learning. Lawrence Erlbaum Associates, Mahwah, NJ (1997)
2. Bernon, C., Gleizes, M., Peyruqueou, S., Picard, G.: Adelfe, a methodology for adaptive multi-agent systems engineering. In: Petta, P., Tolksdorf, R., Zambonelli, F. (eds.) ESAW 2002. LNCS (LNAI), vol. 2577, pp. 156–169. Springer, Heidelberg (2003)
3. Bratus, B.S.: The place of fine literature in the development of a scientific psychology of personality. Soviet Psychology XXV(2), 91–103 (1986)
4. Cossentino, M.: From Requirements to Code with the PASSI Methodology. In: Henderson-Sellers, B., Giorgini, P. (eds.) Agent-Oriented Methodologies, ch. IV, pp. 79–106. Idea Group Publishing, USA (2005)
5. Fuentes, R., Gómez-Sanz, J.J., Pavón, J.: Managing Conflicts between Individuals and Societies in Multi-Agent Systems. In: Gleizes, M.-P., Omicini, A., Zambonelli, F. (eds.) ESAW 2004. LNCS (LNAI), vol. 3451, pp. 106–118. Springer, Heidelberg (2005)

6. Fuentes-Fernández, R., Gómez-Sanz, J.J., Pavón, J.: Integrating Agent-Oriented Methodologies with UML-AT. In: Proceedings of the 5th International Joint Conference on Autonomous Agents and Multiagent Systems (AAMAS–2006), Hakodate, Japan, May 2006, pp. 1303–1310. ACM Press, New York (2006)
7. Gamma, E., Helm, R., Johnson, R., Vlissides, J.: Design Patterns: Elements of Reusable Object-Oriented Software. Addison Wesley Professional Computing Series. Addison-Wesley, London, UK (1995)
8. Gracanin, D., Bohner, S.A., Hinchey, M.: Towards a Model-Driven Architecture for Autonomic Systems. In: Proceedings of the 11th IEEE International Conference and Workshop on the Engineering of Computer-Based Systems, ECBS'04 (2004), Brno, Czech Republic, May 2004, pp. 500–505. IEEE Press, Los Alamitos (2004)
9. Henderson-Sellers, B., Giorgini, P. (eds.): Agent-Oriented Methodologies. Idea Group Publishing, London, UK (2005)
10. Kleppe, A., Warmer, J., Bast, W.: MDA Explained: The Model Driven Architecture-Practice and Promise. Addison-Wesley, London, UK (2003)
11. Leontiev, A.N.: Activity, Consciousness, and Personality. Prentice-Hall, Englewood Cliffs (1978)
12. Newell, A.: The knowledge level. Artificial Intelligence 18, 87–127 (1982)
13. OMG: Unified Modelling Language Specification. Version 2.0 (2005), http://www.omg.org
14. OMG: Object Constraint Language Specification. Version 2.0 (2006), http://www.omg.org
15. Pavón, J., Gómez-Sanz, J.J., Fuentes, R.: The INGENIAS Methodology and Tools. In: Henderson-Sellers, B., Giorgini, P. (eds.) Agent-Oriented Methodologies, ch. IX, pp. 236–276. Idea Group Publishing, USA (2005)
16. Pavón, J., Gómez-Sanz, J.J., Fuentes, R.: Model Driven Development of Multi-Agent Systems. In: Rensink, A., Warmer, J. (eds.) ECMDA-FA 2006. LNCS, vol. 4066, pp. 284–298. Springer, Heidelberg (2006)
17. Sykara, K.P.: Multiagent systems. AI Magazine 19(2), 79–92 (1998)
18. Vygotsky, L.S.: Mind and Society: Development of Higher Psychological Processes. Harvard University Press, Cambridge (1978)

A Norm-Governed Systems Perspective of Ad Hoc Networks

Alexander Artikis[1], Lloyd Kamara[2], and Jeremy Pitt[2]

[1] Institute of Informatics & Telecommunications,
NCSR "Demokritos", Athens, 15310, Greece
[2] Intelligent Systems & Networks Group,
Electrical & Electronic Engineering Department,
Imperial College London, SW7 2BT, UK
a.artikis@acm.org, {l.kamara,j.pitt}@imperial.ac.uk

Abstract. Ad hoc networks are a type of computational system whose members may fail to, or choose not to, comply with the laws governing their behaviour. We are investigating to what extent ad hoc networks can usefully be described in terms of permissions, obligations and other more complex normative relations, based on our previous work on modelling norm-governed multi-agent systems. We propose to employ our existing framework for the specification of the laws governing ad hoc networks. Moreover, we discuss a software infrastructure that executes such specifications for the benefit of ad hoc network members, informing them of their normative relations. We have been developing a sample node architecture as a basis for norm-governed ad hoc network simulations. Nodes based on this architecture consider the network's laws in their decision-making, and can be individually configured to exhibit distinct behaviour. We present run-time configurations of norm-governed ad hoc networks and indicate design choices that need to be made in order to fully realise such networks.

1 Introduction

An *Ad Hoc Network* (*AHN*) is a transient association of network nodes which inter-operate largely independently of any fixed support infrastructure [30]. An AHN is typically based on wireless technology and may be short-lived, supporting spontaneous rather than long-term interoperation [31]. Example AHNs are formed by the devices of consumers entering and leaving an 802.11 wireless hot spot covering a shopping mall (for buying/selling goods consumer-to-consumer style by matching potential buyers and sellers); by participants in a workshop or project meeting (for sharing and co-authoring documents); or by emergency or disaster relief workers, where the usual static support infrastructure is unavailable.

An AHN may be visualised as a continuously changing graph [30]: connection and disconnection may be controlled by the physical proximity of the nodes or it may be controlled by the nodes' continued willingness to cooperate for the formation, and maintenance, of a cohesive (but potentially transient) community.

An issue that typically needs to be addressed when managing and maintaining an AHN is that of routing. AHNs are not usually fully connected; participating nodes are often required to act as routers to assist in the transport of a message (packet) between sender and receiver nodes. Resource-sharing is another challenge that needs to be addressed during the life-time of an AHN; the participating nodes compete over a set of limited resources such as bandwidth. AHNs are often specifically set up for sharing a resource such as broadband Internet access, processor cycles, file storage, or a document in the project meeting example mentioned above.

The nodes of an AHN are programmed by different parties — moreover, there is no direct access to a node's internal state and so one may only make inferences about that state. It is possible, even likely, that the nodes of an AHN will fail to behave as they ought to — for example, a node acting as a router may move out of communication range or run out of power and may therefore be unable to forward packets. Furthermore, system components may intentionally mis-behave in order to seek unfair advantage over others. A 'selfish' node, for instance, may refuse to forward packets for other nodes while gaining services from these nodes [42]. The controller of a resource may grant access to the limited resource based on personal preferences rather than an agreed metric. Moreover, due to the (typically) wireless nature of AHNs, a participating node should be prepared to counteract against rogue peers. For all of these reasons, an AHN needs to be 'adaptable' — that is, it should be able to deal with 'exceptions' such as the ones mentioned above.

Within the EPSRC-funded Programmable Networks Initiative, we are investigating to what extent adaptability can be enhanced by viewing AHNs as instances of *norm-governed* systems, that is, systems in which the actual behaviour of the members is not always ideal and, thus, it is necessary to express what is permitted, prohibited, obligatory, and possibly other more complex normative relations that may exist between the members [18]. We have developed a framework for an executable specification of norm-governed multi-agent systems (ngMAS) that defines the social laws governing the behaviour of the members of such systems [2, 1, 3] (we will use the terms 'social law' and 'norm' interchangeably). We propose to use this framework as an infrastructure for the realisation of norm-governed AHNs (ngAHN)s.

The remainder of this paper is organised as follows. First, we give an overview of ngAHNs. More precisely, we review our work on specifying ngMAS and propose ways to apply this work to AHNs. Second, we discuss a software infrastructure for the realisation of ngAHNs. This infrastructure executes the specifications of ngAHNs to inform members of their permissions, obligations, etc, at any point in time. Third, to simulate ngAHNs, we discuss a sample node architecture. Nodes based on this architecture consider the network's social laws in their decision-making, and can be individually configured to exhibit distinct behaviour. Finally, we summarise the presented work and outline our current research directions.

2 Norm-Governed Ad Hoc Networks

In previous work [2, 1, 3] we presented a theoretical framework for specifying ngMAS in terms of concepts stemming from the study of legal and social systems. The behaviour of the members of a ngMAS is regulated by social laws expressing their:

(i) physical capabilities (that is, what actions members can perform in their 'environment');
(ii) *institutional powers* [38, 19], a characteristic feature of norm-governed systems, whereby designated participants have the institutional power (or are empowered) by the system to create/modify facts of special significance within the system, *institutional facts*, usually by performing a specified kind of act (such as when an agent awards a contract and thereby creates a set of normative relations between the contracting parties);
(iii) permissions, prohibitions and obligations;
(iv) sanctions, that is strategies countering the performance of forbidden actions and non-compliance with obligations.

Note that there is no standard, fixed relationship between physical capability, institutional power and permission. For example, being empowered to perform an action A does not necessarily imply being permitted to perform A or being capable of A. (For further discussion and references to the literature see [25, 19].) The laws comprising level (ii) of the specification correspond to the *constitutive norms* that define the meaning of the agents' actions. Levels (i) and (iii), respectively, can be seen as representing the *physical* and *normative* environment within which the agent interactions take place.

We have employed our framework for specifying a protocol used to regulate the control of access to shared resources [1], a typical issue in AHNs. The protocol expresses the conditions under which a node can be said to have the institutional power to request to access the limited resources. Exercising this power causes the node eligible to be granted access to these resources. Whether a node is *permitted* or not to exercise this power is another aspect expressed by the protocol specification. Typically, the performance of a forbidden action leads to a sanction (irrespective of whether the node performing the action was empowered to do so). Similarly, the protocol expresses, among other things, the circumstances in which it is meaningful to say that a resource controller is obliged or simply permitted to grant/revoke access to the resource, and what the consequences are, and the conditions under which it is physically (practically) possible or permitted for a node to manipulate a shared resource, or obligatory to release the resource, and what the consequences are.

While being a member of an AHN typically implies being within communication range with at least one of the remaining members, being a member of a ngAHN additionally implies being governed by the ngAHN's social laws (see Figure 1), that is, having a set of institutional powers, permissions, and obligations. Consider, in the resource sharing example, a node N that is not a member of the ngAHN but is within communication range of the ngAHN's members.

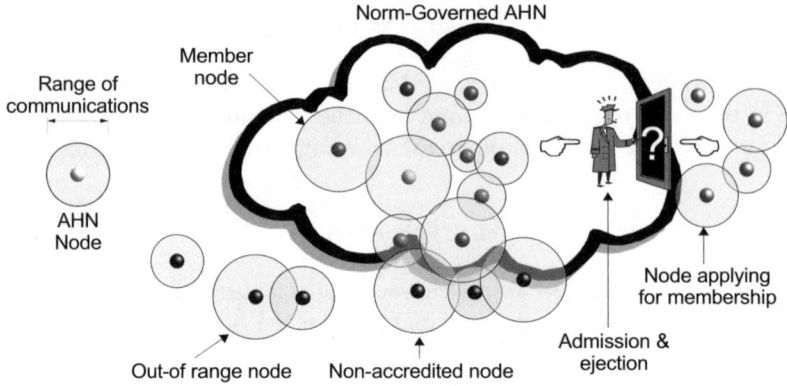

Fig. 1. An AHN Node (left) and a norm-governed AHN

N can communicate a request for access to the shared resources, by means of sending a message of a particular form via a TCP/IP socket connection, for instance, but its request will be ignored, since N is not empowered to request the resources.

In order to become a member of a ngAHN, a node participates in a *role-assignment* protocol. If the node is accepted in the ngAHN then it will occupy a set of roles expressing its institutional powers, permissions and obligations while in that ngAHN. The decision-making procedure for awarding or denying a role is application-specific. Example procedures are *chair-designated* (an elected node assigns roles), *election* (the members of a role-assigning committee comprising existing member nodes or other elected nodes vote on role-assignment), *argumentation* (the members of a role-assigning committee debate on role-assignment) and *lottery scheduling* (role-assignment operates on a probabilistic basis). Formalisations of the first three types of procedure may be found in [1,32,2] respectively. In all cases, the decision-making procedure of the role-assigning authority is informed by, among other things, whether or not the applicant N satisfies the *role conditions* (for example, nodes acting as routers should have broad communication range), whether or not N has been banned from a ngAHN, or how many times it has been disqualified (banning and disqualification are discussed below).

As already mentioned, actuality does not necessarily coincide with ideality in ngAHNs, that is, a node may, inadvertently (due to network conditions or software errors) or maliciously, perform forbidden actions or not comply with its obligations. In this case sanctions may be enforced, not necessarily as a 'punishment', but as a way of adapting the network organisation. Sanctions may come in various forms; for example, a sanctioned node N may be:

- suspended, that is, N loses for a specified time period its institutional powers and permissions. In the resource sharing example, a suspended node loses access to the shared resources as its requests for access will not be serviced (since the node lost the institutional power to request the resources).

- disqualified; N loses its ngAHN membership and thus loses its institutional powers and permissions. However, N may re-apply to enter the ngAHN and, if successful, will regain its institutional powers and permissions.
- banned, that is, N loses its membership and may not re-apply to enter the ngAHN.

Other forms of sanction are possible, such as, for instance, 'bad reputation' (see [9, 16] for a few examples); the choice of a sanction type (when is a node penalised, what is the penalty that it has to face, who applies the penalty, etc) is application-specific.

Another possible enforcement strategy is to try to devise (additional) physical controls that will force nodes to comply with their obligations or prevent them from performing forbidden actions. When competing for hard disk space, for example, a forbidden revocation of the resource access may be physically blocked, in the sense that a node's account on the file server cannot be deleted. The general strategy of designing mechanisms to force compliance and eliminate non-permitted behaviour is what Jones and Sergot [18] referred to as *regimentation*. Regimentation devices have often been employed in order to eliminate 'undesirable' behaviour in computational systems (see, for example, *interagents* [35], *sentinels* [22], *controllers* [26], *guards* and *enforcers* [5]). It has been argued [18], however, that regimentation is rarely desirable (it results in a rigid system that may discourage agents from entering [34]), and not always practical. The practicality of regimentation devices is even more questionable when considering AHNs, due to the transient nature of these networks. In any case, violations may still occur even when regimenting a computational system (consider, for instance, a faulty regimentation device). For all of these reasons, we have to allow for sanctioning and not rely exclusively on regimentation mechanisms.

The process of joining or excluding a node from a ngAHN is a part of a *session control* protocol which further prescribes ways for inviting to join, or withdrawing from a ngAHN, changing the laws of a ngAHN, determining which resources are to be shared, and so on.

In order to realise ngAHNs, we need to provide mechanisms for informing member nodes of the social laws governing their behaviour — a discussion of such mechanisms is presented next.

3 Norm-Aware Nodes

We encode social laws specifications of norm-governed systems in executable action languages. We have shown how two such languages from the field of Artificial Intelligence (AI) may be used to express these specifications: the $\mathcal{C}+$ language [15, 14] and the Event Calculus (EC) [23]. The $\mathcal{C}+$ language, notably when used with its associated software implementation, the Causal Calculator (CCALC), already supports a wide range of computational tasks of the kind that we wish to perform on system specifications. A major attraction of $\mathcal{C}+$ compared with other action languages in AI is its explicit semantics in terms of

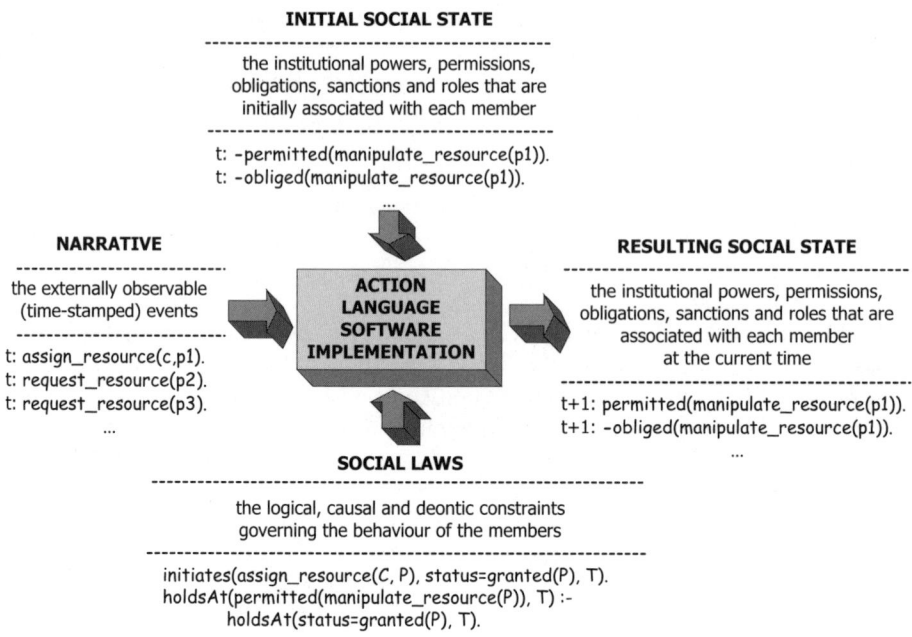

Fig. 2. Computing with Social Laws

labelled transition systems, a familiar structure widely used in logic and computer science. EC, on the other hand, does not have an *explicit* transition system semantics, but has the merits of being simple, flexible, and very easily and efficiently implemented for an important class of computational tasks. It thus provides a practical means of implementing an executable system specification. The Society Visualiser (SV) [3] is a logic programming implementation that supports computational tasks on system specifications formulated in EC. A detailed discussion of both action languages and their software implementations can be found in [3].

Members of a ngAHN should be aware at any point in time of their institutional powers, permissions, obligations and sanctions. Such information may be produced by computing answers to 'prediction' queries on the social laws specification. This type of computational task, which is supported by each action language software implementation (ALSI), that is, CCALC and SV, may be expressed, in the context of ngAHNs, as follows. The input to an ALSI includes an initial *social state* — that is, a description of the institutional powers, permissions, obligations and sanctions that are initially associated with the ngAHN members, and a *narrative*, that is, a description of temporally-sorted externally observable events (actions) of the ngAHN. The outcome of a prediction query (if any) is the current social state — that is, the members' institutional powers, permissions, obligations and sanctions that result from the events described in the narrative.

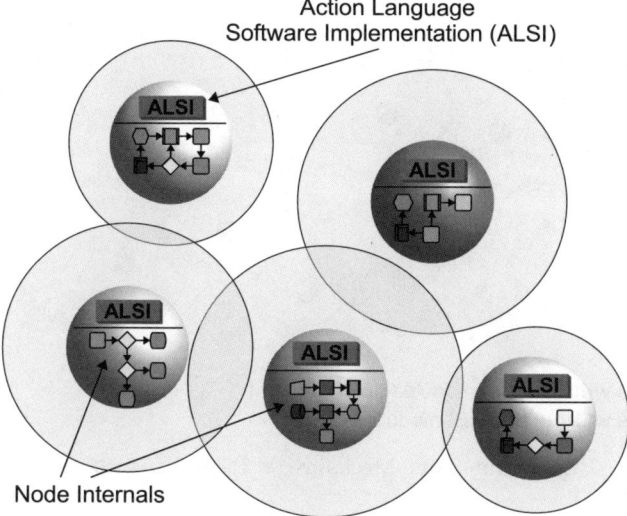

Fig. 3. Run-Time Mechanisms: Total Distribution

Figure 2 illustrates the computation of answer to a prediction query. Example narrative, social laws, initial and resulting social states, expressed in EC pseudocode, concerning the resource sharing example, are presented.

The current social state may be available to (a subset of) the ngAHN members at run-time. Such run-time services may be provided by a central server or, as expected in an AHN, in various distributed configurations. We describe two example configurations below.

Each ngAHN member node (with a private internal architecture) could be equipped with an ALSI module, computing answers to prediction queries for the benefit of the node, informing it of its institutional powers, permissions, obligations and sanctions (see Figure 3). In this configuration the ALSI module of each node may not be aware of all events (that is, the narrative) necessary to compute a prediction query answer. For instance, a node may not be aware of the fact that a resource is no longer available and thus, its ALSI module may compute that the node is still empowered, say, to request access to the resource. A partial narrative could be enriched by 'witnessing' other nodes' actions or requesting from peer nodes to be updated about the events taking place in the ngAHN. In the latter case, however, the node could be, intentionally or not, misinformed which could result in an inaccurate computation of its institutional powers, permissions, and so on.

A design choice that needs to be made in this setting concerns the social laws available to the ALSI module of a node. Clearly a node N's ALSI module should include the laws relevant to the roles N occupies. In some applications, the permissions and obligations of a node need to be private to that node — in this case the ALSI module of each node should contain *only* the laws relevant to its roles. In other applications, a node's ALSI module could contain the complete

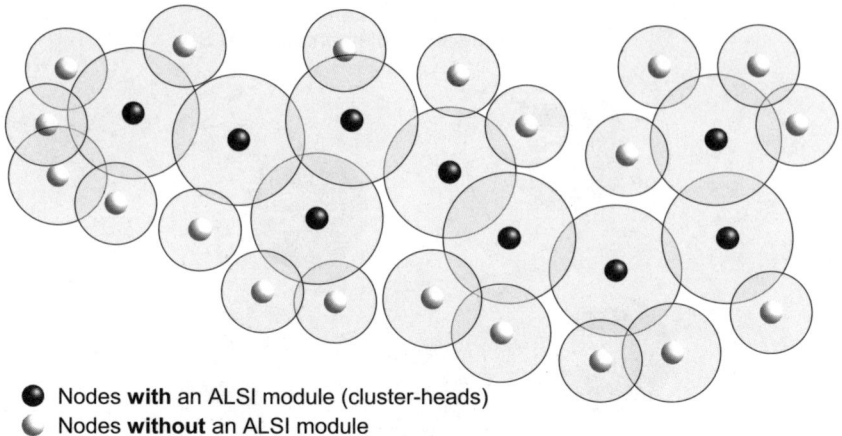

● Nodes **with** an ALSI module (cluster-heads)
◐ Nodes **without** an ALSI module

Fig. 4. Run-Time Mechanisms: Partial Distribution

set of social laws, thus enabling each node to compute information about any other member (provided that a node is aware of the complete ngAHN narrative).

Due to the limited resources (battery, for example) available to the nodes of a ngAHN, it may not be practical or feasible for each node to compute its own institutional powers, permissions and obligations. To address this issue, nodes with rich resources availability could be selected specifically for producing such information and publicising it to the members of a ngAHN. Consider the example topology shown in Figure 4. A network is divided into clusters and each cluster elects a 'cluster-head' (CH), typically a node with longer communication range, larger bandwidth and more power (again, such as battery). A CH's communication range covers all the nodes in the cluster. Links are established to connect CHs to a backbone network. (Such topologies have been proposed in the literature for achieving 'good' routing performance in AHNs — see, for example, [10, 44, 33].) Every message exchange, in this example topology, is carried out via the backbone network; therefore, CHs can compile, in cooperation, the narrative of events of the whole ngAHN. Moreover, each CH is equipped with an ALSI module and the complete set of the ngAHN's social laws. Consequently, a CH is capable of computing a node's institutional powers, permissions and obligations, and detecting non-compliance with obligations and performance of forbidden actions. Such information is publicised to a node upon request (a node requests such information from its CH). Different strategies may be followed for publicising information to nodes — a node's institutional powers could be publicised to all members of a ngAHN, its permissions could be kept private to the node, etc.

It is possible that a CH will not behave as it ought to behave. For instance, it may move out of communication range or run short of resources, and thus be incapable of fullfiling the nodes' requests. Moreover, a CH may intentionally mis-behave — for example, giving incorrect information (informing a node that it is forbidden to perform an action when it is permitted to do so) or disclosing

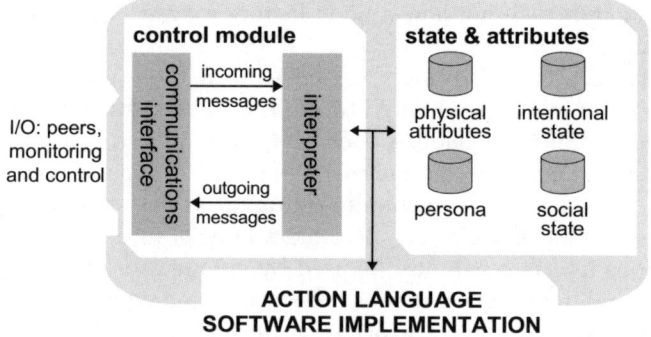

Fig. 5. A Sample Architecture for a Norm-Governed AHN Node

private information (publicising the obligations of a node when these should be private to the node). For these reasons, we may express a CH as role of a ngAHN, specifying the institutional powers, permissions and obligations associated with that role. Violation of these permissions or non-compliance with these obligations will result in 'sanctioning' a node acting as a CH. (Note that the detection of a CH's mis-behaviour may not be straightforward — consider, for example, the disclosure of private information.) A sanctioned CH is replaced by a new one. (Recall that a sanction is not necessarily a penalty; in the case of inadvertent mis-behaviour a sanction can be seen as a way of dealing with the network's exceptions.) The selection of a node as a CH, for the replacement of a 'sanctioned' CH or for the formation of a new cluster, is a typical issue of role-assignment (see Section 2).

Clearly, other topologies are possible for the realisation of a ngAHN. Grizard et al. [16], for instance, discuss an overlay network in which nodes in the overlay, called 'controller agents', monitor the behaviour of the underlying network nodes, called 'application agents', detect whether an application agent violated a norm, and publicise, upon request, details about norm violation (for example, how often each application agent violates a norm). In general, the aforementioned topologies were presented in order to give an indication of the design choices that need to be made for the realisation of a ngAHN.

Apart from mechanisms for informing nodes of their institutional powers, permissions and obligations, the realisation of ngAHNs further requires that nodes actually *consider* such information in their decision-making process. The next section discusses node architectures for ngAHNs.

4 Decision-Making in Norm-Aware Nodes

The functionality of a node within a ngAHN can be likened to that of an agent within a multi-agent system. We have been developing an architecture for a ngAHN node (Figure 5), drawing upon design principles for *agent architectures* [20, 21]. We give an overview of the architecture and focus, in this paper, on the architecture module defining how a node factors a ngAHN's social laws into its decision-making.

The architecture is conceptually organised into three main parts: a *control module*, *state & attributes* and an ALSI module. The control module includes a *communications* interface enabling each node to send and receive its own messages. (We assume error-free communications within simulations, although we note that the introduction of deliberate communications errors and corresponding error-handling mechanisms is a possible avenue of future investigation.) The communications interface can also be used by an operator or manager to externally monitor and instruct the node. That is, the same type of message exchange that occurs between nodes can also occur between a node and an 'external' entity, with emphasis on interrogative and imperative communications. In the simulation platform described later (Section 4.1), we show that in addition to supporting such input/output operations, the same interface can be used to interact with an 'environment'. In the case of a (ng)AHN simulation, the environment process represents the network topology. Communicative acts directed to this process are interpreted as queries or physical acts upon the environment, which can respond accordingly. In this way, the environment process may determine what the nodes perceive of their physical environment. The environment process can also be used to effect exogenous events (such as *time-outs*).

The *interpreter*, a part of the control module, is a focal point of the architecture, linking to all other components. This module is primarily characterised by a processing cycle that enacts, in turn, the perceptive, deliberative and responsive behaviour of the node. These amount to checking for incoming messages, consulting and updating state & attributes accordingly, formulating appropriate (re-)actions and creating optional outgoing messages.

The state & attributes grouping includes elements usually found in deliberative agent architectures. *Physical attributes* include information like a node identifier, location within the environment as well as platform and communications attributes such as battery power and communications range. Note that some of these attributes are beliefs rather than 'facts' as they may be based on (possibly outdated) information obtained from an environment process. The *intentional state* comprises current goals and which, of the methods available, the node in question has selected to achieve them.

While analogues of the preceding architectural elements can be readily found in most deliberative agent architectures, *persona* and *social state* are, perhaps, less likely to have direct counterparts. The former determines a node's social outlook by defining how exactly social laws are factored into its behaviour. The latter is as explained previously in Section 3, with a subjective emphasis — that is, a record of the node's institutional powers, permissions, obligations and sanctions (as opposed to a record of every member's institutional powers, permissions, etc). Such information is produced by consulting an ALSI module, local or remote (for instance, consulting a cluster-head).

We believe that the design approach described above facilitates our modelling and study of ngAHNs, as it allows distinct node behaviours to be straight-forwardly introduced through well-defined adjustment of a single module within a generic architecture (the persona module, for example). This has the outward appearance of introducing a different node architecture without the potential configuration and interoperability issues associated with doing so. In addition, the communications interface 'hides' the internal operation of one node from another, thus capturing the node heterogeneity property of a ngAHN.

4.1 Implementation

We have been developing a simulation base that allows us to model the entities and communications that might take place in a ngAHN [20, 21]. Implemented in *Prolog*, the simulation base enacts the processing cycles of multiple communicating nodes. To do so, it first consults the individual files in which the behaviours of these nodes are specified. These files define the initial configuration of state & attributes elements.

Following initialisation, the simulation base performs the processing cycles of the defined nodes consecutively. The environment cycle is always first to be enacted, while those of the remaining nodes occur in a randomised order. This avoids the development of unfavourable or predictable interaction patterns. Message exchange during processing cycles occurs when data structures, representing node-designated message queues, are consulted and updated.

A processing cycle consists of message exchange as described above, followed by update of the relevant node's state & attributes. During the update, decisions are taken on how to respond to received messages and perceived events — effectively, the deliberative stage of the cycle. This deliberation may produce specific changes to the state & attributes during the update. The realisation that a certain action or event has transpired, or that a particular condition holds, for example, can trigger the adoption of a new goal, attitude or of different means to achieve existing goals.

4.2 Resource-Sharing Protocol

In the following scenario, we provide Prolog-based pseudo-code further illustrating the implementation aspects described above. More precisely, we discuss ways of simulating a ngAHN by specifying different node personas. The scenario is based on an instance of the resource sharing protocol mentioned earlier. In this instance `node1` acts as the resource controller or 'chair', and `node2` and `node3` are the 'subjects', that is, they request access to the shared resource.

An ALSI module expresses, in this example, a logic programming Event Calculus (EC) encoding of the social laws of the resource-sharing scenario. This is used in the first instance to establish the roles occupied by nodes. To summarise the protocol, the chair determines who is the best candidate among the current resource requests and grants resource access accordingly. A grant is for a set amount of time, during which the allocated node derives some utility from the resource and may, at its discretion, either release the resource early or request an extension of the grant period. In addition, the chair can entertain additional resource requests from the other node during the grant period and may insist that (command) the node to whom the resource is currently granted release it immediately. When a violation of social laws occurs (and is detected), a node may be sanctioned. Further consideration of these aspects of the protocol (and variants) is given in [1].

In this example, we characterise `node2` as being willing to violate a social law if the utility of doing so outweighs the expected sanction. In contrast, `node3` will only contemplate social law violation if it has not already done so within an arbitrarily recent time-frame. We consider one potential violation where a node does not comply with the obligation to release a resource. Consider the following example:

```
(1)   process_message( node2, cmd_release_resource( node1, node2 ), T ) :-
(2)      holdsAt( obliged( node2, release_resource( node2 ) ) = true, T ),
(3)      holdsAt( status = granted(node2, TEnd), T ),
(4)      TLeft = TEnd - T,
(5)      TLeft > 0,
(6)      utility( TLeft, U ),
(7)      holdsAt( sanction( obligation_release_resource ) = Cost , T ),
(8)      ( compare(U, Cost) ->
(9)         send_msg( release_resource( node2 ) )
(10)       ;
(11)        true
(12)   ).
```

The `process_message/3` procedure presented above expresses the reaction of a node (in this example, `node2`) to an incoming `cmd_release_resource` ('command to release the resource') message — this message was sent at time `T` by `node1`, in this example. Upon receipt of a `cmd_release_resource` message, `node2` calculates whether or not it is obliged to release the resource (line 2 of the pseudo-code example). A message `cmd_release_resource` creates an obligation for the node holding the resource to release it only when the controller has the institutional power to issue such a command. The controller may not always have the power to command the release of a resource — for example, before the time allocated to the holder ends, the controller may only be empowered to command a release of the resource when it receives an 'urgent' request for the resource from another node. If `node2` is indeed obliged to release the resource, then it recalls the time (`TEnd`) until which access to the resource was originally granted (line 3), and calculates the time difference (`TLeft`) between `TEnd` and the current time `T` (line 4). If `TLeft` is greater than zero, that is, the chair commanded `node2` to release the resource before the allocated time ended, then `node2` computes a subjective utility based on `TLeft` (line 6), that is, it computes the utility it would derive from manipulating the resource until the allocated time ends. It then (line 8) compares the calculated utility with the objective sanction associated with failure to comply with the obligation to release the resource (that is, the sanction expressed by the social laws of the resource sharing protocol). If the comparison favours keeping hold of the resource (in this case the `compare/2` predicate fails), `node2` will ignore the chair's command.

Note, in the above example, that if `TLeft` is less than or equal to zero then `cmd_release_resource` was issued after the time allocated for manipulating the resource ended — this case is dealt by another `process_message/3` procedure.

While we do not provide a specification of the `compare/2` predicate here, we note that it represents a key aspect of a node's (social) decision-making. The predicate provides a subjective assessment of the seemingly objective sanction cost relative to a node's internal utility function. It naturally follows that the value of either complying with or violating a social law is primarily determined by a node's private architecture. A welcome corollary is the ability to nuance node behaviour through the specification of different `compare/2` predicates. A less welcome concern is the additional complexity that such specifications require.

`holdsAt/2` is an EC predicate expressing the social laws of the resource sharing protocol. In other words, the occurrences of `holdsAt/2` that appear in the code above (and below) represent calls to an ALSI module capable of reasoning about social laws. As already mentioned, an invocation of an ALSI module may be 'actual' (in the case where the node in question has its own module) or 'logical' (when the ALSI module is located in another node).

Consider another example node persona:

```
(1)    process_message( node3, cmd_release_resource( node1, node3 ), T ) :-
(2)       holdsAt( obliged( node3, release_resource( node3 ) ) = true, T ),
(3)       last_violation( obligation_release_resource, T2 ),
(4)       compliance_interval( CInt ),
(5)       ( ( CInt < ( T - T2 ) ) ->
(6)           ( retract( last_violation( obligation_release_resource, _ ) ),
(7)             asserta( last_violation( obligation_release_resource, T ) ),
(8)           )
(9)       ;
(10)      send_msg( release_resource( node3 ) )
(11)   ).
```

`node3` maintains a record of the last time (`T2`) it violated the obligation to release the resource. It then compares the time difference `T - T2` (`T` is the current time) with a 'compliance interval'. If the time elapsed since the last such transgression is greater than the compliance interval, `node3` will not comply with the obligation to release the resource.

Clearly, there are other possible attitudes concerning compliance with obligations and performance of forbidden actions. Moreover, we may specify different node personas by expressing the attitude of a node concerning its institutional powers. For instance, a node will not perform an action, say request a resource, if it is not empowered to do so (even if it is permitted to request the resource). Another node N may perform an action, although not empowered to do so, expecting that (some) peer nodes will not be able to tell whether or not N was empowered to perform the action (some peer nodes may not have access to an ALSI module).

By specifying different attitudes to institutional power, permission and obligation — that is, by adjusting the persona module of the proposed node architecture, we may introduce different node behaviours as required for our ngAHN simulations. (Varying node behaviours can be introduced by additionally adjusting the intentional state module of the node architecture.) This is our current research direction.

5 Related Work

There are several approaches in the literature that are related to our work on specifying norm-governed multi-agent systems — [5,27,35,26,45,28,29,39,40,41, 13,12,11,36,4,46] are but a few examples. Generally, work on the specification of multi-agent systems does not distinguish between the constitutive laws and the normative environment within which the agent interactions take place. This is a key difference between our work and related approaches in the literature. Our specification of social laws explicitly represents the institutional powers of the agents, capturing the meaning of the agents' actions, and thus expressing the constitutive laws of a system. Moreover, our specification differentiates between institutional power, permission and physical capability, thus separating the constitutive laws from the normative and the physical environment within which a system is executed. A detailed discussion of research related to the framework for specifying norm-governed computational systems, and an evaluation of this framework, can be found in [3, 2].

Within multi-agent systems research, 'social awareness' can be seen as a development of cooperative behaviour (see, for example [17,43]), which has been primarily based on the use of communication protocols. Castelfranchi et al. [8] note that agents using only fixed protocols are unable to deal with unexpected behaviour on the part of their environment and other agents. The authors consequently identify the need for *autonomous normative agents*: agents with an awareness of social laws and the capacity to both adopt and violate such laws. Castelfranchi et al. position social law-awareness as an influential, but not deterministic, factor in agent behaviour — to them, an agent's ability to choose to violate a recognised social law is equally important as its ability to choose to conform to the same. Our node architecture reflects this consideration by positioning the ALSI as an *advisory* module to whose 'output' the agent can react arbitrarily.

Several researchers have since proposed agent models and associated theory to address the needs identified by Castelfranchi et al. For example, the Beliefs-Obligations-Intentions-Desires (BOID) architecture of Broersen et al. [7,6] introduces mechanisms to resolve conflicts between the representations of cognitive state in a deliberative agent and its obligations. This allows the characterisation of a number of abstract agent social perspectives in terms of the interdependencies, precedences and update strategies of beliefs, obligations, intentions and desires, in which obligations and desires are respectively treated as external (social) and internal motivational attitudes. We adopt a similar first-class view of normative relations in our node architecture from a *logical* perspective but note that the BOID approach differs slightly in its integration of these relations at an *implementation* level. It is through the introduction of a specialised implementation module (the ALSI) to our node architecture that an agent is able to perceive its normative relations.

Lopez et al. [24] characterise social laws in terms of their 'positive and negative effect' on the goals of agents and those of their 'society'. The approach of Lopez et al. is currently a formal model of agent behaviour and remains unimplemented.

Sadri et al. [37] incorporate social decision-making within a deliberative agent model enhanced by EC-based formulations of social laws, noting how these can change over a system's lifetime. They use an abductive logic-based proof procedure to compute an agent's time-constrained *social goals* — expressions encoding what the system expects of an agent. Distinct logic programs encode the *preference policy* of each agent — for example whether an agent puts social goals above its own goals. Social perspectives are identified (for instance, social, anti-social agents) based on the agents' preferences (personal goals versus social goals). Like the other reviewed approaches, the framework of Sadri et al. does not identify the institutional powers of agents, a concept on which we place emphasis.

6 Summary

Malfunctioning, either by intent or by circumstance, is to be expected in AHNs. How to identify and adapt to such situations is essential. By specifying the permissions, obligations, and other more complex normative relations that may exist between the members of an AHN, one may precisely identify 'undesirable' behaviour, such as the performance of forbidden actions (for instance, illicit use of a peer's processor cycles) and non-compliance with obligations (for example, denying access to one's processor cycles). Therefore, it is possible to introduce sanctions and enforcement strategies to adapt to such behaviour (for example, temporarily ejecting a mis-behaving or erroneous node from a network).

We have developed a framework for specifying the laws of norm-governed AHNs and executing these laws for informing the decision-making of the member nodes. We are currently developing such nodes in order to simulate norm-governed AHNs and thus investigate the practicality of, and in general evaluate, the presented approach.

Acknowledgements

This work has been supported by the EPSRC project "Theory and Technology of Norm-Governed Self-Organising Networks" (GR/S74911/01).

References

1. Artikis, A., Kamara, L., Pitt, J., Sergot, M.: A protocol for resource sharing in norm-governed ad hoc networks. In: Leite, J.A., Omicini, A., Torroni, P., Yolum, p. (eds.) DALT 2004. LNCS (LNAI), vol. 3476, pp. 221–238. Springer, Heidelberg (2005)
2. Artikis, A., Sergot, M., Pitt, J.: An executable specification of an argumentation protocol. In: Proceedings of Conference on Artificial Intelligence and Law (ICAIL), pp. 1–11. ACM Press, New York (2003)

3. Artikis, A., Sergot, M., Pitt, J.: Specifying norm-governed computational societies. Technical Report 2006/5, Imperial College London, Department of Computing (2006) Retrieved March 6, 2006, from
   http://www.doc.ic.ac.uk/research/technicalreports/2006/DTR06-5.pdf
4. Bandara, A.: A Formal Approach to Analysis and Refinement of Policies. PhD thesis, Imperial College London (2005)
5. Bradshaw, J., Uszok, A., Jeffers, R., Suri, N., Hayes, P., Burstein, M., Acquisti, A., Benyo, B., Breedy, M., Carvalho, M., Diller, D., Johnson, M., Kulkarni, S., Lott, J., Sierhuis, M., Van Hoof, R.: Representation and reasoning about DAML-based policy and domain services in KAoS. In: Rosenschein, J., Sandholm, T., Wooldridge, M., Yokoo, M. (eds.) Proceedings of Conference on Autonomous Agents and Multi Agent Systems (AAMAS), pp. 835–842. ACM Press, New York (2003)
6. Broersen, J., Dastani, M., Hulstijn, J., Van der Torre, L.: Goal generation in the BOID architecture. Cognitive Science Quarterly 2(3–4), 428–447 (2002)
7. Broersen, J., Dastani, M., Hulstijn, J., Huang, Z., van der Torre, L.: The BOID architecture: conflicts between beliefs, obligations, intentions and desires. In: Proceedings of Conference on Autonomous Agents, pp. 9–16. ACM Press, New York (2001)
8. Castelfranchi, C., Dignum, F., Jonker, C., Treur, J.: Deliberative normative agents: Principles and architecture. In: Jennings, N.R. (ed.) ATAL 1999. LNCS, vol. 1757, pp. 364–378. Springer, Heidelberg (2000)
9. Dellarocas, C.: Reputation mechanisms. In: Hendershott, T. (ed.) Handbook on Economics and Information Systems. Elsevier Publishing, Amsterdam (to appear)
10. Du, X.: QoS routing based on multi-class nodes for mobile ad hoc networks. Ad. Hoc. Networks 2(3), 241–254 (2004)
11. Esteva, M., de la Cruz, D., Sierra, C.: ISLANDER: an electronic institutions. In: Castelfranchi, C., Johnson, L. (eds.) Proceedings of Conference on Autonomous Agents and Multi-Agent Systems (AAMAS), pp. 1045–1052. ACM Press, New York (2002)
12. Esteva, M., Padget, J., Sierra, C.: Formalizing a language for institutions and norms. In: Meyer, J.-J.C., Tambe, M. (eds.) ATAL 2001. LNCS (LNAI), vol. 2333, pp. 348–366. Springer, Heidelberg (2002)
13. Fitoussi, D., Tennenholtz, M.: Choosing social laws for multi-agent systems: minimality and simplicity. Artificial Intelligence 119(1-2), 61–101 (2000)
14. Giunchiglia, E., Lee, J., Lifschitz, V., McCain, N., Turner, H.: Nonmonotonic causal theories. Artificial Intelligence 153(1–2), 49–104 (2004)
15. Giunchiglia, E., Lee, J., Lifschitz, V., Turner, H.: Causal laws and multi-valued fluents. In: Proceedings of Workshop on Nonmonotonic Reasoning, Action and Change (NRAC) (2001)
16. Grizard, A., Vercouter, L., Stratulat, T., Muller, G.: A peer-to-peer normative system to achieve social order. In: Coordination, Organization, Institutions and Norms in Agent Systems II AAMAS 2006 and ECAI 2006 International Workshops, COIN 2006, Hakodate, Japan, May 9, 2006. LNCS, vol. 4386, pp. 274–289. Springer, Heidelberg (2006)
17. Haddadi, A.: Communication and Cooperation in Agent Systems. LNCS, vol. 1056. Springer, Heidelberg (1995)
18. Jones, A., Sergot, M.: On the characterisation of law and computer systems: the normative systems perspective. In: Deontic Logic in Computer Science: Normative System Specification, pp. 275–307. J. Wiley and Sons, West Sussex, England (1993)
19. Jones, A., Sergot, M.: A formal characterisation of institutionalised power. Journal of the IGPL 4(3), 429–445 (1996)

20. Kamara, L., Pitt, J., Sergot, M.: Norm-aware agents for ad hoc networks: A position paper. In: Proceedings of the Ubiquitous Agents Workshop, AAMAS (2004)
21. Kamara, L., Pitt, J., Sergot, M.: Towards norm-governed self-organising networks. In: Proceedings of the NorMAS Symposium, AISB (2005)
22. Klein, M., Rodriguez-Aguilar, J., Dellarocas, C.: Using domain-independent exception handling services to enable robust open multi-agent systems: the case of agent death. Journal of Autonomous Agents and Multi-Agent Systems 7(1–2), 179–189 (2003)
23. Kowalski, R., Sergot, M.: A logic-based calculus of events. New Generation Computing 4(1), 67–96 (1986)
24. Lopez, F., Luck, M., d'Inverno, M.: Normative agent reasoning in dynamic societies. In: Jennings, N., Sierra, C., Sonenberg, L., Tambe, M. (eds.) Proceedings of Conference on Autonomous Agents and Multiagent Systems (AAMAS), pp. 732–739. IEEE Computer Society Press, Los Alamitos (2004)
25. Makinson, D.: On the formal representation of rights relations. Journal of Philosophical Logic 15, 403–425 (1986)
26. Minsky, N., Ungureanu, V.: Law-governed interaction: a coordination and control mechanism for heterogeneous distributed systems. ACM Transactions on Software Engineering and Methodology (TOSEM) 9(3), 273–305 (2000)
27. Moreau, L., Bradshaw, J., Breedy, M., Bunch, L., Hayes, P., Johnson, M., Kulkarni, S., Lott, J., Suri, N., Uszok, A.: Behavioural specification of grid services with the KAoS policy language. Cluster Computing and the Grid 2, 816–823 (2005)
28. Moses, Y., Tennenholtz, M.: On computational aspects of artificial social systems. In: Proceedings of Workshop on Distributed Artificial Intelligence (DAI), pp. 267–284 (1992)
29. Moses, Y., Tennenholtz, M.: Artificial social systems. Computers and Artificial Intelligence 14(6), 533–562 (1995)
30. Murphy, A., Roman, G.-C., Varghese, G.: An exercise in formal reasoning about mobile communications. In: Proceedings of Workshop on Software Specification and Design, pp. 25–33. IEEE Computer Society Press, Los Alamitos (1998)
31. Perkins, C.: Ad Hoc Networking, ch. 1. Addison Wesley Professional (2001)
32. Pitt, J., Kamara, L., Sergot, M., Artikis, A.: Voting in multi-agent systems. Computer Journal 49(2), 156–170 (2006)
33. Pitt, J., Venkataram, P., Mamdani, A.: QoS management in MANETs using norm-governed agent societies. In: Dikenelli, O., Gleizes, M.-P., Ricci, A. (eds.) ESAW 2005. LNCS (LNAI), vol. 3963, pp. 221–240. Springer, Heidelberg (2006)
34. Prakken, H.: Formalising Robert's rules of order. Technical Report 12, GMD – German National Research Center for Information Technology (1998)
35. Rodriguez-Aguilar, J., Martin, F., Noriega, P., Garcia, P., Sierra, C.: Towards a test-bed for trading agents in electronic auction markets. AI Communications 11(1), 5–19 (1998)
36. Rodriguez-Aguilar, J., Sierra, C.: Enabling open agent institutions. In: Dautenhahn, K., Bond, A., Canamero, L., Edmonds, B. (eds.) Socially Intelligent Agents: Creating relationships with computers and robots, pp. 259–266. Kluwer Academic Publishers, Dordrecht (2002)
37. Sadri, F., Stathis, K., Toni, F.: Normative KGP agents. Journal of Computational and Mathematical Organization Theory 12(2-3), 101–126 (2006)
38. Searle, J.: Speech Acts. Cambridge University Press, Cambridge (1969)
39. Shoham, Y., Tennenholtz, M.: On the synthesis of useful social laws for artificial agent societies. In: Swartout, W. (ed.) Proceedings of Conference on Artificial Intelligence (AAAI), pp. 276–281. The AAAI Press/ The MIT Press (1992)

40. Shoham, Y., Tennenholtz, M.: On social laws for artificial agent societies: off-line design. Artificial Intelligence 73(1-2), 231–252 (1995)
41. Tennenholtz, M.: On computational social laws for dynamic non-homogeneous social structures. Journal of Experimental and Theoretical Artificial Intelligence 7, 379–390 (1995)
42. Wang, Y., Giruka, V., Singhal, M.: A fair distribution solution for selfish nodes problem in wireless ad hoc networks. In: Nikolaidis, I., Barbeau, M., Kranakis, E. (eds.) ADHOC-NOW 2004. LNCS, vol. 3158, pp. 211–224. Springer, Heidelberg (2004)
43. Wooldridge, M., Jennings, N.: Formalizing the cooperative problem solving process. In: Readings in Agents, pp. 430–440. Morgan Kaufmann Publishers Inc. San Francisco (1998)
44. Xu, K., Hong, X., Gerla, M.: Landmark routing in ad hoc networks with mobile backbones. Journal of Parallel Distributed Computing 63(2), 110–122 (2003)
45. Yolum, P., Singh, M.: Reasoning about commitments in the event calculus: An approach for specifying and executing protocols. Annals of Mathematics and Artificial Intelligence 42(1–3), 227–253 (2004)
46. Zambonelli, F., Jennings, N., Wooldridge, M.: Developing multiagent systems: The gaia methodology. ACM Transactions on Software Engineering and Methodology (TOSEM) 12(3), 317–370 (2003)

A Definition of Exceptions in Agent-Oriented Computing*

Eric Platon[1,2], Nicolas Sabouret[2], and Shinichi Honiden[1]

[1] National Institute of Informatics, Sokendai,
2-1-2 Hitotsubashi, Chiyoda, 101-8430 Tokyo
[2] Laboratoire d'Informatique de Paris 6,
104, Ave du Pdt Kennedy, 75016 Paris
{platon,honiden}@nii.ac.jp, nicolas.sabouret@lip6.fr

Abstract. The research on exception handling in Multi-Agent Systems has produced some advanced models to deal with 'exceptional situations'. The expression 'agent exception' is however unclear across the literature, as it sometimes refers to extensions of traditional exception models in programming languages, and sometimes to organizational management mechanisms with distinct semantics. In this paper, we propose a definition of 'agent exception' to clarify the notion and justify that specific research is necessary on this theme. We detail properties of this definition, revisit the traditional vocabulary related to exception in software design, propose an adequate agent architecture, and identify some research issues. This work is aimed at federating the endeavors on the question of exception management for Agent-Oriented Computing.

1 Introduction

The notion of exception in Multi-Agent System often refers to usual exceptions in structured and object-oriented programming, defined in the 1970s by the following.

> Of the conditions detected while attempting to perform some operation, exception conditions are those brought to the attention of the operation's invoker. The invoker is then permitted (or required) to respond to the condition [1,2,3].

This definition and the subsequent lineage of exception handling systems are operation-centric approaches [4,5,6]. When an operation is invoked, e.g. by a method of an object, conditions are checked before the actual execution to validate the invocation context. Typical conditions are the correctness of the types and values of the operation input parameters. If a condition is not met, the operation is not executed and an exception is signaled to the invoker to initiate appropriate handling mechanisms.

---
* This research is partially supported by the French Ministry of Foreign Affair under the reference BFE/2006-484446G, Lavoisier grant program.

In the case of agent systems, the usual definition of exception applies as agents are software, but it also misses characteristics of the agent concept. Traditional definitions state that an exception is entirely determined when conditions are violated in the invocation context of an operation. Agents are however free to evaluate whether the *result* of invoking an operation is 'normal' or 'exceptional', due to some form of autonomy and to the loose coupling among agents and resources. Agents can also evaluate differently the *type* of exception they encounter depending on their individual contexts and inner mechanisms. And lastly, the usual definition of exceptions is mostly implemented as specific constructs in a programming language, whereas it is not clear whether a language construct suffices to address the agent case, due to autonomy, openness, and systemic effects [7]. These characteristics actually lead to consider an additional 'agent-centric' approach orthogonal to the notion of exception in programming languages.

In this article, we elaborate a definition of 'agent exception' in MAS, position the definition to traditional ones, and present research issues in Agent-Oriented Computing for the particular problem of exception management.

The structure of this article is as follows. Section 2 reviews related work with earlier attempts for defining exceptions in Multi-Agent Systems. Section 3 presents our proposal to define the term exception and an agent software architecture to support it. Section 4 details research issues for Agent-Oriented Computing with regards to Software Engineering research. Finally, section 5 concludes the paper.

2 Related Work to Defining Agent Exceptions

The literature in MAS tends to follow two approaches to define the term exception. We call them respectively 'continuity' and 'rupture'.

2.1 Continuity: From Traditional to Agent-Oriented Definition

The common ground of much research work relies on the usual intuition of exceptions cultivated in programming languages. Tripathi and Miller thus write that 'an exception is an event that is caused by a program using a language construct, or raised by the underlying virtual machine', where an event deviates the execution of a program to a handler [8,9]. Souchon et al. rely on similar basis that can be also observed in distributed object research [10,11,12,13]. In other words, the first approach to agent exception is to consider it as a natural extension of the usual meaning in programming languages.

The mechanisms related to such definitions are not completely adapted however to the agent paradigm, despite endeavors to match them with some extensions [11]. For example in the above definition from Tripathi and Miller, the usual meaning of exception is applied to agents, without reference to the issues of autonomy and social relationships that can be source of exceptions as 'asocial events'. Autonomy leads agents to decide independently their actions in the environment. The semantics of programming exceptions prevents such independent decisions by imposing the exceptional character of an event, i.e. the 'invoker' agent (as in the definition from Goodenough) is imposed by the 'operation provider' agent to interpret an event as an exception. Preservation of the autonomy requires an appropriate semantics where the invoker can decide autonomously. Social relationships is a specific overlay of MAS, where the 'social fabric' is expected to help structuring the system. Huberman and Hogg showed that social relationships can evolve in various ways, even though a fixed organizational structure has been defined [14]. Evolutions cause the formal organization of the system (defined by rules) to drift to an informal organization that palliates dynamically lacks or deficiencies of the formal one. Interactions in the informal organization are typical sources of exceptions in the formal one, e.g. the violation of interaction protocols due to improper sequences or messages. The usual model of exception in programming languages is unadapted in such situations as exceptions to the formal organization occur while the different components of the system execute *correctly*, in a programming sense.

Finally, this approach based on continuity also assumes agents are single threads and that exceptions make sense in the control flow of one thread only. This structural constraint inherited from the traditional disciplined exception handling facilities does not scale well to the interactive nature of agent societies, where 'exceptions' can concern a group of agent processes, themselves multi-threaded software. Distributed computing research addresses such issues with successful achievements, but most approaches define a framework that allows returning to traditional exception handling, e.g. [10,15]. The advantage is to reuse well-known mechanisms, but this reuse makes the approaches inappropriate to deal with the agent characteristics.

2.2 Rupture: From Agent Paradigm to Agent-Oriented Engineering

Another approach to defining exception is in rupture with traditional ways based on programming. The position of this approach is to consider exceptions as systemic matters, as can be observed in some agent research and component-based development endeavors [16,7,17]. Klein et al. describe intuitively that 'all [...] departures from the "ideal" MAS behavior can be called exceptions' [18]. This definition brings the term exception at the level of the system, as the authors deal with the 'MAS behavior'. The similarity with traditional definitions is the departure from the ideal behavior that recalls the 'deviation from normal behavior' of a program as explained by Parnas and Würges [4]. The difference with traditional definition is however the dimension of the exception that is not constrained to one thread of control. Exceptions become systemic events. This system-wide scope is desired since MAS are open systems that designers need to be able to build and to keep under control. Besides, Klein et al. refer to an *exception management* instead of only exception handling. However, such management and system-level definition must guarantee that agent autonomy and system openness are respected, which is not necessarily the case with such approaches [19].

In addition, this definition does not say how 'departures from the "ideal" MAS behavior' are identified, and more particularly what is the agent point of view in the case of exception. The MAS architecture proposed in the work of Klein et al. addresses the first concern by detailing how normal MAS behaviors are expressed. However, the second question requires a more detailed definition of an agent exception.

Mallya and Singh have addressed this second concern in the case of agents executing commitment protocols [20,21]. From the point of view of agents, *exceptions are abnormal situations compared to what is awaited for in a protocol*. Exception handling is then to appropriately reason on protocols or extending them at runtime when required. This approach clarifies the intuition of an agent exception in the case of interactions based on protocols. Agents do not always interact according to protocols however, and a more general definition would be required. For instance, the work of Bernon et al. introduces an alternative point of view by analyzing 'non-cooperative situations' in the ADELFE methodology, although this work considers exceptions with a more classical sense [22].

3 Definition of Agent Exceptions

In order to clarify and justify the notion of 'agent exception', we propose in this section a series of definitions, first for the term exception, and then for the other technical terms dealing with exceptions. Such terms are diagnosis, raising, propagating, transforming, termination, resumption, handling, and management. In the end, we also propose an agent architecture to manage exceptions.

3.1 Agent Exceptions

The following definition is in our sense in rupture with traditional ones, and we tried to set up proper grounds to exploit the existing work in usual systems, with adapted terminology.

Definition

> An *agent exception* is the evaluation *by* the agent of a perceived event as *unexpected*.

The source of exception is the essential difference with traditional definitions. The source is not the event, which could be related to the call-back from an operation invocation that signals an exception. The source is instead the agent itself. Owing to autonomy, agents can evaluate any percept differently and thus end with either an exceptional or normal execution. An event should be understood in a broad sense of any observable action or state in the system. For example, events are message exchanges or the perceived value of pheromones in stigmergic systems.

In relation to earlier definitions, exceptions are qualified as unexpected events. By unexpected, we mean the agent does not anticipate the arrival of the event in the current execution context (time, resources, value of parameters, etc.).

In other words the agent is not 'ready' to process the event when it occurs. The unexpected characteristic of an event then depends on the kind of agent that evaluates it. For example, the protocol-driven agents of Mallya and Singh consider as unexpected[1] any message that does not belong to the sequences of running protocols [20].

The unexpected character of an event requires an *internal reference* for the agent, such as knowledge of a protocol, that serves as representing the expectations of the agent. 'Purely reactive' agents do not deal with agent exceptions due to the absence of internal reference [23]. Practical reactive models encompass however some built-in implicit representations, so that they deal in fact with agent exceptions but in a less flexible pre-wired manner.

The definition stresses that agent exceptions are based on perception, functional part of the interface between an agent and its environment. This relation to perception has consequences on the type of agent architectures that is required to deal with exception management. In addition, Software engineering practices require the separation of the application logic from the exception handling logic. As for agents, unexpected events lead to using the exception logic. We elaborate more on these points later in this section with the presentation of the agent architecture.

Properties. The first property from the definition is that *exceptions make sense inside an agent.* The meaning of an event is encoded at its source, but the exceptional character depends on the point of view of each receiving agent. For example, an agent can query a database and receive a result. The result can be either a set of tuples that meets the query criteria, or it can be an error message. The querying agent can consider any of these results as an exception, depending on its expectations. The agent can just test the database to confirm a malfunction, so that an error message is not an agent exception in this case. Conversely, the agent can consider the resulting tuples as an exception if the query was a consistency check that turns wrong. A complementary example is when an agent tells another one about an exception. In this case also, the receiving agent evaluates the message (i.e. an event) and decides whether it should be considered as an exception. An exception for the first agent can be a normal event as well as an exception for the second one. This first property is important as it elaborates on the loose coupling between agents. Exceptions make sense inside agents, so that coupling is not increased with exception management. Engineering agent systems or related approaches such as Service-oriented architectures [24] needs careful modeling of interfaces. The present property of agent exceptions implies that no interface extension is necessary for their handling, thus reducing issues related to coupling in agent systems.

The second property emphasizes the original work on exception and recent approaches that claim that *exceptions are not only errors* [1,20]. The dictionary

---

[1] Mallya and Singh use the expression 'unexpected *exception*' to refer to an exception unforeseen by the system designer (at design time). In the present definition, an 'unexpected *event*' is deemed as exception at runtime.

definition of the exception is neutral [25], but usage in programming languages overrides for the major part the original definition to describe 'problematic situations'. Although language constructs often allow to exploit exception mechanisms to deal with non-problematic situations, exceptions are mostly associated with problems. The initial definition from Goodenough suggested however that exceptions could also serve finding better solutions or alternatives, originally by monitoring. Mallya and Singh recently reformulated this initial meaning by distinguishing 'negative exceptions' from desirable ones, named 'opportunities'. In agent systems, the case of opportunities seems particularly interesting for the flexibility of systems [26,27]. In the previous database example, the querying agent can exploit an exception as an opportunity to confirm that a database is in malfunction.

The third property is that *agent exception requires internal representation* (e.g. knowledge). Agents need a reference to determine exceptional situations. When an event is perceived, agents evaluate the situation by considering the event according to their internal representation. Without such representation, agents have no ground to differentiate cases. In the query example, the agent exploits the result or error message in the light of its internal state, for example the state of a plan to test the database system. In the case of reactive agents such as artificial ants, no internal representation is usually integrated explicitly in the agent. However, reactive agents have some form of implicit representation, as they are able to answer to some stimuli and ignore the others. The internal representation is then the type of values accepted in input.

The fourth property is the *asynchronous nature of the exception management mechanisms*. In traditional definitions of programming languages, exceptions are treated synchronously. In agent systems, the decoupled nature of agents makes it impossible to guarantee synchronous management of an exceptional situation. In particular, the case of real-time agent systems is then hardly feasible without specific mechanisms. A corollary issue of asynchrony is that traditional exception propagation mechanisms, which unwind the procedure call stack for a single thread of execution, are not applicable. Propagation in agent systems spans over several agents that may react in a variety of ways where the 'linearity' of traditional approaches is not assumable. Propagation of exceptions in distributed systems is more appropriate in the case of MAS, often involving a third party entity specialized in exception handling [28,18,9].

The fifth and last property identified in this paper is that *agent exceptions can exist without underlying programming exception*. In the previous illustrative examples of this section, we have reviewed several cases where such statement is verified. One typical example is with the protocol-driven agents of Mallya and Singh [20], where exceptions along the protocol can occur while all agents are executing correctly. The converse statement is also relevant: When an agent encounters a programming exception, it incurs an agent exception. For example, a programming fault that causes the termination of the agent provokes a 'death' agent exception as well [18].

3.2 Other Terms for Agent-Oriented Computing

The definition proposed in this paper shifts the source of exception from the invoked operation to the invoker agent. Consequently, the vocabulary used in programming languages should be adjusted to fit the requirements in Agent-Oriented Computing.

Exception diagnosis (or detection) refers to mechanisms to evaluate perceived events and detect exceptions, i.e. unexpected events.

Exception signaling does not seem to need an equivalent in agent systems. Indeed, signaling an exception means traditionally that a signal informs the invoker about an exception. The flow of control is reversed back to the invoker. In MAS, the exceptions are detected by the agents individually, and the need to reverse the control flow disappears, as the agent continues its execution, with potential changes in its behavior.

Exception propagation is the mechanism that describes how agents deal with exception situations they are unable to manage. In such case, an agent can try to find a peer agent for help. The term propagation is used to express that an exception is turned into a message (e.g. a call for support) and propagated to peers that may help. This propagation is from the point of view of the sender. For other agents, this propagation is just an event that may be evaluated as an exception.

Exception transforming is a technique to change the type of an exception, while it is processed. In distributed computing, transformations are used to find a common exception type when several software components detect an exception concurrently [10,9]. In agent systems, the transformation mechanism is done by each agent that evaluates an event as exceptional. The reason for the difference is the loose coupling between agents. Techniques from distributed computing actually assume a close collaboration among processes, which is not always true in open systems like MAS.

Termination refers in usual systems to the end of a program caused by an exception condition ('abnormal termination'). Agent exceptions cannot cause a termination of a MAS due to the loose coupling among agents and their autonomy. Agents are free to choose the consequence of an event (including terminating), and their choices are individual, so that the termination of an agent does not imply the termination of any other.

Resumption is usually defined as the continuation of a program initial execution after the handling of an exception. In agent systems, the definition is the same with different underlying mechanisms, since resumption can concern the activity of several agents simultaneously.

Exception handling is the actual processing of an exceptional situation by an agent. By definition of an exception, handling is the execution of specific tasks defined by a handler, while the execution of other activities of the agent are either unmodified (the exception case is ignored and the execution continues) or suspended (with subsequent termination or resumption).

Exception management refers to all activities involved in the management of exceptions by agents. It encompasses all the previous mechanisms.

In the programming language literature, the aforementioned terms often have formal models of the underlying mechanisms. This work remains to be done for Agent-Oriented Computing. Besides, candidate mechanisms are not necessarily language constructs. Agent exceptions are at the agent level and other 'forms' of mechanisms seem more appropriate. For example, propagation and transformation seem better served by architectural or algorithmic forms than a language construct.

3.3 An Agent Architecture with Exception Management

Basic agent architecture. Agent architectures usually rely on the perception-deliberation-action loop, where the deliberation phase can be minimal in the case of reflex agents, or fairly elaborated in the case of learning agents [29]. Fig. 1 represents this usual loop with architectural details.

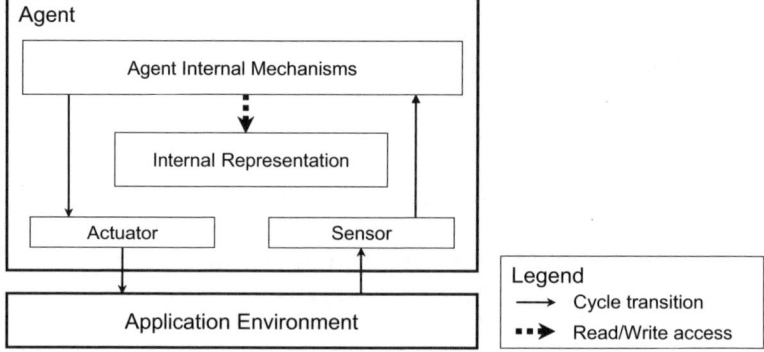

Fig. 1. Basic Agent Architecture

The environment gives the surrounding conditions for the agent to exist [30], and it is the base of the architecture. The agent is composed of four main components with connectors that constitute the aforementioned loop. Events from the environment are received by the agent in the *sensor* component (perception phase), which forwards the percept to the *agent internal mechanisms*. The mechanisms are the central processing of the agent (deliberation phase). Complex architectures can implement these mechanisms with planners, inference engines, or other models such as PRS or MANTA [31]. The mechanisms exploit the *internal representation* of the agent to produce an output in reaction to the percept. The internal representation is an abstract component that refers to any representation type inside the agent. For example, the BDI and KGP architectures are instances where the internal representation is a set of knowledge bases [32,33]. Ant-like agents would have simpler internal representations, such as a set of configuration parameters. The output of the agent internal mechanisms is received by the *actuator* component that commits a corresponding action in the environment (action phase, e.g. sending a message).

Architecture with exception management. The definition of exception in this paper and the subsequent properties lead to reconsider the architecture of an agent to encompass exception awareness. Figure 2 depicts an architecture that integrates necessary and optional architectural elements. The particularity of this architecture pertains to the elaboration of the agent perception and actuation at the architecture level. The novelty is the management of relevance and expectation criteria to classify input events (the 'percepts') and let the agent initiate potential exception management when required. This model can be related to the proposal from Shah et al. for exception diagnosis [34], and also with the work of Weyns et al. on active perception and the notion of focus [35].

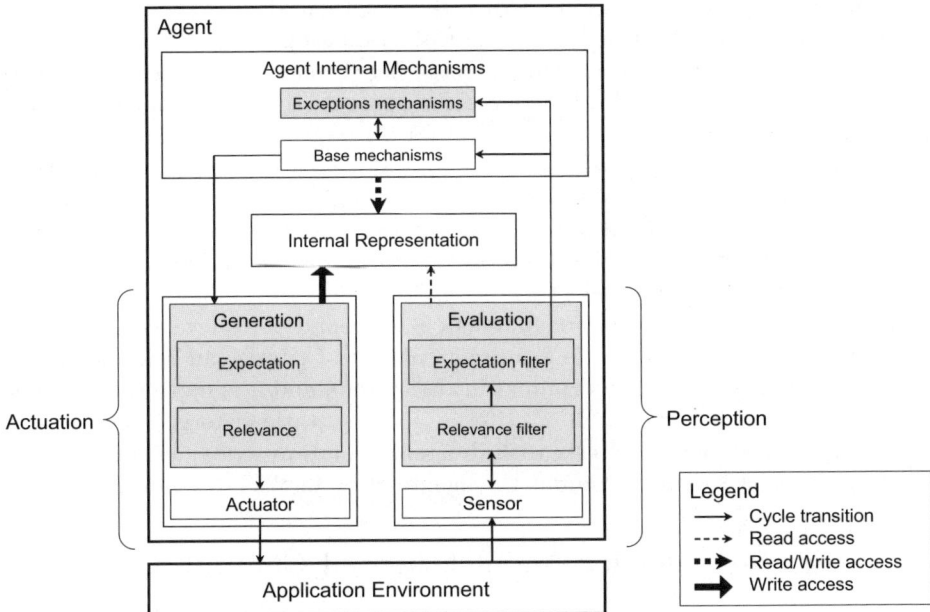

Fig. 2. Agent Architecture with Exception Management

The architecture sets forth necessary components represented in white, and optional components in gray. The necessary components are the ones introduced in the basic agent architecture of Fig. 1. The optional components are the additional mechanisms introduced to manage agent exceptions. This distinction separates the application logic in white, from the exception handling logic in gray, so that designers can choose whether the exception management part is necessary depending on their target application.

The architecture contains the four main components of the basic one, in more elaborated ways. The perception component encompasses the *sensor* and *percept evaluation* functionalities. Sensors receive events from the environment and pass them to the percept evaluation. This latter element is first responsible for distinguishing relevant from irrelevant events. Relevance appears to be essential feature to filter out unnecessary information (potential for exceptions that do not concern the agent in any way) and avoid the high-bandwidth issue in Robotics [36]. The work of Shah et al. and Weyns et al. are instances of mechanisms that fulfill such information selectivity [34,35]. The percept evaluation then identifies unexpected events depending on the criteria of the agent. One example of criteria in planning agents can be that unexpected events are those who are not 'scheduled' in the plan. Such criteria is independent of the architecture, and it is up to the designer to choose one in the development stages, depending on the target application. The percept evaluation uses the *internal representation* as the reference by which the agent can distinguish the events.

Once events are classified by the percept evaluation, they are forwarded to the *agent internal mechanisms* component, where events are processed. The two functional layers presented in this component separate the exception mechanisms from the 'base' that aims at the application logic of the agent. The exception layer introduces appropriate mechanisms to deal with exceptions, and its output should be directed to the base layer, so that the agent can continue its activity despite the occurrence of an unexpected event. The component as a whole manipulates the internal representation and its output is an action passed to the *actuation* for producing an effect in the environment.

The purpose of the *actuation* component is to prepare the relevance and expectation criteria of the agent in its future interactions. Criteria can be dynamically adapted by the agent to fit its context in the system, and it is up to the designer to decide the kind of evolution of criteria, notably static criteria along the life of the agent, or dynamic criteria with different evolution strategies. Finally, the *actuator* element serves to commit the agent action in the environment.

4 Research Issues for Agent-Oriented Computing

The definition of agent exception in this paper raises a number of research issues. The aim of this section is to review these issues and to eventually foster further research on the topic of exception handling in Agent-Oriented Computing.

Disciplined agent exception mechanisms. Agent exceptions require mechanisms adapted to their definition and constraints. Full-fledged MAS are open and have autonomous agents, so that further research is necessary to provide operational models for each mechanism, and eventually to provide implementations.

Time management. Management of agent exceptions is asynchronous, owing to the loose coupling of MAS and agent autonomy. There is no guarantee for when an exception can be handled, especially when an agent needs for the support of some other agents. Time management is therefore necessary to let agents reason on exceptions. One particular issue will be for real-time MAS. This type of systems is particularly significant considering the applications of MAS technologies to multi-robot systems and the financial domain.

Concert exception management. Souchon et al. proposed a mechanism for agents to handle exception 'in concert' in the AGR agency model [11], based on a model of concurrent exception handling [37]. With the present definition of exception, an adaptation of this initial work appears necessary. One perspective of this research issue is for the development of (agent-based) autonomic systems, where agents collaborate —perhaps selfishly— to handle some exceptions and guarantee the adaptability of the system as a whole.

Multi-type exceptions. The present definition is in rupture with exceptions as seen in programming languages. In consequence, we identified two types of exceptions in MAS, one due to the use of programming languages, the other due to the characteristics of MAS. Both types are however linked, i.e. some programming language exceptions have some impacts on agent-level exceptions, and conversely. The links between the two levels need to be modeled and exploited so as to leverage their benefits.

Automatic handler generation. Current work on agent exceptions propose some models of agents that can generate handlers at run-time [20]. Further research is necessary to develop this idea and reach a state where models can be turned into operational mechanisms. This work seems particularly important since it implies more flexibility in MAS and a concrete use of agent autonomy.

The list of issues is not exhaustive at this level of the research on agent exceptions. It sets forth however some of the current concerns that need to be addressed. Further work in these directions should refine the present situation and reveal other issues.

5 Conclusion

In this article, we propose a definition of 'agent exception' based on the current work on this research theme, and based on a model of full-fledged Multi-Agent Systems, i.e. an open system made of autonomous agents in their environment. An agent exception is then the evaluation by the agent of a perceived event as unexpected (application-dependent notion). The particularity of this definition is that exceptions only matter to one agent, whereas other agents are free to consider it as a normal situation. It also advocates that agent exceptions are 'not only errors', but also refers to 'opportunities' for agents to improve their performance. It also shows that exceptions only matter for agents with an internal representation, and that exception management is asynchronous, by opposition to usual mechanisms in programming languages. It finally shows that agent exceptions can occur without any underlying programming exception, which justifies the need for further specific research.

This new definition leads us to define a refined agent architecture to manage exception and to identify a set of research issues. Our future work is to continue our research on agent exceptions addressing these issues.

Acknowledgments

The authors would like to thank Oguz Dikenelli for his valuable comments on this research.

References

1. Goodenough, J.B.: Exception Handling: Issues and a Proposed Notation. Commun. ACM 18(12), 683–696 (1975)
2. Goodenough, J.B.: Structured exception handling. In: POPL '75: Proceedings of the 2nd ACM SIGACT-SIGPLAN symposium on Principles of programming languages, pp. 204–224. ACM Press, New York (1975)
3. Goodenough, J.B.: Exception handling design issues. SIGPLAN Not. 10(7), 41–45 (1975)
4. Parnas, D.L., Würges, H.: Response to undesired events in software systems. In: International Conference on Software Engineering, pp. 437–446 (1976)
5. Stroustrup, B.: The C++ Programming Language. Addison-Wesley, London, UK (2000)
6. Gosling, J., Joy, B., Steele, G., Bracha, G. (eds.): The Java^{TM} Language Specification, 3rd edn. Addison-Wesley, London, UK (2005)
7. Klein, M., Dellarocas, C.: Exception handling in agent systems. In: Agents, pp. 62–68 (1999)
8. Tripathi, A., Miller, R.: Exception handling in agent-oriented systems. [38], pp. 128–146.
9. Miller, R., Tripathi, A.: The Guardian Model and Primitives for Exception Handling in Distributed Systems. IEEE Trans. Software Eng. 30(12), 1008–1022 (2004)
10. Xu, J., Romanovsky, A.B., Randell, B.: Coordinated Exception Handling in Distributed Object Systems: From Model to System Implementation. In: ICDCS, pp. 12–21 (1998)
11. Souchon, F., Dony, C., Urtado, C., Vauttier, S.: Improving Exception Handling in Multi-agent Systems. In: Lucena, C., Garcia, A., Romanovsky, A., Castro, J., Alencar, P.S.C. (eds.) Software Engineering for Multi-Agent Systems II. LNCS, vol. 2940, pp. 167–188. Springer, Heidelberg (2004)
12. Iliasov, A., Romanovsky, A.: Structured Coordination Spaces for Fault Tolerant Mobile Agents. [39], pp. 181–199.
13. Dony, C., Urtado, C., Vauttier, S.: Exception Handling and Asynchronous Active Objects: Issues and Proposal. [39], pp. 81–100.
14. Huberman, B.A., Hogg, T.: Communities of practice: Performance and evolution. Computational and Mathematical Organization Theory 1, 73–92 (1995)
15. Feng, Y.D., Huang, G., Zhu, Y., Mei, H.: Exception Handling in Component Composition with the Support of Middleware. In: Nitto, E.D., Murphy, A.L. (eds.) SEM, pp. 90–97. ACM Press, New York (2005)
16. Dellarocas, C.: Toward Exception Handling Infrastructures in Component-based Software. In: Proceedings of the International Workshop on Component-based Software Engineering (1998)
17. Romanovsky, A.B.: Exception Handling in Component-Based System Development. In: COMPSAC, pp. 580–598. IEEE Computer Society Press, Los Alamitos (2001)

18. Klein, M., Rodríguez-Aguilar, J.A., Dellarocas, C.: Using domain-independent exception handling services to enable robust open multi-agent systems: The case of agent death. Autonomous Agents and Multi-Agent Systems 7(1-2), 179–189 (2003)
19. Platon, E., Sabouret, N., Honiden, S.: Challenges for Exception Handling in Multi-agent Systems. In: Software Engineering for Large-Scale Multi-Agent Systems, pp. 45–50 (2006)
20. Mallya, A.U., Singh, M.P.: Modeling exceptions via commitment protocols. In: Autonomous Agents and Multi-Agent Systems, pp. 122–129. ACM Press, New York (2005)
21. Mallya, A.U.: Modeling and Enacting Business Processes via Commitment Protocols among Agents. PhD thesis, North Carolina State University, Raleigh, United States (2005)
22. Bernon, C., Camps, V., Gleizes, M.P., Picard, G.: Engineering Adaptive Multi-Agent Systems: the ADELFE Methodology. In: Agent-Oriented Methodologies. Whitestein Series in Software Agent Technologies, pp. 172–202. Idea Group Publishing, USA (2005)
23. Brooks, R.: Intelligence without representation. Artificial Intelligence 47(1–3), 139–159 (1991)
24. Singh, M.P., Huhns, M.N.: Service–Oriented Computing: Semantics, Processes, Agents. Wiley, Chichester (2005)
25. Longman, P. (ed.): Dictionary of Contemporary English. Pearson Longman (2003)
26. Platon, E., Sabouret, N., Honiden, S.: Overhearing and direct interactions: Point of view of an active environment. In: Weyns, D., Parunak, H.V.D., Michel, F. (eds.) E4MAS 2005. LNCS (LNAI), vol. 3830, pp. 121–138. Springer, Heidelberg (2006)
27. Platon, E., Sabouret, N., Honiden, S.: Environment Support for Tag Interactions. In: Environment for Multi–Agent Systems (2006)
28. Hägg, S.: A Sentinel Approach to Fault Handling in Multi-Agent Systems. In: Dickson, L., Zhang, C. (eds.) Multi-Agent Systems Methodologies and Applications. LNCS, vol. 1286, pp. 181–195. Springer, Heidelberg (1997)
29. Russell, S., Norvig, P.: Artificial Intelligence: A Modern Approach. Prentice-Hall, Englewood Cliffs (2003)
30. Weyns, D., Omicini, A., Odell, J.: Environment, First-Order Abstraction in Multi-agent Systems. Autonomous Agents and Multi-Agent Systems 14(1), 5–30 (2007)
31. Drogoul, A., Corbara, B., Lalande, S.: MANTA: New Experimental Results on the Emergence of (Artificial) Ant Societies. In: Gilbert, N., Conte, R. (eds.) Artificial Societies: The Computer Simulation of Social Life, pp. 190–211. UCL Press, London (1995)
32. Rao, A.S., Georgeff, M.P.: BDI Agents: From Theory to Practice. Technical report, Australian Artificial Intelligence Institute (1995)
33. Kakas, A.C., Mancarella, P., Sadri, F., Stathis, K., Toni, F.: The KGP model of agency. In: de Mántaras, R.L., Saitta, L. (eds.) ECAI, pp. 33–37. IOS Press, Amsterdam (2004)
34. Shah, N., Chao, K.M., Godwin, N., Younas, M., Laing, C.: Exception Diagnosis in Agent-Based Grid Computing. In: International Conference on Systems, Man and Cybernetics, pp. 3213–3219. IEEE, New York (2004)
35. Weyns, D., Steegmans, E., Holvoet, T.: Towards Active Perception in Situated Multi-Agent Systems. Special Issue of the Journal on Applied Artificial Intelligence 18(8–9) (2004)

36. Kushmerick, N.: Software agents and their bodies. Minds and Machines 7(2), 227–247 (1997)
37. Issarny, V.: Concurrent Exception Handling. [38], pp. 111–127.
38. Romanovsky, A., Dony, C., Knudsen, J.L., Tripathi, A.R. (eds.): Advances in Exception Handling Techniques. LNCS, vol. 2022. Springer, Heidelberg (2001)
39. Dony, C., Knudsen, J.L., Romanovsky, A., Tripathi, A. (eds.): Advanced Topics in Exception Handling Techniques. LNCS, vol. 4119. Springer, Heidelberg (2006)

Toward an Ontology of Regulation: Socially-Based Support for Coordination in Human and Machine Joint Activity

Paul J. Feltovich, Jeffrey M. Bradshaw, William J. Clancey, and Matthew Johnson

Florida Institute for Human & Machine Cognition (IHMC)
40 South Alcaniz Street, Pensacola, FL 32502, USA
{pfeltovich,jbradshaw,wclancey,mjohnson}@ihmc.us

Abstract. In this chapter we explore the role of regulation in joint activity that is conducted among people and how understanding this better can enhance the efforts of researchers seek ing to develop effective means to coordinate the performance of consequential work within mixed teams of humans, agents, and robots. Our analysis reveals challenges to the quality of human-machine mutual understanding; these in turn set upper bounds on the degree of sophistication of human-automation joint activity that can be supported today and point to key areas for further research. These include development of an ontology of regulatory systems that can be utilized within human-agent-robotic teamwork to help with mutual understanding and complex coordination.

Keywords: Coordination, culture, human-agent-robotic systems, joint activity, ontology, policy, predictability, regulation, teamwork.

1 Introduction

One of the most important prerequisites for joint activity among people—and indeed for the functioning of human cultures—is the presence of regulatory systems by which such activity can be coordinated [13]. The kinds of joint activity we have in mind run the gamut of life, including processes as diverse as a conversation, a couple dancing, driving on a busy highway, and military operations.

The essence of joint activity is interdependence—that what party "A" does depends on what party "B" does, and vice versa (e.g., "One if by land, two if by sea" in Longfellow's account of Paul Revere's famous ride). As soon as there is interdependence, there is a need for coordination in time (e.g., timing a live, multi-party phone call) and/or space (e.g., designating a drop-off point), which in turn requires some amount of predictability and order. Such order has been described by Rousseau as the self-crafted bedrock of successful societies:

> ...the social order is a sacred right which serves as the foundation for all other rights. This right, however, since it comes not by nature, must have been built upon conventions."
> [31, p. 170].

In this chapter, we will explore joint activity and the kinds of "conventions," to use Rousseau's term, that serve to bring the order and predictability necessary for coordination among people, although we will call these "regulations" (for reasons that will be explained). By *"regulation,"* we mean any device that serves to constrain or promote behavior in some direction.

We conduct our investigation by identifying and unpacking the many and diverse systems of regulation that humans create and employ for achieving order. In what follows, we first discuss the nature of joint activity (Section 2), and then characterize some of the many culturally-based regulatory systems and their role in joint activity and coordination (Section 3). Finally, sections 4 and 5, respectively, outline some of the research challenges and applications of this work in the context of human-agent-robot teams conducting complex operations, such as space exploration, disaster response, and military operations [3; 5]. In the Appendix, we provide a brief illustration of major categories emerging in our ongoing effort to develop an Ontology of Regulation.

2 Joint Activity

Joint activity, interdependent activity, is important even to animals, which in some ways are better analogues than humans for the limited intellectual capacities of software agents [16]. One important way in which animals accomplish joint activity is through coordination devices, including signaling and display behavior. Animal biologist John Smith has identified ten signal types that he claims are nearly universal to (at least vertebrate) animals, although the specific manner of expression may vary across species [6; 16; 32; 33]. These are simple signals, basic to coordination: e.g., "I am available to take part in joint activity," (or not), "I am going to move," "I am monitoring something important," or "I am under attack." Such signals decouple action from intention and provide opportunity for other parties to join in or stay away.

With the increased complexity of our joint activities, predictability in support of coordination is even more important to humans. Cultural anthropologist Geertz [19] has argued that because of humans' under-determination biologically relative to lower animals, and because of our larger repertory of behavior, we are in even greater need of means of coordination. Notably, for the most part, we are left to fashion these means ourselves. That is, we need to learn and be taught how to live, interact, and control ourselves—hence our relatively long apprenticeships under parents and other educational and training influences. In this view, human *culture* itself is a vast fabrication of *regulatory systems* for guiding and constraining behavior, especially interdependent behavior (see also [20] regarding "guided doings").

Among humans, we distinguish roughly three types of joint activity. These are based on differences in the nature of their points of interdependence, in particular, interdependence among necessary resources only, among actions, and among motivations and goals [10]:

Sharing: This is characterized by interdependence among necessary resources only. Parties have independent goals, and there is no functional coupling of methods. An example is two groups trying to schedule a conference room they both need to use on a certain day. In sharing, constraints on resource allocation require negotiation.

Cooperation: In cooperation, there is interdependence of activities but not of motivations and goals. Often there is also interdependence of resources. Following the last example, two groups trying to conduct their own meetings within the same room at the same time would be a cooperation. So also, interestingly, are competitive games, such as football, where the two teams' actions are clearly interdependent while their aims are not the same and even contrasting.

Collaboration: Shared project objective is the hallmark of collaboration [11]. All parties are trying to achieve the same end (mutually defined), and there is also usually interdependence of actions (often involving different roles) and resources. Team members within *one* team in a football game (or a relay team in track and field) fit this description, as does a group of scholars working together to produce a genuinely multi-authored article on a topic of mutual interest.

2.1 Key Aspects of Joint Activity

We have asserted that one of the major purposes of regulation is the predictability and order it provides to support *coordination* of interdependent activities within joint activity. To make this coordination possible, participating parties need to 1) know some things in common with regard to their activity and, 2) use what they know in common to coordinate their interdependent interactions and moves. Following linguist Herbert Clark, we call the pertinent shared understanding "Common Ground" [14]. Common Ground consists of all the knowledge, beliefs, assumptions, and presuppositions that parties have in common with respect to their joint activity.[1] These knowledge components include the pertinent regulatory systems that apply to their joint activity, as well as related coordination devices that can be used to navigate coordination. Before addressing these directly in Sections 2.2 and 3, we first present other important components of successful joint activity.

We have argued previously that, in addition to adequate Common Ground, joint activity requires a "Basic Compact" that constitutes a level of commitment for all parties to support the process of coordination to achieve group goals [25]. We may say that to coordinate effectively, parties must have the basic resources, including sufficient common knowledge, i.e., have the *ability* to coordinate (Common Ground), and also the willingness to coordinate, the *will* (the Basic Compact). The Basic Compact is an agreement (often tacit) to participate in the joint activity and to carry out the required coordination responsibilities to facilitate group success [25; 26]. We represent as the *ideal* of the Basic Compact to be that parties want to be involved in the joint activity, and they want it to be successful (e.g., relay racers on a track team). Of course, many influences can degrade these ideal conditions (e.g., someone being coerced to participate). Such degradations can affect members' loyalty to the Basic Compact. So also can a member's operational "stance," in terms of having adequate resources, being overloaded or fatigued, being distracted, and so forth.

---

[1] In itself, Common Ground makes no claim about quality; common ground may be well- or poorly-tuned to the joint activity. Parties need to maintain common ground that is good enough to at least keep the joint activity moving forward.

One aspect of the Basic Compact is the commitment to some degree of goal alignment—typically this entails one or more participants relaxing some shorter-term local goals in order to permit group oriented long-term goals to be addressed. These longer-term goals might be shared goals (e.g., a relay team) or individual goals (e.g., drivers wanting to ensure safe journeys). A second aspect of the Basic Compact is a commitment to try to detect and correct any loss of Common Ground that might be disruptive.

We do not view the Basic Compact as a once-and-for-all prerequisite to be satisfied, but rather as a continuously reinforced or renewed agreement. Part of achieving coordination is investing in those things that promote the compact, as well as being sensitive to and counteracting those factors that could degrade loyalty to it.

All parties need to be reasonably confident that they and the others will carry out their responsibilities in the Basic Compact. In addition to repairing Common Ground, these responsibilities include such elements as acknowledging the receipt of signals, transmitting some construal of the meaning of the signal back to the sender, and indicating preparation for consequent acts. The Basic Compact is also a commitment to ensure a reasonable level of interpredictability; that is, agents acting, to the extent they can, so as to be *mutually predictable/understandable and mutually directable* by others.

What is the primary role of a "Basic Compact" in joint activity? We submit that a critical role has to do with *trust and predictability* in the operation of the whole interactive system. For example, when the Basic Compact is strongly in force, we can trust that other people are working on their assignments, are telling the truth about important matters, are going to send an item to another party if they say they will, and so forth. In another example, the Basic Compact requires that if one party intends to drop out of the joint activity, he or she must inform the other parties. Hence, the Basic Compact "washes over" the entire enterprise of the joint activity, largely conferring a trust level in the operations of all the components. When it is functioning at its best, the Basic Compact contributes to the predictability of events within the joint activity, what we have argued is a primary role for regulation in the first place.

A certain way of interacting serves to maintain and even improve Common Ground. This way involves what is called the "joint action ladder" (JAL) [13]. When one party sends a message/signal to another, the second party, in reply, should 1) acknowledge that he has seen the signal arrive [attention], 2) "read" the signal [perception], 3) provide his interpretation of what it means [understanding] and, 4) indicate what he is likely to do as a result [action]. The latter two, in making understandings and intentions public, provide the opportunity for repair of Common Ground, the common understanding among the parties (e.g., through discussion of differences).

2.2 Coordination Devices

People coordinate through signals and more complex messages of many sorts (e.g., face-to-face language, expressions, posture). Human signals are also mediated in many ways—for example, through third parties or through machines such as telephones or computers. Hence, direct and indirect party-to-party communication is one form of a "coordination device," in this instance coordination by *agreement*. For example, a group of scientists working together on a grant proposal, may simply agree, through e-mail exchanges, to set up a subsequent conference call at a specific date and time. There are three other major types of coordination devices that people commonly employ: convention, precedent, and situational salience [13; 25].

Convention: Often prescriptions of various types and degrees of authority apply to how parties should interact. These can range from rules and mandated procedures, to less formal codes of appropriate conduct. These less formal codes include norms of practice in a particular professional community, as well as established practices in a workplace. Convention also often applies to activity devolving from more situationally emergent interactions in which we may engage, e.g., contracts we enter into, and debts and other kinds of obligations we take on. Coordination by convention depends on structures outside of a particular episode of joint activity.

Precedent: Coordination by precedent is like coordination by convention, except that it applies to norms and expectations developed within the ongoing experience of the joint activity. As a process unfolds, decisions are made about the mutually accepted naming and interpretation of things, standards of acceptable behavior and quality, who on the team tends to take the lead, who will enact particular roles, and so forth. As these arise and develop during the course of the activity, they tend to be adopted as devices (or norms) of coordination for the remainder of the activity.

(Situational) Salience: Salience has to do with how the ongoing work arranges the workspace so that next move becomes apparent within the many moves that could conceivably be chosen. Coordination by salience is produced by the very conduct of the joint activity itself. It requires little overt communication and is likely to be the predominant mode of coordination among long-standing, highly practiced teams.

Coordination devices often derive from regulatory systems, as will be discussed in Section 3.3, after further discussion of regulation itself, next.

3 Characterizing Regulation

Culturally-based regulatory systems are many and diverse, and go beyond what we normally construe as "law" or even "rules." In addition to law-like devices, they also include customs, traditions, work-place practices, standards, and even codes for acceptable everyday behavior. In this section we discuss some of what we have discovered in our attempts to characterize regulation.

3.1 Toward an Ontology of Regulation

Paul Wohlmuth, a philosopher of law, once wrote an introductory chapter for a special issue of the *Journal of Contemporary Legal Issues*. This issue focused on the "constitution of authority" [34]. By constitution of authority he meant, roughly, how different kinds of things come to have regulatory power over human activity. He used the example of an automobile traveling a bend in the road to illustrate the ubiquity and diversity of the authoritative forms that can come to bear on human affairs (see also 19; 29). Wohlmuth's full analysis is discussed in [16].

Starting from the numerous examples provided by Wohlmuth, we engaged in an exploration to identify some of the different kinds of devices that serve to govern human conduct. This has been a somewhat informal investigation. It started as a brainstorming effort when we began to build a list of all the different kinds of regulatory concepts we could think of. Then we used dictionaries, thesauruses, synonym finders and the like. The main criterion for inclusion in our list was that a device *constrains or promotes behavior in some direction.*

We needed to choose a word to represent the scope of the entire enterprise, and after considering several plausible candidates we settled on the generic term of "regulation" as one that seemed the most inclusive and accommodating of our purposes. As we continued our effort, rough categories for the numerous basic terms were created and refined. Once we had accumulated a reasonable seed set of regulatory mechanisms, we started circulating them among colleagues for reactions and contributions. As word has spread about the endeavor, friends and colleagues have kindly called in or e-mailed very welcome unsolicited contributions.

We should add that—at least for the present—our central interest has been in the gradual emergence of stable categories—e.g., the difference between a "law-like" device" and an "obligation-like" device—and we have not worried much about the precision of each definition. In fact, up to now nearly all of our rough definitions have come from ordinary dictionaries rather than from careful handcrafting. For this reason, similarities and differences across broad categories seem more stable than distinctions among items within them. As an example, let us examine more closely laws and obligations.

Law-like devices are variants on coded, largely, but not always, written down rules. They characteristically have the power of the State or similar authority behind them. One can be jailed, fined, or otherwise sanctioned for violating them, all "legally." Some examples (including variants within the major category):

Law-like devices (have the power of the State/authority behind them)
- *Law*: All the rules of conduct established and enforced by the authority
- *Statute*: A law passed by a legislative body and set forth in a formal document
- *Bill*: A draft of a law proposed to a lawmaking body
- *Ordinance*: A custom or practice established by usage or authority
- *Mandate*: An authoritative order or command, especially a written one
- *Edict*: An official public proclamation or order issued by an authority; Decree
- *Decree*: An official order, edict, or decision, as of a church or government

While "law" generally develops over time, gets codified, and exists and is enforced over time in a relatively extended process, there is also "fast law," as in a "decree" or "edict." Hence we decided to treat the two kinds as separate categories; that is, within "law," there is both "fast" and "slow" law.

Let us now contrast "law" with "obligation." Obligation generally devolves from position, status, or special group, as in one becoming a parent, priest, Muslim, police officer, or the president of a social club. Some examples (including variants within the major category):

Obligation-like devices (accrue from one's position, role, state)
- *Obligation*: Duty imposed legally or socially. Activity that one is bound to as a result of a contract, promise, moral responsibility, position
- *Duty*: Any action necessary in or appropriate to one's occupation, role, or position. Includes duties to groups of people, e.g., one's elders, one's children
- *Responsibility*: Condition, quality, or fact of being responsible; obligation
- *Requirement*: Something obligatory or demanded as a condition

As in the law example, this category overlaps with some other items such as contracts and agreements—e.g., an agreement, or even a promise, can invoke an obligation—but contracts and agreements seem to have enough unique features that we treat them as a separate category, "Agreement-like things" (see Appendix).

One might wonder about the forms in which regulation exists. The current version of the Ontology contains over 220 regulatory concepts in about forty categories. Given this scope and diversity, one might argue that regulation is just about everywhere, in a myriad of forms. This condition of pervasive regulation has been noted by sociologists and social anthropologists for some time. In addition to Geertz (noted earlier), we cite Erving Goffman, who has claimed that:

> ...one of the consequences of this learning program [socialization, learning the extant systems of regulation] is the transformation of the world into a place that is appreciably governed by, and understandable in terms of, social frameworks. Indeed, adults… may move about through months of their days without once finding themselves out of control of their bodies or unprepared for the impingement of the environment—*the whole of the natural world having been subjugated by public and private means of control* [20, p. 33, emphasis added, our annotations in brackets].

And here is the major link between regulation and coordination: When regulatory systems break down, predictability and order degrade, and coordination becomes impossible. As stated by Goffman:

> If the meaningfulness of everyday activity is similarly dependent on a closed, finite set of [interdependent] rules [and practices]… then one can see that the …the significance of certain deviant acts is that they undermine the intelligibility of everything else we had thought was going on around us, *including all next acts* [predictability], thus generating diffuse *disorder* [20, p. 5, emphasis added, our annotations in brackets].

Given the importance of regulation in successful joint activity, we have, as noted, been developing an ontology of regulatory systems. The current version of the Ontology of Regulation is organized under four main categories:

o *Regulatory Devices*. These are the many forms of regulation themselves that have the power to promote or constrain activity and were exemplified by "law" and "obligation" in our earlier discussions. This is the most highly developed category of the entries within the current Ontology -- with around 200 entries. (Some of the major categories and some examples are provided in the Appendix.)

o *Developmental processes for the Constitution of Authority.* As noted earlier, the constitution of authority refers to the ways in which things come to have regulatory power over human activity. In this category we include three types: The first we refer to as the "Origins and Derivations of Regulatory Devices" themselves. These are ways that regulatory systems come into existence. They include such processes as: force, the process of coming to an agreement, social emergence, legislation, court order, and divine intervention. The second we refer to as "Origins and Derivations of Officiation." These are processes and devices by which people or institutions are initiated into positions of authority to enforce regulation (e.g., a police officer), for example, by election, ritual, testing, or accreditation. The third are "Origins and derivations of Interpretive Prerogative." These are processes and devices by which entities gain the authority to interpret regulation (e.g., judges). Examples are appointment, credentialing, or election. The first two categories listed in this subsection, together, are the second most highly developed categories in the current Ontology, with more than thirty entries.

o *Objects of Regulation.* These are the entities to which regulatory devices apply. They apply to the activities of people--including people in roles, such as a medical doctor, a priest, a certified public accountant, or a citizen of a county--and institutions (e.g., a publicly traded company), as they participate in such processes as a marriage ceremony, audit, or corporate merger.

o *Guardians of Regulation.* These are individuals or institutions empowered to enforce regulatory systems (as they apply to certain groups of people and their activities). Examples are a police officer with regard to the public, a parent with regard to his or her children, or the Securities and Exchange Commission with regard to publicly traded companies. These enforcing entities often gain their regulatory authority by virtue of some process involving the "Origins and Derivations of Officiation" as described above (e.g., by licensure, appointment, or election by a people).

o *Interpreters of Regulation.* Because the relationship between regulations and concrete applications in the world is often not straightforward (e.g., see Section 3.5 on the "Bureaucrat's Dilemma"), there must be people and institutions with authority to adjudicate alternative interpretations (e.g., judges or appeals courts). Such authority is usually granted through processes involving the "Origins and Derivations of Interpretive Prerogative." Examples are appointment of a judge and licensure of an attorney.

The last three categories, "Objects of Regulation," "Guardians of Regulation," and "Interpreters of Regulation, except as illustrated, are largely undeveloped. It should also be noted that, depending on context, the same entity can appear in any of these three categories. For example, a parent may have a variety of regulatory authority over his or her children but, at the same time, be subject to the laws of the land.

A sample of major categories of regulatory devices from the Ontology is given in the Appendix. We should note that many, if not most, of these regulatory devices— e.g., agreements or even contracts— are, themselves, best characterized as processes and joint activities (see [18]).

3.2 Fast and Slow Regulation

We take the "speed" of a regulatory device to have three main dimensions: how quickly it can be enacted and acquire authority, how quickly it can be changed, and how quickly it can be enforced. For good reason, human activity happens at many different speeds, and regulatory devices need to be appropriate to these different paces. We would not want to have to engage the mechanisms of changing constitutional law for guidance and reprieve when someone has a gun to our face, for instance. The distinctions we address below have similarities to distinctions David Woods has made between activity at the "sharp end" (i.e., at the point of connection and impact with the world) and the "blunt end" (i.e., behind-the scenes culture and practice that bear on sharp-end activity) [15]. As a first attempt at addressing the speed of regulation, we have divided the major classes of regulatory devices into three types:

Systemic Schemas, including things such as folkways (e.g., customs, traditions, mores) and elements of "natural law." These are deeply engrained, pervasive guidance systems of a people, perhaps largely implicit, slow to develop, and slow to change. Interestingly, these types of regulation may be quick to enforce. It seems that in these kinds of matters people are inclined and feel authorized to take matters into their own hands when things go awry; that is, they may deal with enforcement personally and on-the-spot. For example, many citizens might be inclined to intervene upon seeing deliberate burning of their nation's flag or the desecration of their prophet's image or holy book.

Organizational Constraints and Allowances, including such things as authorizations, policies, practices, obligations, and codified rules (e.g., laws). These are more special-purpose devices that are explicitly and deliberately enacted and enforced by different configurations of people, within different socially constructed bodies (e.g., a club, agency, nation). While they can generally be enacted and modified more quickly than the Systemic Schemas, they can, in some instances, be slow to change through what may involve complicated processes (e.g., changing a country's constitution).

People tend to put enforcement in this category into the hands of some socially-sanctioned authority. Speed of enforcement varies by the degree of closeness/access of such an authority to the pertinent (regulated) activity. On the highway, for instance, for the *law* to be engaged as regulation requires spotting of the incident by a law enforcement officer and may subsequently involve courts and procedures and judgments by designated individuals. (As an aside, technology is now affecting speed of enforcement in many ways. For instance, cameras and sensors are now enabling on-the-spot detection of traffic violations. This technology can document the violation, gain the driver's identity and address from the license plate, and stuff and mail an envelope for delivery of the ticket and fine.) In contrast, good and bad manners (which are more like systemic schemas) on the highway (e.g., following too closely or cutting in too quickly after a pass) can and often are enforced by the individual participants—by honking horns or giving angry hand gestures—who feel they have authority in this context.

Action Guides. One kind of action guide is what we call "design of affordances." This involves what we make *hard or easy to do* by design, such as in the placing of doorways, sidewalks, streets, and bridges in particular places to channel traffic [28]. Not all design of affordances is physical. For instance, we can make some activities easy or hard to do through the allocation of resources (e.g., restricting gasoline). Another kind of action guide, what we call "transactional utilitarian devices," (see Appendix) are fast acting, often fluid, regulatory devices that we set for ourselves in the process of conducting everyday affairs with others—e.g., making promises, agreements, appointments, gestures, and so forth. Creating and dismantling affordances can be slow or fast (putting up a roadblock vs. building a road). The key is that once they are in place, they have nearly immediate regulatory efficacy. On the other hand, almost by definition, transactional utilitarian devices can be quickly created, dismantled, or enforced at the point of activity.

3.3 Relationships Between Regulatory Systems and the Coordination Devices

We have argued that regulatory systems are created by any social group to increase inter-predictability and order necessary for coordination in joint activity. Hence, it is not surprising that the main types of coordination devices (see previous Section 2.2) people use in the actual conduct of joint activity bear a strong relationship to various categories involved in the Ontology of Regulation (Appendix). Some of these relationships are clear, e.g., in the cases of agreement-like coordination devices, precedent-like devices, convention-like devices, and devices that are like "transient utilitarian devices/salience." Hence, the coordination devices can be thought of as the operational mechanisms of the more abstracted regulatory systems. It is an interesting challenge for further research to investigate more deeply these kinds of correspondences (e.g., identifying additional types of coordination devices by examining categories of the Ontology that do not match well to the four discussed previously, for example a new category of "schedules").

3.4 The Special Status of Norms Among Regulatory Devices

It is common in socio-cultural research to characterize social behavior as being subject to "norms" [12]. In this sense, norm means what Bourdieu called "habitus"—dispositions or schemas relating perception, thought, and action that socially structure experience and behavior, while dialectically structuring the social world [2]. We suggest that this characterization can be further specified. In our present treatment, we take behavior to be subject to regulation of *many and diverse* forms, as exemplified by the entries in our Ontology. Norms are but one these, indicating something about *personal and societal acceptability*. In this regard, norms still do have a special status among regulation types. In particular, there are at least three variations of norms, all of which possess some overlap, but at the same time suggest significant difference. All are socially constructed and enforced:

That which exists now: This norm, call it "Norm1," pertains to *what is* actually in place now (regardless of what the pertinent official rules, laws, etc., may be—see below). Regarding driving a highway, Norm1 refers to the largely self-organized, in-place, ambient traffic speed: maybe 10+ the posted speed limit in clear, dry weather—lower in ice or snow. What is extant is special for a number of reasons. First it has a certain momentum and inertia by way of its development, implementation, and execution, including enforcement and interpretation [34]. "Possession is nine-tenths of the law," so to speak. Think of the intense societal processes associated with the regulation of abortion in the United States. Many people have devoted a good share of their lives (and spent a lot of money) to persuade the public, legislators, and judges to accept their point of view. Moreover, "what is" attracts a following, in particular all those who benefit from the current state of regulation, either in their personal behavior, by favor of their constituent voters (or not), in medical facilities that can or cannot operate, or in related official posts that were created, and people were subsequently able to assume, because of the state of affairs.

That which is socially tolerable (relative, of course, to some reference group): This Norm2 (the "stretched norm") refers to what people of a certain community will *actually still tolerate*, beyond in-place norms. For instance, Norm2 reflects what a driver can get away with without incurring the wrath of other drivers—e.g., their taking down a tag number or calling in the police, honking their horns, or trying to run him of her off the road. While drivers, following the in-place norm, routinely drive faster than the posted speed limit, if somebody drives *too* fast, or even at the *normal* rate in icy and rainy conditions, others may try to take some action. It is clear that there are ranges of public tolerance for deviation from social norms established as "what is." Audience members at the symphony may dislike but tolerate some degree of coughing or whispering, but may chastise louder talking or repeated cell-phone use.

A more formal benchmark or standard by which some thing or process is judged acceptable: These are the more **formal** standards for regulation of behavior or products in a community. They often lie behind norms of the first two kinds, employed only when there are breakdowns in more informal employment and enforcement. In driving the highway, Norm3 refers to such things as the posted speed limit and all the legal machinations behind and entwined within it (e.g., the motor vehicle code, the courts). Norms of type three include laws, ordinances, posted job rules, rules-of-the-game, and so forth. As an example, let us look at the three kinds of norms as they might apply to a junior lawyer in the (hypothetical) law firm of Dewey, Cheatem, & Howe:

o   Norm1 refers to what the young lawyer does, pretty much like the other junior lawyers do, on a day-to-day basis, when everything is going uneventfully, within routine.
o   Norm2 refers to what deviations people around him will tolerate without reprisal (e.g., taking off at noon each Friday to get a jump on the weekend,

being persistently behind in logging his billable hours, doing a great deal of travel away from the office).
o   Norm3 refers to the formalized rules and regulations pertaining to "being a junior partner at the Dewey firm."

So, what is special about norms as a regulatory device, compared to all the others? We propose that as benchmarks and criteria for acceptability, essentially for necessary quality, norms apply in the execution *of all the other* regulatory devices. These other regulatory devices can be carried out, applied in ways that are themselves subject to extant norms of all three kinds: the ways that are currently honored in a social group, the deviations that are tolerated, and in the formal regulations that can be brought to bear under challenge. One can easily see how these might play out in the execution of an agreement, even more clearly with regard to a contract, but even with regard to such activities as engaging in a promise, giving and responding to a command, executing one's duties and obligations, or just "obeying the law."

Norms also apply to the internal workings of any joint activity itself, in the cues that signal the need for and initiation of the joint activity, in the coordination devices and reciprocal actions utilized by the parties within the execution of the joint activity, and in the signals that indicate the joint activity is coming to a close. In a simple example, in the activity of a customer checking out at the counter of a convenience store, the customer generally initiates the interchange by walking up to the sales clerk's counter and placing an item on it. It is not the norm, for this purpose, for the customer to continue clutching the item—which may in fact be a cue to the clerk that the customer is approaching to ask a question ("Where's the restroom?"). One can envision how norms might be involved in the other components of this seemingly simple, everyday transaction, including the coordination devices that are employed.

3.5 Qualifying Regulation: The Bureaucrat's Dilemma

Regulations may be implicitly or explicitly qualified in order to take account of context. I promise to drive my friend to the airport on Friday—*if* my car is out of the repair shop by then. It is okay to break the speed limit—*if* one is driving an accident victim to the hospital. When "authorities" do not tie the hands of "enforcers" too rigidly, the latter can recognize and appropriately adapt to such circumstances. However, because of their limited perceptual and reasoning abilities and their difficulty in grounding and adapting to context, this is a daunting challenge for software agents.

Consider this seemingly simple rule: *Congress shall make no law abridging freedom of speech.* Then we start thinking about the circumstances in which one would not want this rule to be honored—for example, the famous crime of (falsely) crying "Fire!" in a crowded theater. So, we further qualify: *Congress shall make no law abridging freedom of benign speech.* But what is benign? (or "speech" for that matter?).

For every qualification made, one is left with a choice. Either a judgment needs to be made (e.g., about what is "benign") or there needs to be yet another qualification (e.g., defining what constitutes "benign"). In human affairs, this can lead to absurdity if there is an attempt to expunge all human judgment from such interpretations—to specify so extremely that criteria are mechanically analyzable. An example has been presented in the literature about national workplace rules that require that all handrails be 28 inches high, whether the users are professional basketball players or midgets [24]. Revisiting speech, and our example, such specifications might take the form: *Congress shall make no law abridging freedom of speech that lasts less than one-half hour,* a degree of specification that enables mechanical and reliable interpretation by artificial agents, even by relatively stupid ones, but which can lead to absurd rigidity and reversals of intended effects.

4 How Can We Make Software Agents Better Team Players?

A key challenge in any kind of joint activity is to successfully navigate points of interdependency among the participants. In this section, we will focus on two challenges for software agents to become better team players: 1) participation in the Basic Compact, and 2) the crucial "handoff" operation embodied in the "joint action ladder" (JAL) as it contributes to mutual understanding.

Being part of the Basic Compact means that participants want to participate in the joint activity and want the group's goals to succeed. This is an ideal, and variants often occur (e.g., forced participation). This affects trust and confidence throughout the enterprise, for example, the belief that participants will function and report appropriately.

Of course, except for rare software agents that can reason about the relative utility of their participation in competing tasks, the issue of an agent's dedication to the group work is moot. More germane to the subject for most agents is their responsibility to uphold certain kinds of standards in reporting and functioning. For instance, if an agent reaches a failure point and will not be able to complete its task, this needs to be reported to others, and agents can often do this. On the other hand, more nuanced kinds of "appraisals of progress" can pose daunting challenges for software agents. In contrast, highly expert human teams have ways of understanding and reporting about "how things are going" that can anticipatorily *foreshadow* success or failure.

With respect to the first rung of the JAL, *attending*, agents need to share a common repertoire of watched-for signals that call for entry (and exit) into interdependency cycles—for example, signals such as a customer placing a store item on the check-out counter at a convenience store. The signals (and more complex messages) that the agents send and receive ideally would be expressed at many levels of overtness and abstraction, depending on the situation. These could range from simple beeps and buzzes (e.g., the warning of a big truck backing up) to complex appraisals that would tax the best human experts (e.g., "The battle here is going badly").

In critical situations, agents cannot afford to miss important signals. This is a matter of attention management, both in one's self and in others. Sometimes attention is directed appropriately by the nature of the ongoing activity itself, through the coordination device of salience. However, in many cases where practice is less of a routine, the need for attention focus calls for the ability of agents to direct each other to help ensure that messages are not missed. Operative regulatory frameworks can help, as in certain circumstances cues to the start of joint activity may be mandated, conventional, part of accepted practice, and so forth (e.g., the reaching out of a hand for hand shaking).

Once arrival of signals/messages has been detected, the JAL prescribes that they must be *read* and *understood*. Analysis must reveal their "meaning in context." This is largely a matter of adequate Common Ground, the pertinent (to the activity) knowledge that agents either bring with them or gain in the course of ongoing participation, including knowledge of the regulatory mechanisms and coordination devices that apply. Knowledge of pertinent regulatory devices can help by constraining the space of possible meanings for both sender and receiver (e.g., conversation policies [21; 23]). A major challenge, however, is for the parties to construct mutually a model of *which* regulatory systems *apply*. This is particularly challenging because the question is highly context sensitive; that is, it depends on how the parties conceptualize "what they are doing now." Consider the pertinent forms of regulation that would be taken to apply if a couple (or an observer watching them) conceived that they were "inspecting a house" for possible purchase, versus "casing a house" for a robbery, or a group thinks it is "conducting a holy war" versus "committing criminal murder" [1; 9]. Fluidly adopting different perspectives is difficult even for humans [17]. A major challenge for software agent participation in joint activity is the sophistication with which agents can adopt such "points of view" regarding their activities or changing roles.

When an agent has created a candidate understanding, it must broadcast this understanding for public scrutiny, in particular, for appraisal by those others with whom the agent's actions are most interdependent. This provides opportunity for the detection of failures and slips of Common Ground among the parties. Additionally, the final *action* rung of the JAL, involving broadcasting the next steps one is likely to take on the basis of this current, tentative understanding, not only provides additional information about understanding, but it also decouples intention to act from action itself, allowing for corrective intervention before actual moves are made. Both of these mechanisms contribute to increasing the predictability and transparency (e.g., of intent, stance, and state) of interdependent agents to each other and, hence, contribute to productive coordination.

Finally, the parties need to participate in repair of their understandings, as needed, to enable successful further progress in the joint activity. Making mutual understandings visible aids this repair process, and, hence, improves the quality of Common Ground. Aspects of discordance among the parties' understandings can vary in transparency and abstraction, from those involving simple facts and observable states, to those involving more complex levels of interpretation and reasoning that may tax agents, e.g., relations among things, inferences, interpolations, and appraisals of progress [7]. In any event, discrepancies in the agents' understandings require that they be capable of some forms of *negotiation*. As with determination of meanings in general, knowledge of the regulatory systems under which the agents are operating may help to delimit the range of discrepancies that occur.

The sketch just presented of the nature of understanding and interaction needed for successful joint activity is a caricature, intended to overly pull apart and concretize operations that in practice are often more blended and abbreviated, especially as groups work together over time and learn about their activities and each other. That is, in reality, things are often different. One example is "downward evidence" in the JAL [13], by which higher levels of response are taken as indications of lower ones (e.g., a response that conveys sensible understanding of a message also conveys that it was noticed and read). Other examples of real complexities that belie the simplifications presented here involve qualitative changes of state in items such as coordination devices and even regulations. For example, agreements over time can become precedents, at which point they no longer require operational deliberation. Procedures conducted with explicit reference to plans, over time can come to be directed mostly by situational salience. Practice over time can be transformed into law.

The treatment above, though compromised, is intended to accentuate the kinds of understandings and operations that are required to conduct successful joint activity in its rawest form, and hence to point to the most important places where software agent capability must be considered (and possibly improved over time) in creating successful, mixed human-agent working groups.

5 Applications to Human-Agent-Robotic Teamwork

We are applying the ideas presented in this chapter to facilitate joint activity in mixed human-agent-robot teams. We are doing this in two primary ways. First we are implementing regulatory systems that help coordinate joint activity through constraints of authorizations and obligations that we call "policies" [4]. In this we utilize KAoS Policy and Domain Services, a framework that can reliably and flexibly specify, analyze, enforce, and adapt policy. Applications include modeling a point-of-view, such as in "adversarial modeling," in which we attempt to model actions and reactions to events by groups of people sharing a culture greatly different from our own. Another is cross-domain information exchange (CDIX), in which we attempt to facilitate the sharing of information across different governmental and relief agencies, perhaps highly different in their rules, procedures, and organizational "cultures." We are also applying this work in the context of human-agent-robot teams within complex operations, such as space exploration, disaster response, and military operations [3; 5].

A second direction is to enhance our understanding of the regulatory systems operative in human social behavior and to develop our "Ontology of Regulation" that can be implemented to support our human-agent modeling systems. The ontology is a work in progress. We are currently formalizing its concepts and relations within an OWL ontology using the Cmap Ontology Editor (COE; [22]). (For a current version, contact Feltovich).

We believe that the unusual approach of modeling social behavior with more attention to operative regulatory structures, relative to individual behavior and cognition, will complement other modeling approaches in important ways. For example, much of a people's culture may well be more stable than individual behavior and cognition, and may well be more amenable to being represented in advance, when, for example, there is need to quickly ramp-up a model of a brand-new hostile social group.

Acknowledgements

Our thanks to the following individuals for their contributions: Beth Adelson, Guy Boy, Kathleen Bradshaw, Judith Donath, Andy Feltovich, Anne Feltovich, Ellen Feltovich, Joan Feltovich, Robert Hoffman, Gary Klein, Donald Norman, Paolo Petta, Jeremy Pitt, Gene Santos, Rand Spiro, Maarten Sierhuis, Niranjan Suri, Bob Weber, Gloria Valencia-Webber, Paul Wohlmuth (deceased), David Woods, and the Editors of this volume. The research was supported in part by grants and contracts from the U.S. Air Force Office of Scientific Research to Dartmouth College through subcontract (5-36195.5710), the Office of Naval Research (N00014-06-1-0775), and the U.S. Army Research Laboratory through the University of Central Florida under Cooperative Agreement Number W911NF-06-2-0041. The article does not necessarily reflect the views of any of these agencies.

References

[1] Anderson, R.C., Pichert, J.W.: Recall of previously unrecallable information following a shift of perspective. Journal of Verbal Learning and Verbal Behavior 17, 1–12 (1978)
[2] Bourdieu, P.: Outline of a theory of practice (Trans. by R. Nice). Cambridge University Press, Cambridge, NY (1977)
[3] Bradshaw, J.M., Acquisti, A., Allen, J., Breedy, M.R., Bunch, L., Chambers, N., Feltovich, P., Galescu, L., Goodrich, M.A., Jeffers, R., Johnson, M., Jung, H., Lott, J., Olsen Jr., D.R., Sierhuis, M., Suri, N., Taysom, W., Tonti, G., Uszok, A.: Teamwork-centered autonomy for extended human-agent interaction in space applications. In: AAAI 2004 Spring Symposium. AAAI Press, Stanford University, CA (2004)
[4] Bradshaw, J.M., Feltovich, P.J., Jung, H., Kulkarni, S., Allen, J., Bunch, L., Chambers, N., Galescu, L., Jeffers, R., Johnson, M., Sierhuis, M., Taysom, W., Uszok, A., Van Hoof, R.: Policy-based coordination in joint human-agent activity. In: Proceedings of the IEEE International Conference on Systems, Man, and Cybernetics. The Hague, Netherlands (2004)
[5] Bradshaw, J.M., Jung, H., Kulkarni, S., Johnson, M., Feltovich, P., Allen, J., Bunch, L., Chambers, N., Galescu, L., Jeffers, R., Suri, N., Taysom, W., Uszok, A.: Toward trustworthy adjustable autonomy in KAoS. In: Falcone, R. (ed.) Trusting Agents for Trustworthy Electronic Societies. Springer, Heidelberg 2005 (in press)
[6] Cheney, D.L., Seyfarth, R.M.: Precis of how monkeys see the world. Behavioral and Brain Sciences 15(1), 135–182 (1992)
[7] Chi, M.T.H., Feltovich, P.J., Glaser, R.: Categorization and representation of physics problems by experts and novices. Cognitive Science 5(2), 121–152 (1981)
[8] Clancey, W.J.: Situated Cognition: On Human Knowledge and Computer Representations. Cambridge University Press, Cambridge (1997)
[9] Clancey, W.J.: Conceptual coordination: How the mind orders experience in time. Lawrence Erlbaum, Hillsdale, NJ (1999)
[10] Clancey, W.J.: Automating Capcom: Pragmatic Operations and Technology Research for Human Exploration of Mars. In: Cockell, C. (ed.) Martian Expedition Planning. AAS Science and Technology Series, vol. 107, pp. 411–430 (2004)
[11] Clancey, W.J.: Roles for agent assistants in field science: Personal projects and collaboration. IEEE Transactions on Systems, Man, and Cybernetics, Part C: Applications and Reviews 34(2), 125–137 (2004)

[12] Clancey, W.J., Sierhuis, M., Damer, B., Brodsky, B.: Cognitive modeling of social behavior. In: Sun, R. (ed.) Cognition and Multi-Agent Interaction: From Cognitive Modeling to Social Simulation, Cambridge University Press, New York (2005)
[13] Clark, H.H.: Using Language. Cambridge University Press, Cambridge, UK (1996)
[14] Clark, H.H., Brennan, S.E.: Grounding in communication. In: Resnick, L.B., Levine, J.M., Teasley, S.D. (eds.) Perspectives on Socially Shared Cognition, American Psychological Association, Washington, D.C. (1991)
[15] Cook, R.I., Woods, D.D.: Operating at the sharp end: The complexity of human error. In: Bogner, S.U. (ed.) Human Error in Medicine. Lawrence Erlbaum, Hillsdale, NJ (1994)
[16] Feltovich, P., Bradshaw, J.M., Jeffers, R., Suri, N., Uszok, A.: Social order and adaptability in animal and human cultures as an analogue for agent communities: Toward a policy-based approach. In: Omicini, A., Petta, P., Pitt, J. (eds.) ESAW 2003. LNCS (LNAI), vol. 3071, pp. 21–48. Springer, Heidelberg (2004)
[17] Feltovich, P.J., Hoffman, R.R., Woods, D., Roesler, A.: Keeping it too simple: How the reductive tendency affects cognitive engineering. IEEE Intelligent Systems, 90–94 (May-June 2004)
[18] Feltovich, P.J., Spiro, R.J., Coulson, R.L., Myers-Kelson, A.: The reductive bias and the crisis of text (in the law). Journal of Contemporary Legal Issues 6(1), 187–212 (1995)
[19] Geertz, C.: The Interpretation of Cultures. Basic Books, New York (1973)
[20] Goffman, E.: Frame Analaysis. Harper and Row, New York (1974)
[21] Greaves, M., Holmback, H., Bradshaw, J.M.: What is a conversation policy? In: Greaves, M., Bradshaw, J.M. (eds.) Proceedings of the Autonomous Agents '99 Workshop on Specifying and Implementing Conversation Policies, Seattle, WA, pp. 1–9 (1999)
[22] Hayes, P.J.: Speaking informally. In: Clark, P., Schreiber, G. (eds.) Proceedings of the Third International Conference on Knowledge Capture (K-CAP 2005), Banff, Alberta, Canada. ACM, New York (2005)
[23] Holmback, H., Greaves, M., Bradshaw, J.M.: A pragmatic principle for agent communication. In: Bradshaw, J.M., Etzioni, O., Mueller, J. (eds.) Proceedings of Autonomous Agents '99, Seattle, WA, pp. 368–369. ACM Press, New York (1999)
[24] Howard, P.: The Death of Common Sense: How Law is Suffocating America. Random House, New York (1994)
[25] Klein, G., Feltovich, P.J., Bradshaw, J.M., Woods, D.D.: Common ground and coordination in joint activity. In: Rouse, W.B., Boff, K.R. (eds.) Organizational Simulation, pp. 139–184. John Wiley, New York (2004)
[26] Klein, G., Woods, D.D., Bradshaw, J.M., Hoffman, R., Feltovich, P.: Ten challenges for making automation a "team player" in joint human-agent activity. IEEE Intelligent Systems 19(6), 91–95 (2004)
[27] Locke, J.: An essay concerning the true original, extent, and end of civil government. In: Barker, E. (ed.) Social Contract, pp. 1–143. Oxford University Press, New York (1689/1962)
[28] Norman, D.A.: The Psychology of Everyday Things. Basic Books, New York (1988)
[29] Norman, D.A.: Cognitive artifacts. In: Carroll, J.M. (ed.) Designing Interaction: Psychology at the Human-Computer Interface, pp. 17–38. Cambridge University Press, Cambridge (1992)
[30] Norman, D.A.: Turn signals are the facial expressions of automobiles. In: Turn Signals Are the Facial Expressions of Automobiles, pp. 117–134. Addison-Wesley, Reading, MA (1992)
[31] Rousseau, J.J.: The social contract. In: Barker, E. (ed.) Social Contract, pp. 167–307. Oxford University Press, New York (1762/1962)

[32] Smith, W.J.: The Behavior of Communicating. Harvard University Press, Cambridge, MA (1977)
[33] Smith, W.J.: The biological bases of social attunement. Journal of Contemporary Legal Issues 6, 361–371 (1995)
[34] Wohlmuth, P.C.: Traveling the highway: Sources of momentum in behavioral regulation. Journal of Contemporary Legal Issues 6, 1–9 (1995)

Appendix: Some Major Categories from the Ontology of Regulatory Devices:

- Coercion (various uses of force, pressure, intimidation)
- Agreements (including contracts, promises, etc.)
- Precedents (guidance from pertinent past cases, decisions, events)
- Plans (recipe-like things, including schedules, designs, forms)
- Standards/norms (benchmark kinds of things—what is contextually acceptable)
- Fashion: (the current style or mode of dress, speech, conduct, etc.)
- Exceptions (various kinds of suspensions of rules, e.g., zoning variances, waivers)
- Commands (orders by an authority)
- Permissions (allowances to conduct certain actions)
- Folkways (including practices, taboos, customs, ceremonies, myths, rituals...)
- Lessons (ways of operating, acting learned from study or, often, unfortunate, experience—"That surely taught me a lesson.")
- Codified rules (things similar to laws)
- Obligations (accrue from positions held or assumed)
- Authorizations (means of granting authority)
- Incentives (which implicates *disincentives,* i.e., classes of enforcement and punishment)
- Design of Affordances (by what we make *hard or easy to do* by design, even physical design, such as the placement of doorways, sidewalks, streets, bridges—cf. Norman [28])
- Transactional utilitarian devices (fast acting, often fluid, regulatory devices that we set for ourselves in the process of conducting everyday affairs—e.g., making promises, agreements, appointments, pledges, gesture, expression…
- "Natural Law:" (for example, rights to life, liberty, property, …rights to punish violations of natural law, etc. "Social Contract" theorists, e.g., Locke [27], Hume, Rousseau [31])
- Physical and physiological law
- Ideology/Belief/philosophy (idea systems of a people, e.g., Confucianism, democracy)
- Honor Codes (including "face," codes of ethics, and duties)

An Algorithm for Conflict Resolution in Regulated Compound Activities

Andrés García-Camino, Pablo Noriega, and Juan-Antonio Rodríguez-Aguilar

IIIA-CSIC, Campus UAB, 08193 Bellaterra, Spain
{andres,pablo,jar}@iiia.csic.es

Abstract. The use of norms is a well-known technique of co-ordination in multi-agent systems (MAS) adopted from human societies. A normative position is the "social burden" associated with individual agents, that is, their obligations, permissions and prohibitions. Compound activities may be regulated by means of normative positions. However, conflicts may appear among normative positions of activities and sub-activities. Recently several computational approaches have appeared to make norms operational in MAS but they do not cope with compound activities. In this paper, we propose an algorithm to determine the set of applicable normative positions, i.e., the largest set of normative positions without conflicts in the state of an activity, and propagate them to the sub-activities.

1 Introduction

Society has frequently come across the need of coordinating interactions among individuals and one way of addressing that need has been to establish restrictive environments where the interactions are constrained to only those participants and those interactions that are meant to be. For analogous reasons, the MAS community has proposed regulated environments where agents –human or software– interact as [1,2,3].

The environments we will have in mind in this paper are regulated environments where agent interactions are structured as repetitive interactions –that we shall call *activities*– and the whole environment is the result of the composition of many such activities. These activities are subject to explicit sets of conventions that prescribe how the actions of agents that participate in a given activity establish or fulfil commitments that affect the participants of that activity and of subordinate activities. For lack of a better term we will refer to such environments as regulated compound activities.

Many real world societies conform to this type of regulated environments and virtual counterparts are easy to conceive. For instance, Figure 1 describes the example of an on-line commodities trading market that has different price-fixing conventions which may have different simultaneous enactments (different auctions to buy, say, wholesale fruit and poultry; one-to-many negotiations for supermarkets to stock their weekly supply, direct purchasing for scarce quality

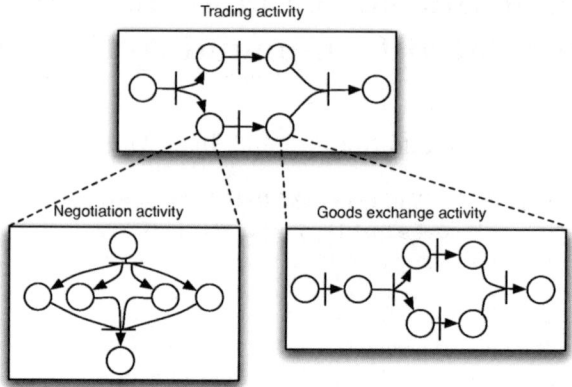

Fig. 1. Example of compound activities

goods like spices). These price-fixing activities are, on one hand, preceded by activities whose purpose is to set the grounds for the day's trading (e.g., activities to introduce whatever is to be exchanged during the day, or the accreditation of buyers and their credit lines, etc.) and, on the other hand, they may be followed by other activities like delivery contracting, temporary warehousing or packaging, which in-turn may be compound activities on their own, etc. Other examples of compound activities that naturally come to mind are hospital operation, the football world association (FIFA) activities, or the execution of everyday local government activities.

The conventions that regulate activities, as the examples show, usually have both a procedural component and a declarative one. The conventions may be expressed in different ways although the most familiar ones are commitment-based interaction protocols (e.g., [1,4,3]) and logical (and logic-based) systems (e.g., [5,6]), or as a combination of both [7]. Some of these approaches have elegant conceptual frameworks behind and a few have also an operationalisation that is amenable to be implemented and still a few have been able to integrate the three previous types of convention representation. This last family is what we aim at in our proposal.

In advancing conceptual or implementation frameworks for compound activities, one of the main problems to address —from a social perspective— is to keep track of the commitments that are being established and fulfilled dynamically anywhere in the (compound) society while the society is active. This is a particularly significant problem in societies where truly autonomous entities intervene. The actual problem, then, is to keep an appropriate record of the commitments that are being made and their follow-up, to make sure that the commitments are consistent. This entails the need to make the problem operational, state it in such a way that formal and implementations are feasible and practical.

In this paper, consequently, we want to make headway towards a proposal of a framework for commitment management in regulated environments formed by compound activities. For this purpose we adopt a social perspective to the problem and take a simplified and unconventional approach along the following lines: We abstract the notion of commitment and commitment management by focusing only on the prohibitions, permissions and obligations associated to actions, what we will call the *normative positions*. We also abstract the way normative positions propagate in the society by having a directed graph linking the activities that inherit normative positions and assuming that the graph is *predefined* and to that extent independent of the way activities are actually connected in formal and implementational ways. The management of commitments is also abstracted in this paper by focusing only in the resolution of conflicts among normative positions and using conventional priority criteria to choose between conflicting positions and only then propagate normative positions to the subordinate activities. Since we are interested in making our proposal operational we also present an algorithm that implements these ideas and keeps track of the evolution of normative positions in acyclic compound activities and maintains a conflict-free normative positions base.

The rest of the paper is structured as follows. In section 2, normative positions, deontic conflicts and criteria for conflict resolution are introduced. Compound activities and their deontic conflicts are defined in section 3. In this section, properties of normative consistency in regulated compound activities, as e.g. strong and weak conflict-freedom, are also introduced. In section 4, an algorithm to resolve deontic conflicts in compound activities is proposed. In section 5 we present an example of how the algorithm works. Finally, conclusions and future work are outlined in the last section.

2 Normative Positions

A normative position is the "social burden" associated with individual agents, that is, their obligations, permissions and prohibitions (cf. [5]). Depending on what agents do, their normative positions may change – for instance, permissions/prohibitions can be revoked or obligations, once fulfilled, may be removed.

In regulated actions, the change of normative positions maybe determined by rules that, for example, are time-dependent. For instance, a permission to lend a book is enabled every week day at 9:00a.m. and disabled every week day at 9:00p.m. Action performances can enable normative positions that can be subsequently fulfilled or cancelled. For example, an obligation to pay for a good is enabled if that good is received (generation). This obligation can be disabled either by paying for the good (fulfilment) or by returning it (cancellation).

Deontic conflicts may appear when norms enable new normative positions that are incompatible with the normative positions already enabled. Traditionally, three principles have been used to resolve deontic conflicts: *legis posterior*, *legis specialis* and *legis superior*. These principles order the norms to avoid conflicts following three criteria: a *chronological* criterion (*lex posterior*), a *speciality* criterion (*lex specialis*) by which a specific law prevails over a general law and a *source* criterion, where preference is linked to the rank of the issuing authority (*lex superior*). We extend these criteria with an extra criterion, the *salience* criterion where we may capture whatever other notions of pertinence or relevance may be of use in a specific activity.

Although such criteria are used to resolve deontic conflicts, in some occasions two or more criteria may need to be sequentially applied to achieve that goal. In those circumstances the criteria involved need to be totally ordered. An example of this meta-ordering is when the source criterion prevails over the speciality, salience and the chronological ones, the salience criterion prevails over the speciality and the chronological ones, and the speciality criterion prevails over the chronological one.

In this paper, the chronological and salience criteria will have a straightforward operationalisation. The speciality criterion will correspond to the hierarchical dependence of an activity and its subactivities. Establishing certain agents as sources of law, defining the ordering of sources of law, and ordering norms using the source criterion are left for future work.

We may now illustrate these ideas and state them in a more precise way. We mentioned that a normative position is a permission, a prohibition or an obligation to perform a specific action. Since we are concerned with resolving conflicts between normative positions, we find useful to associate to every normative position a time stamp that corresponds to the moment it becomes effective (enabled). For the same reason we find useful to associate to normative positions an argument that stands for its salience, although it is beyond the scope of this paper how values may be assigned to that parameter. More precisely:

Definition 1. *Let $\delta \in \{per, prh, obl\}$ be a label for the "social burden" of performing an action identified by a, a salience constant $s \in \mathbb{N}$ and a time-stamp $t \in \mathbb{N}$, the formula $\delta(a, s, t)$ stands for a normative position that states that at time t, and with priority s, action a becomes permitted, obligatory or prohibited.*

Examples of normative positions may be: $per(bid_{ag_1}, 0, 0)$, $prh(bid_{ag_1}, 2, 1)$, etc. The former normative position intuitively states that agent ag_1 is permitted to bid since time 0 and this normative position has priority 0. The latter normative position intuitively means that agent ag_1 is prohibited to bid since time 1 and this normative position has priority 2.

As mentioned above, deontic conflicts among normative positions can arise as agent interactions progress. We will say that two normative positions are in conflict if one is a permission or obligation and the other is a prohibition over the same action than the former, regardless of their corresponding salience and enabling times. That is:

Definition 2. *Given two normative positions np, np' such that $np = \delta(a, s, t)$ and $np' = \delta'(a', s', t')$; np, np' are in conflict, denoted $np \ggg np'$, iff:*

1. $\{\delta\} \cup \{\delta'\} = \{per, prh\}$, $a = a'$; or
2. $\{\delta\} \cup \{\delta'\} = \{obl, prh\}$, $a = a'$.

In the previous example, $per(bid_{ag_1}, 0, 0)$ and $prh(bid_{ag_1}, 2, 1)$ are in conflict by the first condition of definition 2.

We take care of the other two parameters of normative positions with the following definitions.

Definition 3. *A normative position* $np = \delta(e,s,t)$ *is a chronological successor of* $np' = \delta'(e',s',t')$, *written as* $np \succ_T np'$, *iff* $t > t'$.

Definition 4. *A normative position* $np = \delta(e,s,t)$ *is more salient than* $np' = \delta'(e',s',t')$, *written as* $np \succ_P np'$, *iff* $s > s'$.

In our example, $prh(bid_{ag_1}, 2, 1)$ is more salient and a chronological successor of $per(bid_{ag_1}, 0, 0)$.

3 Regulated Compound Activities

In this paper we have in mind that a performance of a set of actions —subject to some regulation— constitute an activity and that an activity may be composed of several sub-activities which in turn may also be decomposed into other sub-activities. Take the example of a clearinghouse involving different activities outlined in Figure 1. The main activity, trading, involves subactivities —like auctioning, one-to-many negotiation or direct purchasing— that serve the purpose of fixing the conditions for purchasing goods and other activities required for payment and delivery of goods.

Several models and methodologies for MAS (*e.g.*, [1,2,3]) have looked into the notion of compound activity using different names such as performative structure, missions or simply interaction. For our purposes we only need to look into those aspects that relate to the evolution of normative positions within an activity and how these are propagated in the hierarchy.

Each activity behaves like a transition system: it has a state, represented by a set of grounded terms, that changes with the performance of the actions of the agents in the activity. This transition function is partial since not all the actions may occur in all the states of an activity. Norms establish what actions are permitted, forbidden or obligatory (and their effects) in a given state of the activity, defining the transition function of the activity. Normative positions are part of the state of an activity and they also change by the performance of actions and the application of the transition function. Note also that since different activities may be connected —in the sense that what happens in one has effect on the other— when those normative positions in the first change, the normative positions in the other may also change. We will say that the scope of a normative position is the activity where it becomes enabled and all the sub-activities associated with that activity. Finally, recall that conflicts among normative positions could be avoided through the sequential use of criteria (chronological, salience, speciality) that order normative positions. Once these conflicts are resolved, we obtain the set of normative positions that will be applied and propagated to the sub-activities.

All these elements are part of the following definition:

Definition 5. *An* activity state *is a tuple* $q = \langle Ag, Ac, I, N, N^{in}, N^{out}, \tau, O, \Omega \rangle$ *where*

- *Ag is a finite, non-empty set of agent identifiers;*
- *Ac is a finite, non-empty set of action labels;*
- *I is a finite subset of grounded terms that describe the current value of the parameters involved in the activity;*
- *N is the set of normative positions of an activity state q;*
- N^{in} *is the set of normative positions propagated from a state of a super-activity to activity state q;*
- N^{out} *is the set of applicable normative positions propagated by activity state q to the sub-activities;*
- $\tau : Q \times 2^{Ac} \to Q$ *is a partial transition function from the set of activity states and sets of actions to the set of activity states Q, which defines the state $\tau(q, ac)$ that would result by the performance of actions $ac = \{ac_1, \ldots, ac_n\}$ from state q – note that, as this function is partial, not all the set of actions are possible in all the states;*
- *O is a finite, non-empty set of partial order relations $\succeq_i$ in the set of normative positions; and*
- Ω *is a total order relation over O.*

Henceforth, we will respectively denote with a q subscript the components of activity state q.

N_q^{in} is the set of normative positions propagated to a activity state q. Algorithm 1, presented in section 4, calculates N^{in} of the super-activities of q prior to use Algorithm 2 to calculate N_q^{out}. Intuitively, the set of inherited normative positions is the union of applicable normative positions of the super-states.

N_q^{out} is the set of applicable normative positions in an activity state q that will be propagated to the sub-activities. It is obtained by removing conflicting normative positions from the union of those that are inherited from super-activities and those that arise from the transition that produces state q. For the removal of conflicting normative positions we use the ordering criteria in O in the sequence established by Ω. To calculate N_q^{out}, we use Algorithm 2 presented in section 4 that applies the meta-ordering Ω of activity state q in $N_q \cup N_q^{in}$ in order to remove the less priority, conflicting normative positions.

Figure 2 shows an example of state of an auctioning activity with 3 agents: an auctioneer, and two buyer agents. Auctioneer made an offer and the buyer can bid for that offer. Buyers have a credit that is decreased when they win an auction. If an agent performs an unsupported bid a sanction of 10 is applied. The set of inherited normative positions (N^{in}) is empty since we assume that auctioning is not part of other activity. The set of applicable normative positions is equal to the set of associated normative positions since there is no conflict. Partial function τ is defined using $\cup_C$ and $\setminus_C$ operators that respectively adds and removes formulae from a set C of the tuple defining an activity state. For instance, $q \cup_N \{obl(pay_{ag_1}, 0, t)\}$ intuitively means that the obligation is added to the set of normative positions N of activity state q. Notice that τ checks the set

$q = \langle Ag, Ac, I, N, N^{in}, N^{out}, \tau, O, \Omega \rangle$ where

- $Ag = \{auct, ag_1, ag_2\}$;
- $Ac = \{offer_{auct}, bid_{ag_1}, bid_{ag_2}, pay_{ag_1}, pay_{ag_2}\}$;
- $I = \left\{ \begin{array}{l} credit(ag_1, 100), credit(ag_2, 50), item(it1), price(it1, 100), \\ decrement(it1, 10), reserveprice(it1, 30) \end{array} \right\}$;
- $N = \{per(bid_{ag_1}, 0, 0), per(bid_{ag_2}, 0, 0)\}$;
- $N^{in} = \emptyset$;
- $N^{out} = \{per(bid_{ag_1}, 0, 0), per(bid_{ag_2}, 0, 0)\}$;

- $\tau(q, ac) \begin{cases} q \cup_N \{obl(pay_{ag_1}, 0, t)\} & \begin{array}{l} if\ ac = \{bid_{ag_1}\}\ and \\ price(i, p), credit(ag_1, x) \in I_q\ and \\ x \geq p\ and\ time(t) \end{array} \\ \vdots \\ \begin{array}{l}(q \setminus_I \{credit(ag_1, x)\}) \cup_I \\ \{credit(ag_1, x - 10)\}\end{array} & \begin{array}{l} if\ ac = \{bid_{ag_1}\}\ and \\ price(i, p), credit(ag_1, x) \in I_q\ and \\ x < p\ and\ time(t) \end{array} \\ \vdots \\ \begin{array}{l}((q \setminus_I \{credit(ag_1, x)\}) \setminus_N \\ \{obl(pay_{ag_1}, s, t)\}) \cup_I \\ \{credit(ag_1, x - p)\}\end{array} & \begin{array}{l} if\ ac = \{pay_{ag_1}\}\ and \\ price(i, p), credit(ag_1, x) \in I_q\ and \\ x \geq p\ and\ obl(ac, s, t) \in N_q^{out} \end{array} \\ \vdots \\ q \cup_I \{collision\} & if\ ac = \{bid_{ag_1}, bid_{ag_2}\} \\ \vdots \end{cases}$

- $O = \{\succ_T, \succ_P\}$; and
- $\Omega = \{\langle \succ_P, \succ_T \rangle\}$.

Fig. 2. Example of activity state

of applicable normative positions, since N_q^{out} is supposed to be conflict-free, but changes the set of existing normative positions N_q. Furthermore, there are the two order relations introduced in section 2, as mentioned above salience criterion is preferred over the chronological one.

The scope of norms is established by the scope of the normative positions that the norms enable (or disable). Their scope is the activity (and sub-activities) where their normative positions are associated. When a compound activity has a normative position associated, it is propagated to its sub-activities.

Definition 6. *If an activity A is a sub-activity of activity B, denoted $A \ll B$, then there exists at least one state q' of activity B such that its applicable normative positions are a subset of the normative positions propagated to each state q of A. Formally, $A \ll B \implies \forall q \in A, \exists q' \in B : N_{q'}^{out} \subseteq NP_q^{in}$.*

An activity structure defines which sub-activities compose it and which sub-activities compose the former sub-activities by relating the states of activities.

Definition 7. *An activity structure is an acyclic directed graph $H = \langle Q, E \rangle$ where Q is the finite, non-empty set of activity states; and E is a finite, non-empty set of edges $\langle q, q' \rangle$. If $\langle q, q' \rangle \in E$ we say that q is a sub-state of q' and q' is a super-state of q.*

Henceforth, we will denote with a subscript H the components of activity structure H.

Definition 8. *Given two activities A and B, A is a sub-activity of B iff all the states of A are a sub-state of, at least, one state of B. Formally, $A \ll B \iff \forall q \in A, \exists q' \in B : \langle q, q' \rangle \in E$.*

3.1 Deontic Conflicts in Compound Activities

Deontic conflicts can appear either between the set N of normative positions generated in an activity state by the transition function or the set N^{in} of normative positions inherited from other activity states. As expressed in definition 2, two normative positions of the same activity state are in conflict if one is a permission or obligation and the other is a prohibition over the same action than the former. Two normative positions from different activity states are in conflict if one is a permission or obligation and the other is a prohibition over the same action than the former and one of them is associated to a sub-state of the other.

Definition 9. *Given two normative positions np and np', respectively pertaining to activity states q and q', such that $np = \delta(a, s, t)$, $np' = \delta'(a', s', t')$ and $q, q' \in Q_H$; np, np' are in conflict in an activity structure H, denoted $np \overset{H}{\gg\!\!\!\ll} np'$, iff $np \gg\!\!\!\ll np'$ and $q = q'$; or $np \gg\!\!\!\ll np'$ and exists a path between q and q' in E.*

By relating activity states in an activity structure, we can adapt the speciality ordering criterion introduced in section 2 to our definition of activity state:

Definition 10. *A normative position np is more specific than np' in activity state q, written as $np \succ_S np'$, if $np \in N_q$ and $np' \in N_q^{in}$.*

Given the example of a trading activity composed of two auctioning sub-activities and the activity structure introduced above, we have:

- $N_{trading}^{in} = \emptyset$ since *trading* has no super-state.
- $N_{trading} = \{prh(bid_{ag_1}, 1, 1)\}$ if agent ag_1 made an unsupported bid.
- $N_{trading}^{out} = \{prh(bid_{ag_1}, 1, 1)\}$ since there is no conflict in $N_{trading} \cup N_{trading}^{in}$.
- $N_{auction1}^{in} = \{prh(bid_{ag_1}, 1, 1)\}$ since *auction1* activity state has only *trading* as super-state and $N_{trading}^{out} = \{prh(bid_{ag_1}, 1, 1)\}$.
- $N_{auction1} = \{per(bid_{ag_1}, 0, 0)\}$, for example, as agents are permitted to bid in auction houses.

Thus, there is a conflict between $N_{auction1}$ and $N_{auction1}^{in}$. Normative position $per(bid_{ag_1}, 0, 0)$ is more specific than $prh(bid_{ag_1}, 1, 1)$ because the former belongs to $N_{auction1}$ and the latter to $N_{auction1}^{in}$. In order to calculate the set of normative positions without conflict, we may use the speciality criterion to remove less priority, conflicting normative positions resulting $N^{out}_{auction1} = \{per(bid_a, 0, 0)\}$.

When the normative positions in an activity state are not conflicting, we will say that this activity state is conflict-free:

Definition 11. *An activity state q is* conflict-free *if there is no conflict among its normative positions (from $N^{in}_q \cup N_q$).*

In the example above, the *trading* activity state is conflict-free since $N^{in}_{trading} \cup N_{trading}$ has no deontic conflict.

When the applicable normative positions in an activity state q (N^{out}_q) are not conflicting, we will say that activity state q is Ω-conflict-free. Ω is the meta-ordering of activity state q used to resolve any deontic conflict.

Definition 12. *An activity state q is Ω-conflict-free if the set of applicable normative positions in the activity state (N^{out}_q) is the largest set of normative positions $N^{out}_q \subseteq N^{in}_q \cup N_q$ that is conflict-free after applying meta-ordering Ω of activity state q to resolve deontic conflicts.*

In the example above, *auction1* activity state is Ω-conflict-free.

On the one hand, there are activity structures without conflicts before applying any method of conflict resolution, we call them *strongly conflict-free*. An activity structure H is strongly conflict-free if there is no conflict in the normative positions of any activity state.

Definition 13. *An activity structure H is* strongly conflict-free *if $\forall q \in Q_H$, $\forall np, np' \in N_q \cup N^{in}_q : \neg(np \overset{H}{\gg} np')$.*

On the other hand, there are activity structures without conflicts after applying a method of conflict resolution, we call them *weakly conflict-free*. This property holds when the method of conflict resolution is effective. An activity structure is weakly conflict-free if there is no conflict in the set of applicable normative positions (N^{out}) of any activity state.

Definition 14. *An activity structure H is* weakly conflict-free *if $\forall q \in Q_H$, $\forall np, np' \in N^{out}_q : \neg(np \overset{H}{\gg} np')$.*

In the example above, the activity structure that constitutes the trading activity composed of two auctioning activities is not strongly conflict-free because there is a conflict between two normative positions in *auction1* activity state (in $N_{auction1} \cup N^{in}_{auction1}$). However, it is weakly conflict-free because after applying the Ω meta-ordering, the conflict is resolved and $N^{out}_{auction1}$ includes no conflicting normative positions.

4 Maintenance OF Ω-Conflict-Freedom

After the norms are applied in each activity state using τ, the set of normative positions associated in each activity state (N) changes. Since the scope of normative positions also include the sub-states of the activity state where the normative position is associated, the set of applicable normative positions should be calculated and propagated to the sub-states.

For that purpose, we introduce in this section PROPAGATE-NPS algorithm that takes as input a list of the leaves of the trees defined by the activity structure and ensures that each activity state in the activity structure is Ω-conflict-free.

Algorithm 1. PROPAGATE-NPS($leaves, h$)

Require: $leaves$ is a list of the leaves of each tree of the activity structure $h = \langle Q, E \rangle$
Ensure: Every N^{out} is conflict-free
1. **for all** $leaf \in leaves$ **do**
2.   **if** $N^{out}_{leaf} = null$ **then** {First execution for activity state $leaf$}
3. $parents \leftarrow \{q' \mid \langle leaf, q' \rangle \in E\}$
4. $l \leftarrow \emptyset$
5.   **if** $parents \neq \emptyset$ **then**
6.   PROPAGATE-NPS($parents, h$)
7.   **for all** $parent \in parents$ **do** {Append N^{out} of parents}
8. $l \leftarrow l \cup N^{out}_{parent}$
9.   **end for**
10.   **end if**
11. $N^{in}_{leaf} \leftarrow l$
12. $N^{out}_{leaf} \leftarrow$ GET-CONFLICTFREENPS($leaf$)
13.   **end if**
14. **end for**
15. **return** $leaves$

In algorithm 1, for each leaf of the activity structure, if N^{out} has not been calculated yet (line 2), we apply recursively PROPAGATE-NPS to the parents (if they exist) (line 6) and we gather the list of all the normative positions inherited by the current activity state and update N^{in} (line 12). We set to N^{out}, the result of the algorithm GET-CONFLICTFREENPS applied to the updated activity state (line 13).

When normative positions are propagated, the set of applicable normative positions of an activity state should be calculated. For that purpose, Algorithm 2 returns this set ensuring that it is conflict-free. In algorithm 2, for each normative position np, we gather in s, by calling GET-SETINCONFLICT, a list of normative positions including np and the ones in conflict with np (line 4). If there is at least one normative position in conflict, then we call GET-PRIORITYNP to resolve the conflict and get the highest priority normative position (line 6). This normative position is added to the result list anp (line 8). Otherwise, np is added to the result list in anp (line 11).

Although different activity states can share super-states, Algorithm 1 calculates N^{out} of each activity state only once. Addition and removal of normative positions from a activity state require that N^{out} for that activity state and its sub-states are recursively set to null using Algorithm 3.

Algorithm 2. GET-CONFLICTFREENPS(q)

Require: $q = \langle Ag_q, Ac_q, I_q, N_q, N_q^{in}, N_q^{out}, \tau_q, O_q, \Omega_q \rangle$
Ensure: anp is the list of applicable normative positions
1. $anp \leftarrow \emptyset$
2. $nps \leftarrow N_q \cup N_q^{in}$
3. **for all** $np \in nps$ **do**
4. $s \leftarrow$ GET-SETINCONFLICT(np, nps)
5.   **if** LENGTH(s) > 1 **then**
6. $pnp \leftarrow$ GET-PRIORITYNP(Ω_q, s)
7.   **if** $pnp \not\subseteq anp$ **then**
8. $anp \leftarrow anp \cup pnp$
9.   **end if**
10.   **else if** LENGTH(s) $= 1$ **then**
11. $anp \leftarrow anp \cup np$
12.   **end if**
13. **end for**
14. **return** anp

Algorithm 3. CLEAR-ANP(q, h)

Require: $q = \langle Ag_q, Ac_q, I_q, N_q, N_q^{in}, N_q^{out}, \tau_q, O_q, \Omega_q \rangle$ and $h = \langle Q, E \rangle$
Ensure: For q and its sub-states, $N^{out} = null$
1. $N_q^{out} \leftarrow null$
2. $children \leftarrow \{q' \mid \langle q', q \rangle \in E\}$
3. **for all** $child \in children$ **do**
4.   CLEAR-ANP($child$)
5. **end for**

5 Example

In this section, we introduce an example of a regulated compound activity. Picture a set of auctioning activities, regulated by their own norms, that constitute a trading activity. At the trading activity level there is a norm stating that a buyer that makes an unsupported bid in a auctioning activity will be banned to bid in any of the auctioning activities except in the auctioning activity called *auction2*. The meta-ordering to be applied will be $(\succ_P) \prec (\succ_S) \prec (\succ_T)$: the salience criterion prevails over the speciality and the chronological ones, while the speciality criterion prevails over the chronological one.

Figure 3 shows an example of execution for a trading activity compound of two auctioning activities and a buyer agent. Figure 3(a) illustrates the state of the activities prior to any unsupported bid. Activity state *trading* has no normative position associated. Thus, its set of applicable normative positions is empty. Activity state *auction1* only has associated a permission with normal priority (salience 0). Since N^{out} of the super-state is empty, its N^{out} only contains the associated permission. Activity state *auction2* only has associated a permission

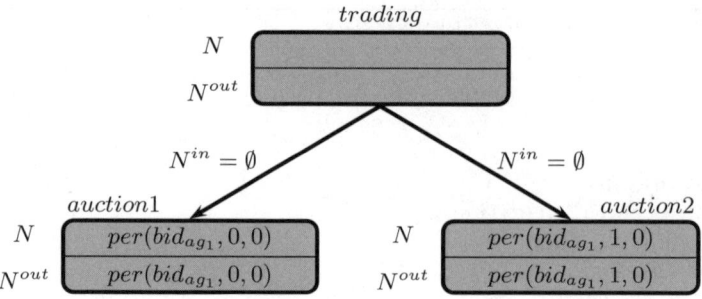

(a) activity states before an unsupported bid

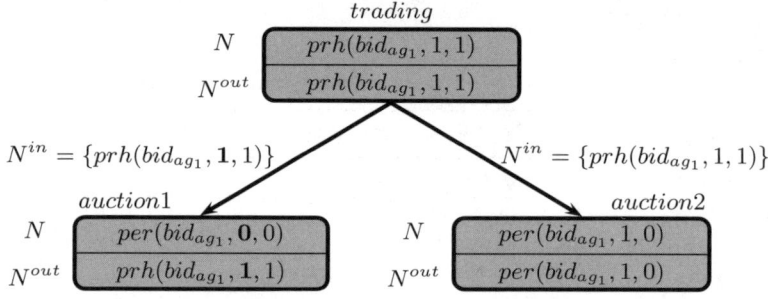

(b) activity states after an unsupported bid

Fig. 3. Example of the normative state of the activity structure

with higher priority (salience 1). Since the N^{out} of the super-state is empty, its N^{out} only contains the associated permission.

Figure 3(b) shows the state of the activities after an unsupported bid performed by agent ag_1. Activity state $trading$ has associated the prohibition (with salience 1) for ag_1 to bid. Since the $trading$ state has no super-states, its N^{out} is equal to its associated normative positions, i.e., the prohibition. Recall that activity state $auction1$ only has associated a permission with normal priority (salience 0). Since the prohibition in N^{in} (inherited from $trading$ activity state) has higher salience, it will belong to the N^{out} of $auction1$ state. Recall that $auction2$ state only has associated a permission with salience 1. Since the prohibition in N^{in} (inherited from $trading$ state) has the same salience, the speciality criterion is applied. The permission will belong to the N^{out} of $auction2$ state because it belongs to N, i.e., it is associated to the sub-state.

6 Related Work

There are many works in deontic conflicts (*e.g.*, [8,9,10,11,12]) from a logical point of view but there are few works that implement computational strategies to solve deontic conflicts in the area of multi-agent systems. Two developments are related quite directly with the problems we face in our proposal [13,14].

In [13], the authors present an agent architecture where obligations, permissions and prohibitions can be added to the agents' plans. Contrary to our work, in which deontic conflicts appear when defining sub-activities, deontic conflicts appear in this work when the agent has to adopt non-hierarchical norms. They (explicitly or implicitly) associate norms to Instantiation Graphs which represent an action or state declaration as a hierarchy of all its possible forms of partial instantiation of variables. Thus, norms are also ordered: explicit norms override implicit norms; and new norms override old norms. Although they consider deontic conflicts, they only adopt an agent-centred stance.

In [14], the author proposes the use of *RuleML* for representing business contracts. The underpinning of the proposal is the use of Defeasible Logic (DL) as the inferential mechanism for *RuleML*. The primary use of DL in that work is the resolution of conflicts that might arise from clauses of a contract. DL analyses the conditions laid down by each rule in the contract, identifies the possible conflicts that may be triggered and uses the priorities defined over the rules to eventually resolve a conflict. By using DL, a normative position receive different priority depending on the antecedents of the rule and not on the normative position by itself. In contrast, our priorities are defined at the normative position level, i.e., we assign a priority to each normative position.

7 Final Remarks

We took an unconventional approach to a complex problem and in this paper we made many simplifying assumptions that we intend to relax in future work. For the moment we wanted to keep our framework simple so that we could explore the main components of the problem. We also wanted to keep it concrete enough so that it could be applied to real organisations and because of that aim we wanted to profit from implemented systems that are already available to deal with regulated simple activities.

In spite of the austerity of this proposal, it is evident that most intuitions we have explored here are prone to a serious logical treatment. Conflict resolution in this paper has been limited to a total ordering of normative positions. It is true that this type of resolution may be adequate in some conflicts and some activities. However, we realise that this issue may be treated in other interesting ways accommodating other culturally accepted conflict resolution mechanisms like negotiation, arbitration or argumentation. Likewise, the current conflict resolution components of the framework could be revisited to incorporate other pertinent normative aspects like peer to peer conflict settlements, contract breaches and blame assignment, enforcement policies, etc.

To define normative positions, in this paper we use an action identifier a that hides too much information. Here we only wanted to be able to decide whether two actions are the same or not. However, that makes the crude handling of normative positions to be simplistic stand-in for commitments. In a system that does serious commitment management, action identifiers will very likely be terms that adequately reflect the state the activity and the organisation at the time of establishing the commitment: in particular, who are the agents involved in the commitment and what are the specific current values of the parameters involved in the definition of the commitment and its fulfilment or up-keeping.

In a similar vein, we have assumed a somewhat obscure notion of activity state because we have hidden many significant elements inside the "transition function". One option we have at hand to make our notion of transition functions clear is to use the notion of performative structures [1]. In that light, the transition function correspond to their speech-acts labelled finite-states machines of a scene together with the scene transitions. Another option is to think of activities as logical theories with a deduction mechanism. In that case, transitions occur when a new action (or possibly concurrent actions) is added to the theory and its consequences deducted. In both cases, performative structures and logical theories, for each transition we still need to keep track of the starting and terminating states of the activity. For that purpose, we may take from [1] the notion of scene state and institutional state and extend them by including the state of the normative positions in the activity and in the compound activities. As part of the state of an activity we need to take into account those commitments that are active, but also their relevance to the actions that take place so that their propagation is correct (sound and complete).

The propagation of normative positions that we discuss in this paper is hand-wired and fix on top of the activities hierarchy. The paths of commitment propagation should be an inherent outcome of the way activities are regulated and combined to constitute the regulated environment. Furthermore, although it may be rather natural to assume nested hierarchies of activities, the express assumption of acyclicity is questionable from an applications point of view, no matter how convenient it may be for formal and algorithmic reasons. In this respect, it looks attractive to distinguish between propagation links among activities that are established through the flow transitions designed into the compound activities on one hand and propagation between states whenever an action takes place anywhere in the organisation and for that purpose a notion of pertinence would be welcome.

We have taken care to make our proposal compatible with the model from [1] and as such it constitutes an extension of that model. In that language we can say that our activities correspond to scenes, performative structures, and nested normative structures. Because of this last possibility, we need an extended notion of transition to handle connections between all of these that produces non-nested compound activities. In addition, this richer activity composition leads us to consider an enriched notion of transition that deals in a smooth way with commitment propagation. We are also interested in profiting from expressive extensions to that model so that conventions may be stated not only as transition graphs but as normative expressions of good expressive power endowed also with a deduction mechanism. Our proposal, as it stands, should work properly with our recent production rules extensions [15].

In this prospective paper, we have followed an unconventional approach to the problem of managing dynamic commitment-making in regulated agent systems. We think that the problem is a fundamental one for regulated environments where the autonomy of participants is an essential ingredient. Although we are aware that what we propose here is far from being a solution to that problem, we acknowledge that the approach has brought to light many challenging questions that we believe deserve further analysis.

Acknowledgements

This work was partially funded by the Spanish Science and Technology Ministry as part of the Web-i-2 project (TIC-2003-08763-C02-00). García-Camino enjoys an I3P grant from the Spanish Scientific Research Council (CSIC).

References

1. Arcos, J.L., Esteva, M., Noriega, P., Rodríguez-Aguilar, J.A., Sierra, C.: Engineering open environments with electronic institutions. Engineering Applications of Artificial Intelligence 18(2), 191–204 (2005)
2. Hannoun, M., Boissier, O., Sichman, J.S., Sayettat, C.: MOISE: An organizational model for multi-agent systems. In: Monard, M.C., Sichman, J.S. (eds.) SBIA 2000 and IBERAMIA 2000. LNCS (LNAI), vol. 1952. Springer, Heidelberg (2000)
3. Dignum, V.: A model for organizational interaction: based on agents, founded in logic. PhD thesis, Utrecht University (2003)
4. Yolum, P., Singh, M.: Flexible protocol specification and execution: Applying event calculus planning using commitments. In: Proceedings of the First International Joint Conference on Autonomous Agents and Multiagent Systems (AAMAS'02) (2002)
5. Sergot, M.: A Computational Theory of Normative Positions. ACM Trans. Comput. Logic 2(4), 581–622 (2001)
6. Pacheco, O., Carmo, J.: A role based model for the normative specification of organized collective agency and agents interaction. Autonomus Agents and Multi-Agent Systems 6, 145–184 (2003)
7. García-Camino, A., Rodríguez-Aguilar, J.A., Sierra, C., Vasconcelos, W.: Norm-Oriented Programming of Electronic Institutions: A Rule-based Approach. In: Coordination, Organization, Institutions and Norms in agent systems (COIN 2006) in Fifth International Joint Conference on Autonomous Agents and Multiagent Systems (AAMAS 2006). LNCS, vol. 4386, pp. 177–193. Springer, Heidelberg (2006)
8. Sartor, G.: Normative conflicts in legal reasoning. Artificial Intelligence and Law 1(2-3), 209–235 (1992)
9. International Workshop in Deontic Logic in Computer Science: DEON (1991-2006)
10. Cholvy, L., Cuppens, F.: Solving normative conflicts by merging roles. In: Fifth International Conference on Artificial Intelligence and Law, Washington, USA (1995)
11. Cholvy, L., Cuppens, F.: Reasoning about norms provided by conflicting regulations. In: McNamara, P., Prakken, H. (eds.) Norms, Logics and Information Systems. New Studies in Deontic Logic and Computer Science. Frontiers in Artificial Intelligence and Applications, vol. 49, pp. 247–262, Washington, USA. IOS Press, Amsterdam (1999)

12. Boella, G., van der Torre, L.: Permission and Obligations in Hierarchical Normative Systems. In: Procs. 8th Int'l Conf. in AI & Law (ICAIL'03), Edinburgh. ACM, New York (2003)
13. Kollingbaum, M., Norman, T.: Strategies for resolving norm conflict in practical reasoning. In: ECAI Workshop Coordination in Emergent Agent Societies 2004 (2004)
14. Governatori, G.: Representing business contracts in *RuleML*. International Journal of Cooperative Information Systems 14(2-3), 181–216 (2005)
15. García-Camino, A., Noriega, P., Rodríguez-Aguilar, J.A.: Implementing Norms in Electronic Institutions. In: Fourth International Joint Conference on Autonomous Agents and Multiagent Systems (AAMAS'05), Utrecht, The Nederlands, pp. 667–673 (2005)

Modeling the Interaction Between Semantic Agents and Semantic Web Services Using MDA Approach

Geylani Kardas[1], Arda Goknil[2], Oguz Dikenelli[3], and N. Yasemin Topaloglu[3]

[1] Ege University, International Computer Institute, 35100 Bornova, Izmir, Turkey
geylani.kardas@ege.edu.tr
[2] Software Engineering Group, University of Twente, 7500 AE, Enschede, The Netherlands
a.goknil@ewi.utwente.nl
[3] Ege University, Department of Computer Engineering, 35100 Bornova, Izmir, Turkey
{oguz.dikenelli,yasemin.topaloglu}@ege.edu.tr

Abstract. In this paper, we present our metamodeling approach for integrating semantic web services and semantic web enabled agents under *Model Driven Architecture (MDA)* view which defines a conceptual framework to realize model driven development. We believe that agents must have well designed environment specific capabilities to fully utilize the power of semantic web environment. Hence, we first define a conceptual architecture for semantic web enabled agents and then discuss how this conceptual architecture can form the basis of a metamodel that can be used in the development of semantic web enabled agents with a model driven approach. We then zoom into the specific part of the metamodel that defines the interactions between semantic web enabled agents and semantic web services since it is not possible to cover all the aspects of the metamodel at one time. So we extend the metamodel of the conceptual architecture from the point of entity aspect for the interaction between semantic agents and semantic web services. Finally, we discuss the mappings between the entities of this extended metamodel and the implemented entities of SEAGENT framework.

1 Introduction

Recently, model driven approaches have been recognized and become one of the major research topics in agent oriented software engineering community [2] [17] [27]. Model driven development is considered as the most promising generational shift in programming technology [28] and even has been characterized as a paradigm shift [6] by several researchers. Model driven development aims to change the focus of software development from code to models. This would increase the level of abstraction in development. Therefore software products would be less affected from the changes in the technological advancements and also the productivity of software developers would be improved [1]. To work in a higher abstraction level is of critical importance for the development of Multi-agent Systems (MAS) since it is almost impossible to observe code level details of MAS due to their internal complexity, distributedness and openness.

The key activity in model driven development is model transformation [29] and model transformation requires syntactical and semantical definitions of models which are provided by metamodels. Various metamodels have been proposed for specific MAS methodologies like Gaia, Adelfe, PASSI [5] and SODA [21]. These metamodels have been generally used for presenting concepts and only recently they are being considered as a foundation for MAS development tools [26].

Collaborating with Object Management Group's (OMG) Agent SIG, the FIPA Modeling Technical Committee proposes a metamodel called Agent Class Superstructure Metamodel (ACSM) [24] which is based on – and extends – UML 2.0 superstructure [25]. The metamodel presents a formal proposal for agent organizations considering the agent, group and role concepts and their relations. In fact, representing the MAS structure with these main meta-entities is not new and formerly proposed in AALAADIN MAS metamodel [12] but not as formal as FIPA Modeling TC's work.

On the other hand, MetaDIMA [15] is a metamodeling project which aims at bridging the gap between existing agent architectures with their development tools and agent-based methodologies, inspired by the Model Driven Architecture. It deals with metamodeling and transformations for agents. However, the project is currently in its preliminary phase.

In [26], Pavon et al reformulates their agent-oriented methodology called INGENIAS in terms of the Model Driven Development paradigm. This reformulation increases the relevance of the model creation, definition and transformation in the context of multi-agent systems.

However, we believe that a significant deficiency exists in above mentioned agent metamodeling and model-driven MAS development studies when we consider modeling of agent systems working on Semantic Web [4] environment. Near future's agent systems will doubtlessly work in this environment and agents in these systems will have capabilities to interact with other semantic entities such as semantic web services.

In this study, we present our approach for integrating semantic web services and semantic web enabled agents under a model driven view. The primary focus of our work is the semantic web environment. We believe that agents must have well designed environment specific capabilities to fully utilize the power of semantic web environment. Hence in this paper, we first define a conceptual architecture for semantic web enabled agents and then discuss how this conceptual architecture can form the basis of a metamodel that can be used in the development of semantic web enabled agents with a model driven approach.

Model driven architecture (MDA) [23] defines a conceptual framework to realize model driven development. MDA is based on developing Platform Independent Models (PIMs) and then converting these PIMs to Platform Specific Models (PSMs) by model transformation. Therefore definitions of PIM and PSM are required for the development of semantic web enabled MAS with the MDA approach. In this paper, we zoom into the specific part of the metamodel that defines the interactions between semantic web enabled agents and semantic web services since it is not possible to cover all the aspects of the metamodel at one time. So we extend the metamodel of the conceptual architecture from the point of *entity aspect* for the interaction between semantic agents and semantic web services. We model the agents and the relation between these agents and semantic web services.

The paper is organized as follows: In Section 2, we introduce the proposed approach for Semantic Web enabled MAS modeling. Our conceptual architecture for Semantic Web enabled MASs is discussed within this section. Section 3 introduces our metamodel that extends ACSM. This metamodel is the first step to incorporate a model driven approach to the development of MASs. So, in section 4, we model the interaction between the semantic agents and semantic web services using MDA approach from the entity view. We also discuss a model transformation example for agent plans within this section. Conclusion and future work are given in Section 5.

2 Proposed Approach for Semantic Web Enabled MAS Modeling

The basic entities of a Semantic Web enabled Multiagent System must be defined in order to apply model driven approaches for development of these systems. We believe that these entities can be derived from the conceptual architecture of Semantic Web enabled MASs. These conceptual entities derived from the conceptual architecture will constitute the key point for application models which are defined within the context of model driven software development. For this reason, we introduce the conceptual architecture of Semantic Web enabled MASs in the first following subsection and discuss the use of these conceptual entities and components within the context of model driven approach in the second subsection.

2.1 A Conceptual Architecture for Semantic Web Enabled MASs

As it is mentioned in Berners-Lee et al's study [4], the real power of the Semantic Web will be realized when programs are created that collect Web content from diverse sources, process the information and exchange the results with other programs. The computer programs in question are software agents and their effectiveness will increase exponentially as more machine-readable Web content and automated services (including other agents) become available. First of all, we need to define a conceptual architecture for semantic web enabled MASs to realize this vision. In this MAS architecture, autonomous agents can also evaluate semantic data and collaborate with semantically defined entities such as semantic web services by using content languages.

Our proposed conceptual architecture for Semantic Web enabled MASs is given in Figure 1. The architecture defines three layers: *Architectural Service Layer*, *Agency Layer* and *Communication Infrastructure Layer*. A group of system agents provides services defined in the Architectural Service Layer. Every agent in the system has an inner agent architecture described in the Agency Layer and they communicate with each other according to the protocols defined in the Communication Infrastructure.

Semantic web agents are agents which are initiated by using the platform architecture and able to use semantic services within the service layer. In Architectural Service Layer, services (and/or roles) of semantic web agents inside the platform are described. All services in the Architectural Service Layer use the capability of the Agency Layer. Besides domain specific agent services, yellow page and mediator services should also be provided.

Agent Registry is a system facilitator in which capabilities of agents are semantically defined and advertised for other platform members. We also define a conceptual entity called *Semantic Service Registry* in the proposed architecture in order to provide semantic service discovery and execution for platform agents by advertising semantic capabilities of services. *Ontology Mediator* is another architectural service in which translation and mapping of different ontologies are performed to support interoperability of different agent organizations using different ontologies.

The middle layer of the architecture is the Agency which includes inner structural components of Semantic Web enabled agents. Every agent in the system has a *Semantic Knowledgebase* which stores the agent's local ontologies. Those ontologies are used by the agent during his interaction with other platform agents and semantic web services. Evaluation of the ontologies and primitive inference are realized by the *Reasoner*. *Semantic Knowledge Wrapper* within the Agency provides utilization of above mentioned ontologies by upper-level Agency components.

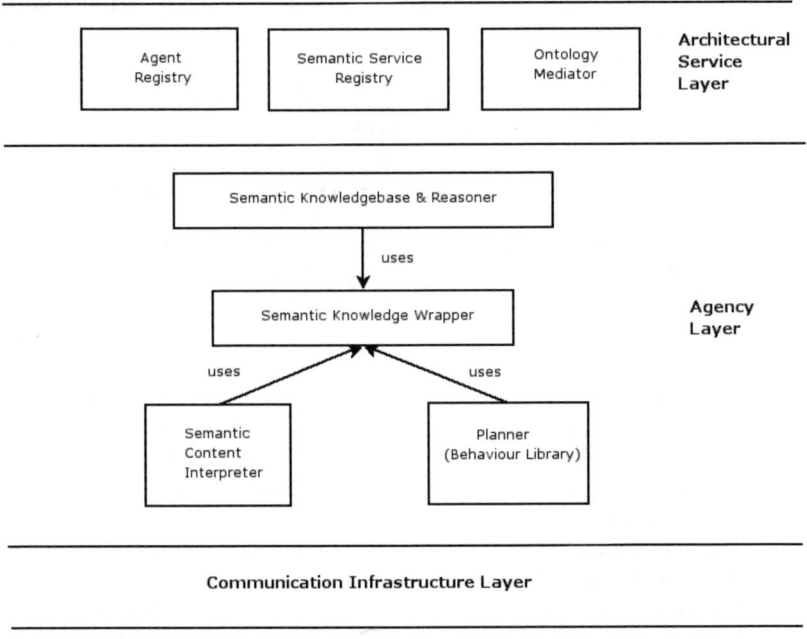

Fig. 1. The conceptual architecture for Semantic Web enabled MASs

The *Planner* of the Agency Layer includes necessary reusable plans with their related behavior libraries. On the other hand, the *Semantic Content Interpreter* module uses the logical foundation of semantic web, ontology and knowledge interpretation in order to check content validity and interpretation of the message during agent communications.

The bottom layer of the architecture is responsible of abstracting the architecture's communication infrastructure implementation. More detailed discussion of this proposed conceptual architecture can be found in [20].

2.2 Model Driven Engineering Approach for Semantic Web Enabled MAS

The implementation of methods and tools for the development of semantic web enabled multi agent systems based on the conceptual architecture discussed in Section 2.1 can be addressed by *Model Driven Engineering (MDE)*. MDE [6] is a recent approach that aims to increase the abstractness level in software development by using models in different phases and therefore by freeing the developers from the code level details. Each conceptual part of the architecture should be analyzed and designed while developing a semantic web enabled multiagent system. Using a model driven approach will enable us to reuse the components of the architecture and to generate the source code of the system from high level abstraction models.

Model Driven Architecture (MDA) [23] is one of the realizations of MDE to support the relations between platform independent and various platform dependent software artifacts. MDA defines several model transformations which are based on the Meta-Object-Facility [22] framework. These transformations are structured in a three-layered architecture: *the Computation Independent Model (CIM), the Platform Independent Model (PIM)*, and *the Platform Specific Model (PSM)*. A CIM is a view of a system from the computation independent viewpoint [23]. Such a model is sometimes called a domain model or a business model. CIM requirements should be traceable to the PIM and PSM constructs by marking the proper elements in CIM. For instance, although the CIM does not have any information about agents and web services, the entities in the CIM are marked in an appropriate notation to trace the agents and semantic web services in the PIM of the semantic web enabled MAS. Bauer and Odell [2] discuss which aspects of a MAS could be considered at CIM and PIM.

The PIM specifies a degree of platform independency to be suitable for use with a number of different platforms of similar type [23]. In our perspective, the PIM of a semantic web enabled MAS should define the main entities and interactions which are derived from the conceptual architecture in Section 2.1. Also, the PIM of semantic web enabled MAS should have different aspects where specific concerns can be addressed. The PIM for service-oriented architecture discussed in [3] identifies four aspects: *information aspect, service aspect, process aspect* and *Quality of Service aspect*. In our approach, the PIM of semantic web enabled MAS can have mainly two aspects: *entity aspect* and *interaction aspect* in order to avoid decomposing the system into too many views. While the entity aspect combines the information aspect and service aspect defined in [3], the interaction aspect is similar with the process aspect and describes a set of interactions between agents and semantic services in terms of message exchange.

On the other hand, the PSM combines the PIM with the additional details of the platform implementation. The platform independent entities in the PIM of semantic web agents are transformed to the PSM of an implemented semantic web enabled agent framework like SEAGENT [8]. The flexible part of this approach is that the

PIM enables to generate different PSMs of semantic web enabled agent frameworks automatically. These PSMs can be considered as the realizations of our conceptual architecture.

The metamodel proposed in [20] defines the general concepts and entities of the proposed conceptual architecture. This metamodel provides the key point for customizing the entities for PIM and PSM metamodels in the MDA based development of this agent system. However, the current metamodel could not be considered as a complete PIM for semantic web enabled MAS. For instance, Semantic Web Service meta-entity and its related entities such as Service Ontology should be detailed. We believe that new entities for agent - semantic service interaction will be needed to add into the metamodel to provide this metamodel as a PIM for modeling the interaction in question.

Obviously, a Semantic Web Service encapsulates a service interface and a service process mechanism for its discovery and execution by semantic web agents. This interface and the semantic process should also be represented by appropriate entities in the metamodel in order to constitute the PSM of such MAS or directly generate semantic web enabled agent platform source code. Although there are ongoing efforts e.g. OWL-S [31] and WSMO [33] which aim to describe web services semantically, there is currently no platform independent standard for representation of these web services in order to be used in the semantic web environment. Due to the lack of this standard, representation of the service interface and the process mechanism constructs in the metamodel are difficult. Hence, it is not possible to define a PIM metamodel for semantic web enabled MAS without including these appropriate entities in the metamodel.

Another part of the semantic web enabled MAS architecture that must be detailed in the metamodel is the behavior library (planner). One of the implementation of reusable plans in the behavior library is the Hierarchical Task Network (HTN) planning [11] which is an AI planning methodology that creates plans by task decomposition. From the point of MDA based development, this HTN or other realization techniques of planning is defined in PSM level. For instance, SEAGENT which is a semantic web enabled MAS framework is based on the HTN planning framework presented by Sycara et al. [30] and the DECAF architecture [13]. If we consider the metamodel of SEAGENT framework as PSM, the HTN and other specific entities of the SEAGENT framework are defined in this metamodel as PSM entities. In PIM level, the general concepts of planning mechanism should be modeled and any specific component of HTN or other planning mechanisms should not be considered for platform independence.

In this study, we model the planning mechanism and the relation between this planning mechanism and semantic web service from the point of entity aspect. That is why we use the Class diagram to represent the model of this relation. In semantic web enabled MAS architecture, planner mechanism has the capability of executing plans consisting of special tasks for semantic service agents in a way described in [16]. The agents in the system can discover the appropriate service and invoke this service through the planning mechanism. The metamodel of the semantic web enabled MAS should consider the general entities of planner mechanism, semantic web service profile parameters and the relation between these entities. While the PIM metamodel does not have any platform specific entities like HTN, OWL-S or WSMO, the implementation of this mechanism in SEAGENT could be considered as platform specific realization.

3 A Metamodel for Semantic Web Enabled MASs

In [20], we introduced a core agent metamodel superstructure to define elements and their relationships of a Semantic Web enabled MAS depending on the previously discussed conceptual architecture. However, the metamodel in question was improper to be used in model transformations and it was too primitive to support widely-accepted software modeling tools due to its arbitrary formalism. Therefore, in this study, we present one representation of the above metamodel by extending FIPA Modeling TC's Agent Class Superstructure Metamodel (ACSM) [24]. Although ACSM is currently also in its preliminary phase, we believe that it neatly presents an appropriate superstructure specification that defines the user-level constructs required to model agents, their roles and their groups. By extending this superstructure we do not need to re-define basic entities of the agent domain. Also, ACSM models assignment of agents to roles by taking into consideration of group context. Hence, extending ACSM clarifies relatively blurred associations between Semantic Organization, Semantic Agent and Role concepts in our metamodel by appropriate inclusion of ACSM's Agent Role Assignment entity. However, ACSM extension is not sufficient and we provide new constructs for our metamodel by extending UML 2.0 Superstructure and Ontology UML Profile which is defined by Djuric [9].

Before discussing our metamodel, ACSM is briefly mentioned below. More information about ACSM can be found in [24] and [25]. ACSM has a specification which is based on –and extends- UML superstructure. It proposes a superstructure for modeling agents, agent roles and agent groups. Its class model is illustrated in Figure 2.

ACSM utilizes the distinction between UML Classifier and UML Class. The agent classification in the model is based on an extension of Classifier. This provides omitting features of object-orientation (such as object-based messaging and polymorphism) which are troublesome for agents.

An Agent Classifier in the model defines various ways in which agents will be classified. It has two subclasses: Agent Physical Classifier which defines the primitive or basic classes describing core requirements of an agent and Agent Role Classifier which classifies agents by the various kinds of roles agents may play.

The Agent class defines the set of all agents that populate a system. Each instance of an Agent is associated with one or more Agent Classifiers that define its necessary features.

Group is defined as a set of agents which have been collected together for some reason. Within a group, its member agents interact according to the roles that they play. Groups are partitioned into Agentified Groups and Non-Agentified Groups according to whether or not they are addressable as an agent and can act as an agent in their own right.

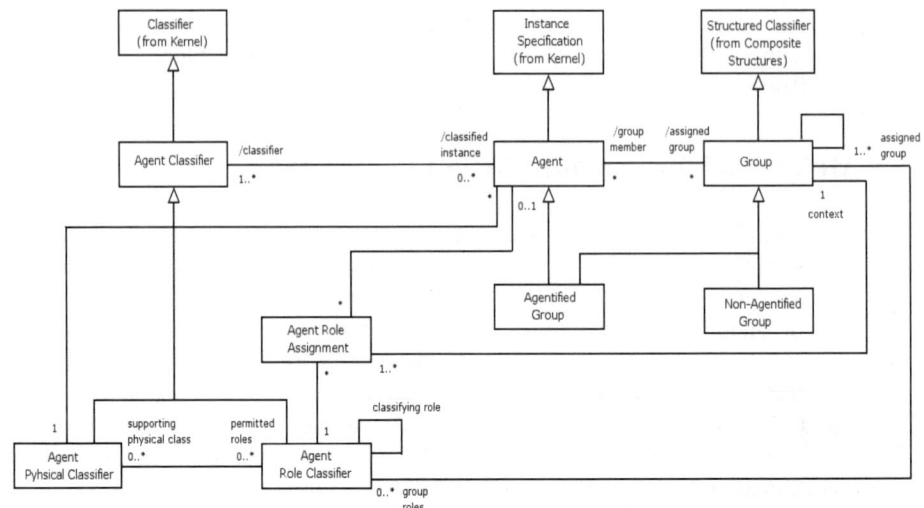

Fig. 2. FIPA Modeling TC's Agent Class Superstructure Metamodel [24]

Besides UML Classifier utilization, another noteworthy feature of the ACSM is modeling Agent – Role assignment as a ternary association. The fact that assignment of Agents to Roles is dynamic, required association is modeled by the Agent Role Assignment entity. A Role assignment between an agent and its role must be qualified by a group context. Hence, an Agent Role Assignment is a Class in the model whose associated instances associate Roles, Groups and Agents. Each instance of the ternary Agent Role Assignment associates a role, a group and an agent.

The Semantic Web enabled MAS metamodel being proposed in this study is given in Figure 3. The model extends FIPA Modeling TC's Agent Class, UML 2.0 superstructures and Ontology UML Profile.

As given in [20] a *Semantic Web Agent* is an autonomous entity which is capable of interaction with both other agents and semantic web services within the environment. It is a special form of the ACSM's Agent class due to its entity capabilities. It includes new features in addition to Agent classified instance.

Roles provide both the building blocks for agent social systems and the requirements by which agents interact as it has been remarked in [25]. We believe that the same is true for roles played in Semantic Web enabled agent environments. However, this general model entity should be specialized in the metamodel according to task definitions of architectural and domain based roles: An *Architectural Role* defines a mandatory Semantic Web enabled MAS role that should be played at least one agent inside the platform regardless of the organization context whereas a *Domain Role* completely depends on the requirements and task definitions of a specific Semantic Organization created for a specific business domain.

The Role concept in the metamodel is an extension of Agent Role Classifier due to its classification for roles the semantic agents are capable of playing at a given time. This conforms to the Agent – Agent Role Classifier association defined in ACSM

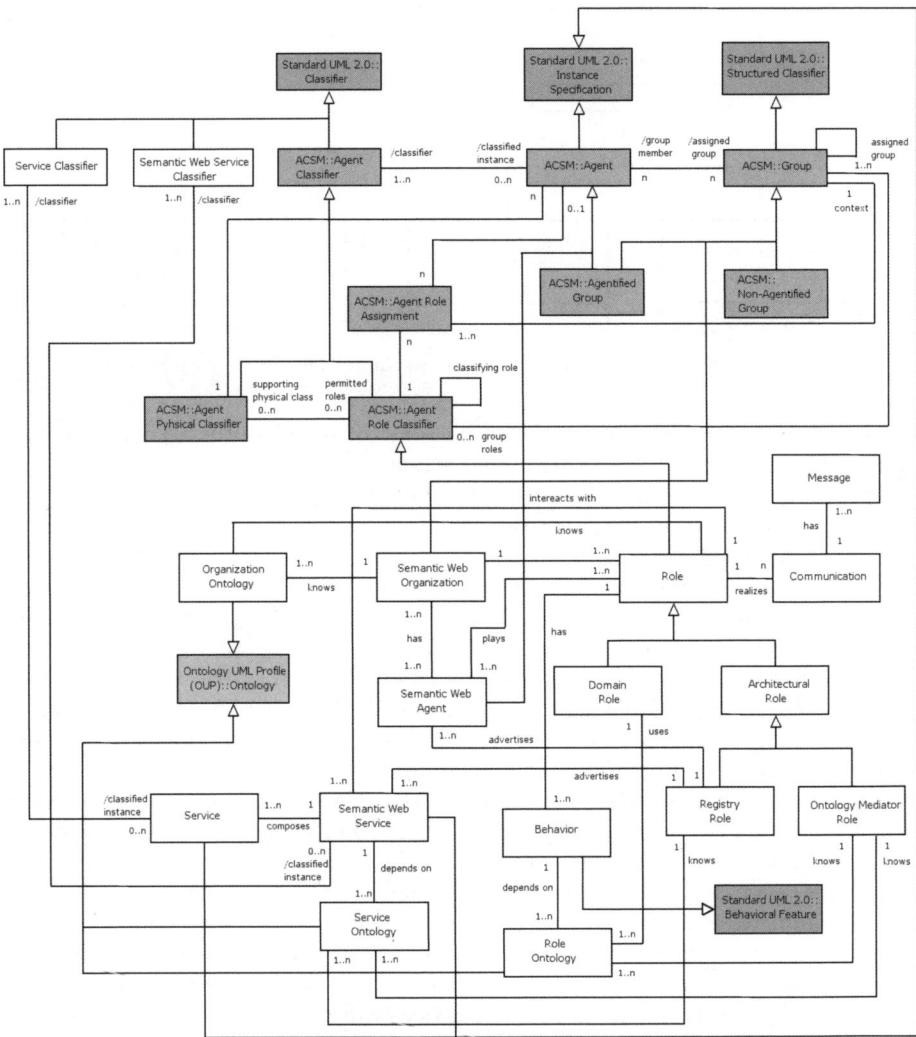

Fig. 3. The metamodel for Semantic Web enabled MASs which extends FIPA Modeling TC's Agent Class, UML 2.0 Superstructure and Ontology UML Profile

[25]: *Semantic Web Agents can be associated with more than one Role (which is also an Agent Role Classifier) at the same point in time (multiple classification) and can change roles over time (dynamic classification).*

Agent Role Classifiers form a generalization hierarchy. This is also valid for Semantic environment's Role elements. For example, in Figure 4, a hierarchy of Architectural Roles in SEAGENT [8] MAS framework is given. Due to its FIPA compliancy, related framework also defines a Registry Role called Directory Facilitator (DF). However, it also includes a service role called Semantic Service

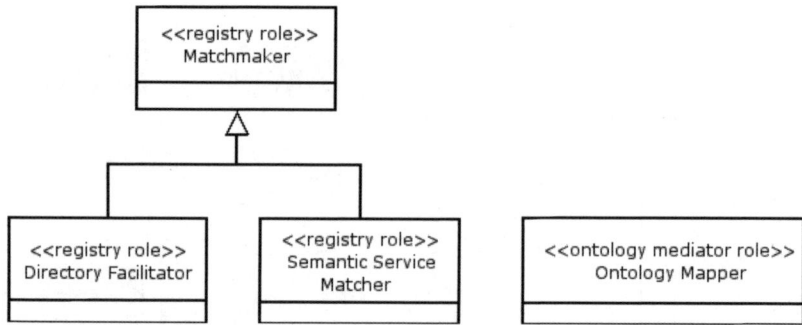

Fig. 4. A generalization hierarchy of Architectural Roles in SEAGENT MAS

Matcher (SSM) which should be played by some of the platform agents in order to realize Semantic Web Service – Agent interaction.

On the other hand, *Semantic Web Organization* is defined as a specialization of the ACSM's Group entity in the proposed model because it should be implemented as only a composition of Semantic Web Agents. However, a Semantic Web Organization may or may not behave as a Semantic Web Agent in overall manner. Hence, it shouldn't be defined neither as Agentified nor Non-Agentified Group. It is a direct extension of the Group Composite Structure.

Above discussed ACSM extensions provide clarification of the relations between Semantic Web Agent, Role and Semantic Web Organization in our model by presenting practicability of ACSM's Agent Role Assignment ternary association between Agent, Agent Role Classifier and Group.

The metamodel is also based on – and extends – UML 2.0 Superstructure to define meta-elements of the Semantic Web environment. For example, we have defined a first-class entity called Semantic Web Service Classifier in our core model. This entity is defined in the final model as a UML 2.0 Classifier extension.

A *Semantic Web Service* represents any service (except agent services) whose capabilities and interactions are semantically described within a Semantic Web enabled MAS. A Semantic Web Service composes one or more *Service* entities. Each service may be a web service or another service with predefined invocation protocol in real-life implementation. But they should have a semantic web interface to be used by autonomous agents of the platform.

Like agents, semantic web services have also capabilities and features which could not be just based on object-oriented paradigm. Hence, we define new Classifiers and their related Instance Specifications in the metamodel to encapsulate semantic web entities. We have applied classifier – classified instance association between *Semantic Web Service Classifier* and *Semantic Web Service*. Same is valid for *Service Classifier – Service* relationship.

Ontology entities (*Organization Ontology*, *Service Ontology* and *Role Ontology*) are defined as extensions of the *Ontology* element of the Ontology UML Profile (OUP) defined in [9]. OUP captures ontology concepts with properties and relationships and provides a set of UML elements available to use as semantic types in our metamodel. By deriving the semantic concepts from OUP, there will be already-defined UML elements to use as semantic concepts within the metamodel.

One Role is composed of one or more *Behaviors*. Task definitions and related task execution processes of Semantic Web agents are modeled inside Behavior entities. The Behavior entity is defined in the metamodel as a UML 2.0 Behavioral Feature because it refers to a dynamic feature of a Semantic Web Agent (e.g. an agent task which realizes agent interaction with other agents).

According to played roles, agents inevitably communicate with other agents to perform desired tasks. Each *Communication* entity defines a specific interaction between two agents of the platform which takes place in proper to predefined agent interaction protocol. One Communication is composed of one or more *Message*s whose content can be expressed in a RDF based semantic content language.

Figure 5 portrays an example semantic role assignment considering a MAS working in Tourism domain.

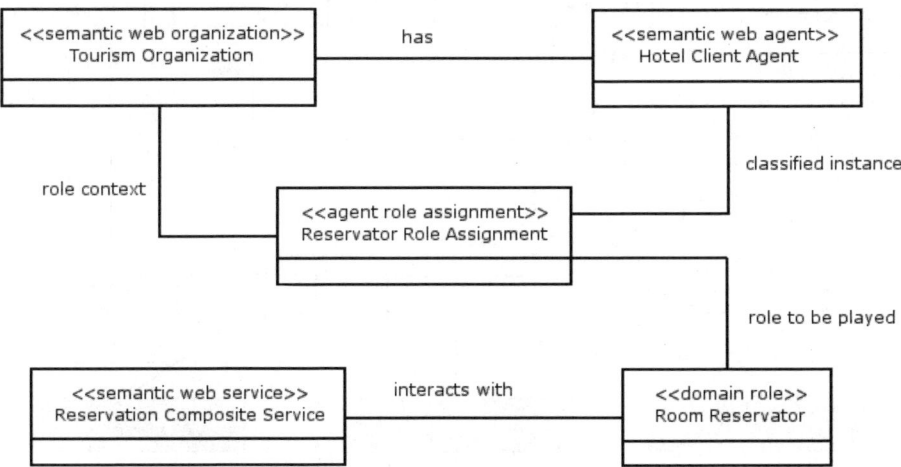

Fig. 5. A Semantic Role Assignment for a MAS working in Tourism domain

In this system, there exists an Agent Role Assignment Class called "Reservator Role Assignment" which represents the three-way association between a Hotel Client Agent, the Room Reservator Role and Tourism Organization. Hotel Client is a Semantic Web Agent which reserves hotel rooms on behalf of its human users. Within the Semantic Web Organization called Tourism Organization, the semantic web agent plays a Room Reservator Role. The related role includes a semantic web service interaction during its task execution: Hotel Client Agent uses Reservation Composite semantic web service which may be a composition of discovery, engagement and invocation services for hotel room reservation.

4   Elaboration of the Metamodel by Considering the Interaction Between Semantic Agents and Semantic Web Services

The metamodel discussed in the previous section defines required meta-entities and entity relations of a Semantic Web enabled MAS architecture. However, interaction between semantic agents and external services needs to be studied in more detail in order to realize model transformations during system development. Such a study also provides a practical evaluation of the proposed metamodel. The extended model given in Figure 6 elaborates the agent – service interaction from the point of entity aspect.

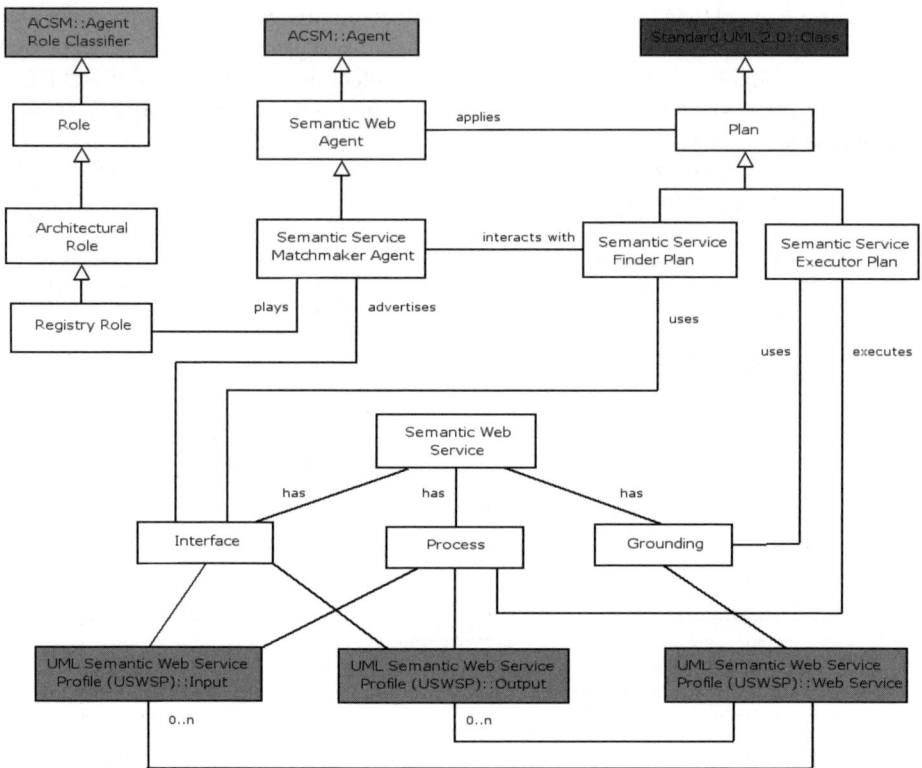

Fig. 6. The extended metamodel of the interaction between Semantic Agents and Semantic Web Services

Semantic Web Agents have *Plan*s to discover and execute Semantic Web Services dynamically. In order to discover service capabilities, agents need to communicate with a service registry. For this reason, the model includes a specialized agent entity, called *Semantic Service Matchmaker Agent*. This meta-entity represents matchmaker agents which store capability advertisements of semantic web services within a MAS and match those capabilities with service requirements sent by the other platform agents.

When we consider various semantic web service modeling languages such as OWL-S [31] and WSMO [33], it is clear that services are represented by three semantic documents: *Service Interface*, *Process Model* and *Physical Grounding*. Service Interface is the capability representation of the service in which service inputs, outputs and any other necessary service descriptions are listed. Process Model describes internal composition and execution dynamics of the service. Finally

Physical Grounding defines invocation protocol of the web service. These Semantic Web Service components are given in the metamodel with *Interface*, *Process* and *Grounding* entities respectively. Semantic input, output and web service definitions used by those service components are exported from the UML Semantic Web Service Profile proposed in [14].

Semantic Web Agents have two consecutive plans to interact with Semantic Web Services. *Semantic Service Finder Plan* is a Plan in which discovery of candidate semantic web services takes place. During this plan execution, the agent communicates with the service matchmaker of the platform to determine proper semantic services. After service discovery, the agent applies the *Semantic Service Executor Plan* in order to execute appropriate semantic web services. Process model and grounding mechanism of the service are used within the plan. An instance model of the above metamodel is given in Figure 7 for the interaction between a Hotel Client Agent and a Reservation Service within a MAS working in Tourism domain.

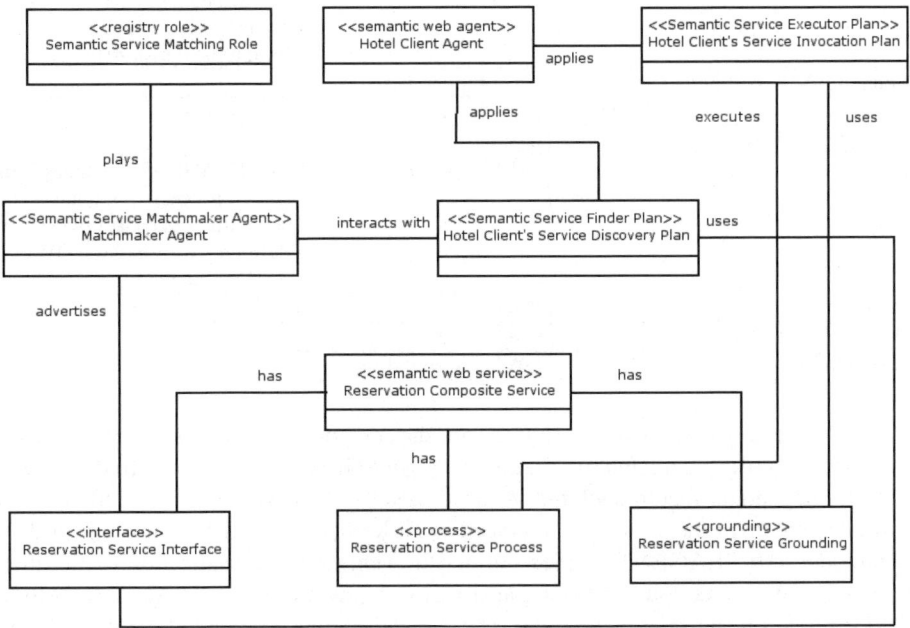

Fig. 7. An instance model for the agent – service interaction within a MAS working in Tourism domain

As previously mentioned, the client agent is a Semantic Web Agent which reserves hotel rooms on behalf of its human users. During its task execution, it needs to interact with a semantic web service called Reservation Composite Service. Matchmaker Agent is the service matcher of the related agent platform.

When we consider the metamodel as a PIM of the agent – service interaction, we should give the corresponding PSM entities in an implemented Semantic Web enabled MAS environment. As previously mentioned, SEAGENT [8] is a MAS development framework which provides built-in components for Semantic Web enabled MASs. Hence, in Table 1, we give the mappings between entities of our proposed metamodel and SEAGENT framework. These mappings precede the transformation between PIM and PSMs of such kind of MASs according to MDA approach.

Table 1. Mappings between the metamodel and SEAGENT framework entities

Metamodel Entity	SEAGENT Entity	Explanation
Registry Role Semantic Service Matchmaker Agent (SSMA)	Semantic Service Matcher (SSM)	Both Registry role and SSMA in the metamodel corresponds to the SSM in SEAGENT.
Plan	HTN Plan	In SEAGENT MAS, agent plans are designed as hierarchical task networks.
Semantic Service Finder Plan	HTN Finder Task	
Semantic Service Executor Plan	HTN Executor Task	
Semantic Web Agent	Agent	
Semantic Web Service	OWL-S Service	In SEAGENT, capabilities and process models of semantic web services are defined by using OWL-S markup language.
Interface	OWL-S Profile	
Process	OWL-S Process	
Grounding	OWL-S Grounding	

To derive a transformation (based on the mappings listed in Table 1) from metamodel entities depicted in Figure 6 to SEAGENT entities, we first define a platform dependent metamodel and instance models of SEAGENT. Since SEAGENT is implemented in Java, we do not need a customized metamodel instead of Java metamodel. All SEAGENT entities defined in Table 1 are a realization of the meta classes in Java metamodel. In our transformation we use a Kernel MetaMetaModel (KM3) based on the Java metamodel which is defined in a metamodel zoo [32]. All the metamodels available in this zoo [32] are expressed in KM3 [18] metamodel format and can be injected to an "ecore" file which is an Eclipse Modeling Framework (EMF) [10] format.

SEAGENT dependent plan and semantic web agent models based on this Java metamodel for the interaction between semantic web agents and semantic web services contain the components of SEAGENT plan structure. Gürcan et al [16] define a software platform which fulfills fundamental requirements of Semantic Web Services Architecture's (SWSA) [7] conceptual model including all its sub-processes and a planner that has the capability of reusable plans in which these sub-processes are modeled for development of semantic service agents. This plan structure [16] is similar to the frameworks presented by Sycara et al [30] and the DECAF architecture [13]. As a requirement of HTN, tasks might be either complex (called behaviors) or primitive (called actions). Each plan consists of a complex root task consisting of sub-tasks to achieve a predefined goal.

Components of our plan structure are shown in Figure 8. Tasks have a name describing what they are supposed to do and have zero or more provisions (information needs) and outcomes (execution results). The provision information is supplied dynamically during plan execution. Tasks are ready, and thus eligible for execution, when there is a value for each of its provisions. Related control is done via *isAllProvisionsAreSet()* method. The more detailed information about SEAGENT plan structure can be found in [16].

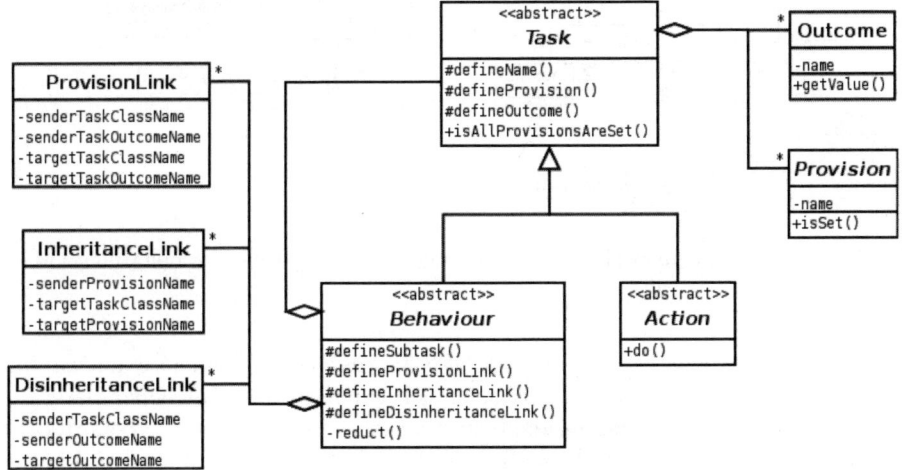

Fig. 8. Components of SEAGENT Plan Structure [16]

According to the plan structure depicted in Figure 8, we define an instance SEAGENT plan model based on Java metamodel for the agent – service interaction within a MAS working in Tourism domain whose platform independent model is shown in Figure 7.

Figure 9 shows an instance plan model in SEAGENT for the agent – service interaction within a MAS working in Tourism domain. This is the corresponding platform dependent model of the plan part of the platform independent model depicted in Figure 7. Every entity in this model is a Java class which is defined as a meta class in the Java metamodel. In this plan, *FindaHotel*, *FindaRoom*, and *MakeRoomReservation* sub-tasks of the plan are concrete realizations of *ExecuteService* task. They are connected with their provisions and outcome slots, and because they are domain dependent plans they know what input parameters they will take. Since the realization of a Plan from another plan is done through inheritance

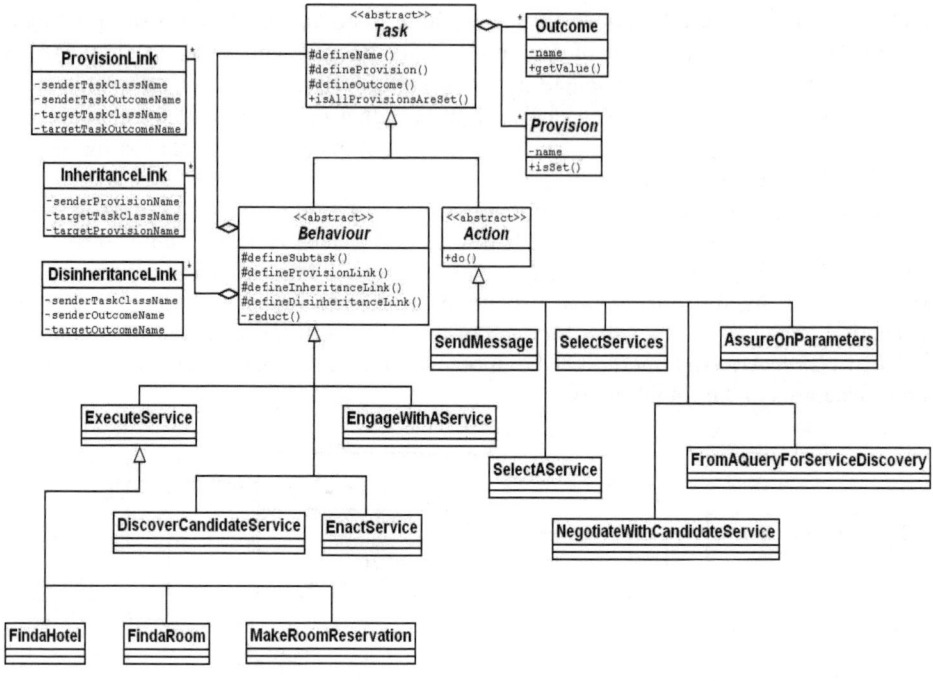

Fig. 9. An instance plan model in SEAGENT for the agent – service interaction within a MAS working in Tourism domain

relations between Java classes in SEAGENT, *FindaHotel* class is extended from *ExecuteService* class and *ExecuteService* class is extended from *Behaviour* class.

Currently, we are working on implementing the transformations derived from the mappings given in this study to realize the model driven development of Semantic Web enabled MASs using MDA approach. For this purpose, we use ATLAS INRIA & LINA research group's ATL (Atlas Transformation Language) which is a model transformation language specified as both a metamodel and a textual concrete syntax [19].

Figure 10 summarizes the full model transformation process. A model M_a, conforming to a metamodel MM_a, is here transformed into a model M_b that conforms to a metamodel MM_b. The transformation is defined by the model transformation model M_t which itself conforms to a model transformation metamodel MM_t. This last metamodel, along with the MM_a and MM_b metamodels, has to conform to a metametamodel MMM such as MOF (Meta Object Facility) or Ecore [10].

In our transformation case, MMM is Ecore and MM_t is ATL. Our source model (M_a) is the model given in Figure 7 which conforms to metamodel (MM_a) given in Figure 6. When we apply a transformation into our source model, we aim to obtain the platform specific destination model (M_b) which conforms to metamodel of the SEAGENT planner depicted in Figure 9.

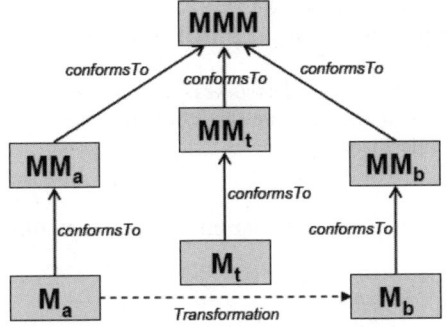

Fig. 10. An overview of model transformation [19]

Consider the simple example in which we transform a Semantic Service Finder Plan (in Figure 6) into its corresponding SEAGENT plan which is DiscoverCandidateService (in Figure 9) within the ATL environment. To do this we have to create EMF encodings -.ecore files- of both models and use them in ATL transformation.

EMF provides its own file format (.ecore) for model and metamodel encoding. However the manual edition of Ecore metamodels is particularly difficult with EMF. In order to make this common kind of editions easier, the ATL Development Tools (ADT) include a simple textual notation dedicated to metamodel edition: the Kernel MetaMetaModel (KM3) [18]. This textual notation eases the edition of metamodels. Once edited, KM3 metamodels can be injected into the Ecore format using ADT integrated injectors. More information about KM3 and Ecore injection can be found in [18, 19].

Following is the part of the KM3 file in which Semantic Service Finder Plan is represented:

```
package SemanticServiceFinderPlan {
    class SemanticServiceFinderPlan {
        attribute plan_name : String;
        reference desiredServiceInterface : Interface;
    }
    class Interface {
        attribute input: String;
        attribute output: String;
        attribute precondition: String;
        attribute effect: String;
    }
}

package PrimitiveTypes {      datatype String;       }
```

Notice that service interface metamodel definition in here is extremely simplified for the demonstration purposes. In a real transformation, IOPE (Input, Output, Precondition and Effect) attributes of a semantic service interface would have complex types. The ecore model conforming to above metamodel includes the following model instance which will be given into transformation process:

```xml
<?xml version="1.0" encoding="ISO-8859-1"?>
<xmi:XMI   xmlns:xmi="http://www.omg.org/XMI" xmlns="SemanticServiceFinderPlan">
    <SemanticServiceFinderPlan>
        <name>Hotel Client's Service Discovery Plan</name>
        <Interface>ReservationServiceInterface</Interface>
    </SemanticServiceFinderPlan>
</xmi:XMI>
```

The KM3 representation of the destination model's metamodel is given below:

```
package DiscoverCandidateService {
    class DiscoverCandidateService {
        attribute name : String;
        attribute candidateServiceInputList : String;
        attribute candidateServiceOutputList : String;
    }
}

package PrimitiveTypes {      datatype String;      }
```

Finally, here is the transformation rule written in ATL which will be used by the ATL engine in order to generate the model conforming to DiscoverCandidateService's metamodel:

module SemanticServiceFinderPlan2DiscoverCandidateService;
create OUT : DiscoverCandidateService **from** IN : SemanticServiceFinderPlan;
rule SemanticServiceFinderPlan {
  from
    ssfp : SemanticServiceFinderPlan!SemanticServiceFinderPlan
  to
    dcs : DiscoverCandidateService!DiscoverCandidateService (
      name <- ssfp.name,
      candidateServiceInputList <- ssfp.desiredServiceInterface.input,
      candidateServiceOutputList <- ssfp. desiredServiceInterface.output
)
}

The engine applies the above rule in order to transform "Hotel Client's Service Discovery Plan" model which conforms to SemanticServiceFinderPlan metamodel into a model instance that can be used within the SEAGENT environment conforming to plan metamodel of DiscoverCandidateService.

5 Conclusion and Future Work

A metamodel for Semantic Web enabled MASs and the extended part of this metamodel for the interaction between semantic agents and semantic web services are introduced in this paper. This extended metamodel can be considered as a part of *Platform Independent Model* within the context of MDA approach. This PIM models the planning mechanism and the relation between this planning mechanism and semantic web service from the point of entity aspect. The agents in the system can discover the appropriate semantic services and invoke these services through the planning mechanism. General entities of the planner mechanism, semantic web service profile parameters and the relation between these entities are considered. While the PIM does not have any platform specific entities of HTN, OWL-S or WSMO, the implementation of these mechanisms in SEAGENT could be considered as platform specific realization. The mappings between the entities of the metamodel and the implemented entities of SEAGENT framework in Section 4 show the practical relevance of the metamodel.

In our future work, we aim to define interaction aspect of this extended metamodel at first. Meanwhile, we also intend to improve mappings and model transformations introduced in this study. The metamodel in here is only extended for interaction between semantic agents and semantic web services. Hence, as our further work, we plan to extend other parts of the metamodel according to the components of the layered conceptual architecture and provide tool support for the proposed metamodel.

References

[1] Atkinson, C., Kühne, T.: Model-Driven Development: A Metamodeling Foundation. IEEE Software 20, 36–41 (2003)
[2] Bauer, B., Odell, J.: UML 2.0 and Agents: How to Build Agent-based Systems with the New UML Standard. Journal of Engineering Applications of Artificial Intelligence 18(2), 141–157 (2005)
[3] Benguria, G., Larrucea, X., Elvesaeter, B., Neple, T., Beardsmore, A., Winchester, M.: A Platform Independent Model for Service Oriented Architectures. In: Interoperability for Enterprise Software and Applications Conference (I-ESA'06), Bordeaux, France (2006)
[4] Berners-Lee, T., Hendler, J., Lassila, O.: The Semantic Web. Scientific American 284(5), 34–43 (2001)
[5] Bernon, C., Cossentino, M., Gleizes, M., Turci, P., Zambonelli, F.: A Study of some Multi-Agent Meta-Models. In: Odell, J.J., Giorgini, P., Müller, J.P. (eds.) AOSE 2004. LNCS, vol. 3382. Springer, Heidelberg (2005)
[6] Bezivin, J.: Model Driven Engineering: Principles, Scope, Deployment and Applicability. In: Proceedings of 2005 Summer School on Generative and Transformational Techniques in Software Engineering (July 2005)
[7] Burstein, M., Bussler, C., Zaremba, M., Finin, T., Huhns, M., Paolucci, M., Sheth, A., Williams, S.: A semantic web services architecture. IEEE Internet Computing 9(5), 72–81 (2005)
[8] Dikenelli, O., Erdur, R.C., Kardas, G., Gümüs, O., Seylan, I., Gurcan, O., Tiryaki, A.M., Ekinci, E.E.: Developing Multi Agent Systems on Semantic Web Environment using SEAGENT Platform. In: Dikenelli, O., Gleizes, M.-P., Ricci, A. (eds.) ESAW 2005. LNCS (LNAI), vol. 3963, pp. 1–13. Springer, Heidelberg (2006)
[9] Djuric, D.: MDA-based Ontology Infrastructure. Computer Science Information Systems (ComSIS) 1(1), 91–116 (2004)
[10] Eclipse Modeling Framework (2006) available at: http://www.eclipse.org/emf
[11] Erol, K., Hendler, J.A., Nau, D.S.: Complexity Results for HTN Planning. Ann. Math. Artif. Intell. (1996)
[12] Ferber, J., Gutknecht, O.: A Meta-Model for the Analysis and Design of Organizations in Multi-Agent Systems. In: Proc. 3rd International Conference on Multi-Agent Systems, pp. 128–135. IEEE Computer Society Press, Los Alamitos (1998)
[13] Graham, J.R., Decker, K., Mersic, M.: DECAF – a flexible multi agent system architecture. Autonomous Agents and Multi-Agent Systems (2003)

[14] Gronmo, R., Jaeger, M.C., Hoff, H.: Transformations between UML and OWL-S. In: Hartman, A., Kreische, D. (eds.) ECMDA-FA 2005. LNCS, vol. 3748, pp. 269–283. Springer, Heidelberg (2005)
[15] Guessoum, Z., Thiefaine, A., Perrot, J., Blain, G.: META-DIMA: a Model-Driven Architecture for Multi-Agent Systems, (last accessed: 2006), http://www-poleia.lip6.fr/%7Eguessoum/MetaDima.html
[16] Gürcan, Ö., Kardas, G., Gümüs, Ö., Ekinci, E.E., Dikenelli, O.: A Planner for Implementing Semantic Service Agents based on Semantic Web Services Initiative Architecture. In: Fourth European Workshop on Multi-Agent Systems, Lisbon, Portugal (2006)
[17] Jayatilleke, G.B., Padgham, L., Winikoff, M.: Towards a Component Based Development Framework for Agents. In: Lindemann, G., Denzinger, J., Timm, I.J., Unland, R. (eds.) MATES 2004. LNCS (LNAI), vol. 3187, pp. 183–197. Springer, Heidelberg (2004)
[18] Jouault, F., Bezivin, J.: KM3: A DSL for Metamodel Specification. In: Gorrieri, R., Wehrheim, H. (eds.) FMOODS 2006. LNCS, vol. 4037, pp. 171–185. Springer, Heidelberg (2006)
[19] Jouault, F., Kurtev, I.: Transforming Models with ATL. In: Bruel, J.-M. (ed.) MoDELS 2005. LNCS, vol. 3844, pp. 128–138. Springer, Heidelberg (2006)
[20] Kardas, G., Goknil, A., Dikenelli, O., Topaloglu, N.Y.: Metamodeling of Semantic Web Enabled Multiagent Systems. In: Multiagent Systems and Software Architecture (MASSA), Erfurt, Germany, pp. 79–86 (2006)
[21] Molesini, A., Denti, E., Omicini, A.: MAS Meta-models on Test: UML vs. OPM in the SODA Case Study. In: Pĕchouček, M., Petta, P., Varga, L.Z. (eds.) CEEMAS 2005. LNCS (LNAI), vol. 3690. Springer, Heidelberg (2005)
[22] Object Management Group (OMG): Meta Object Facility (MOF) Specification. OMG Document AD/97-08-14, (September 1997)
[23] Object Management Group (OMG): MDA Guide Version 1.0.1. Document Number: omg/2003-06-01 (2003)
[24] Odell, J., Levy, R., Nodine, M.: FIPA Modeling TC: Agent Class Superstructure Metamodel (2004), available at: http://www.omg.org/docs/agent/04-12-02.pdf
[25] Odell, J., Nodine, M., Levy, R.: A Metamodel for Agents, Roles and Groups. In: Odell, J.J., Giorgini, P., Müller, J.P. (eds.) AOSE 2004. LNCS, vol. 3382. Springer, Heidelberg (2005)
[26] Pavon, J., Gomez, J., Fuentes, R.: Model Driven Development of Multi-Agent Systems. In: Rensink, A., Warmer, J. (eds.) ECMDA-FA 2006. LNCS, vol. 4066, pp. 284–298. Springer, Heidelberg (2006)
[27] Perini, A., Susi, A.: Automating Model Transformations in Agent-Oriented Modeling. In: Odell, J.J., Giorgini, P., Müller, J.P. (eds.) AOSE 2004. LNCS, vol. 3382. Springer, Heidelberg (2005)
[28] Selic, B.: The Pragmatics of Model-Driven Development. IEEE Software 20, 19–25 (2003)
[29] Sendall, S., Kozaczynski, W.: Model Transformation – the Heart and Soul of Model-Driven Software Development. IEEE Software 20, 42–45 (2003)
[30] Sycara, K., Williamson, M., Decker, K.: Unified information and control workflow in hierarchical task networks. In: Working Notes of the AAAI-96 workshop 'Theories of Action, Planning, and Control' (1996)
[31] The OWL Services Coalition: Semantic Markup for Web Services (OWL-S) (2004), http://www.daml.org/services/owl-s/1.1/
[32] The Atlantic Zoo: Metamodels expressed in KM3 (2006), available at: http://www.eclipse.org/gmt/am3/zoos/atlanticZoo/
[33] Web Service Modeling Ontology (2005), http://www.wsmo.org/

Formal Modelling of a Coordination System: From Practice to Theory, and Back Again[*]

Eloy J. Mata[1], Pedro Álvarez[2], José A. Bañares[2], and Julio Rubio[1]

[1] Departamento de Matemáticas y Computación, Universidad de La Rioja,
Edificio Vives, Luis de Ulloa s/n, E-26004 Logroño (La Rioja, Spain)
{eloy.mata,julio.rubio}@dmc.unirioja.es
[2] Departamento de Informática e Ingeniería de Sistemas, Universidad de Zaragoza,
María de Luna 3, E-50015 Zaragoza, Spain
{alvaper,banares}@unizar.es

Abstract. In this work, we report an experience that illustrates the interplay between formal methods and real software development. Starting from a Web-enable Coordination Service (WCS) based on JavaSpaces technology which had been successfully used in an industrial project, we built a formal model for the system in order to study its properties; specifically, our aim was to prove that Linda semantics was preserved in several layers of complex mappings from XML documents to Java objects. Once this objective was achieved (at least in a simplified, idealistic version), we observed several possibilities of extending the coordination system at the model level. In particular, we identified that it was possible to enhance the formal model with transactional capabilities, taking advantage of the similarity of our model to rule-based systems. At present, we are working on the translation of this theoretical result to practice, in order to improve our Web Coordination Service.

Keywords: Web services, coordination, formal methods, Linda.

1 Introduction

Both *multi-agent systems* and *service oriented architectures* have an underlying communication infrastructure. The most widespread standard for agents is based on FIPA-ACL to achieve interoperability between heterogeneous agent-based systems. In service-oriented computing, Web services are the basic building blocks to create new applications. Once a standardized way for accessing them has been defined, research needs focus on service coordination and composition, namely, coordination and composition middlewares to weave those services together and subsequently expose the resulting artifacts themselves as a Web service. In this research context, in our opinion, it is important to note that: 1) service composition is an aspect that is mostly internal to the implementation

---

[*] Partially supported by Comunidad Autónoma de La Rioja, project ANGI2005/19, and Ministerio de Educación y Ciencia, projects MTM2006-06513, TIN2006-13301.

of the service that composes other Web services, whereas protocols for service coordination are the properties required for the external interactions between Web services [1]; 2) coordination middleware is the cornerstone of more complex coordination protocols (transactions, security, reliable messaging, etc.) and Web service composition; and 3) successful experiences to avoid Web interface complexities should be revisited to solve coordination and composition problems, for example, applying a simple and generic interface to a broad range of distributed applications and delegating the complexity of interaction to the exchanged data based on XML-formats [2] (more precisely, at an Internet scale the success of HTTP as an application protocol was due to its reduced set of operations -GET, POST, PUT, and DELETE- and the standardization of the data exchanged by Web-based applications).

The key element around which to construct agent, service or middleware infrastructure is a *mediator software*. Mediation includes the mediation of different technologies as well as different interaction styles (e.g. *Message Brokers, Enterprise Service Bus, Normalized Message Router, XML Bus*). Therefore, according to these previous remarks it seems interesting to propose first a coordination framework based on a reduced set of basic operations and data-driven coordination.

The core of our proposal is the definition of a pure coordination model inspired by the Blackboard architectural pattern [3]. This model is the conceptual basis of a Web Coordination Service (WCS) [4] which plays the Web service coordinator role as a message broker (this type of broker has been considered essential in Web-based application-integration platforms [1]). We did not consider the possibility of creating a new coordination model from scratch because there were some proposed solutions that could be used as a starting point, for example, the Linda model [5,6]. Linda is based on Generative Communication and allows a collection of independent services to work cooperatively on a common data structure, or blackboard, using a shared vocabulary. These services use a reduced set of communication coordination primitives to put data messages into the common structure, which can be retrieved later on by other services asking for a certain template message. The use of Linda in open environments is promising because it allows for an uncoupled cooperation in space and time and a flexible modelling of interactions among services without adapting or announcing themselves.

One of the benefits of using Linda is its structural simplicity that allows the modeller, in particular, to get clear semantics insights. Nevertheless, there is always a trade off between simplicity and (real or industrial) practice. In our case, the Linda coordination ideas were materialized in a Java program written on top of *JavaSpaces* [7], a Java implementation of Linda. Thus, a main concern was to ensure that Linda's semantic behavior was preserved in our system. To this aim, our method implied establishing a formal model of the system (more precisely, of a simplified version of the system) and then, proving a theorem asserting that the model satisfies Linda semantics.

It is worth noting that the nature of the problem considered led us to choose a particular mathematical machinery. Since our goal was to analyze an already-in-use system, we selected a transition system approach for Linda [8], instead of a more abstract approach based on process algebra techniques like in [9] and [10]. This decision allowed us to set particular models of tuple organization which faithfully reflected the shared memory in our coordination system.

This paper is organized as follows. In Sections 2 and 3 generalities, on both Web services coordination and our particular proposal, are presented. The formal treatment and our main theoretical results appear in Section 4. The potential benefits of this formal modelling to enhance the real software system are briefly explained in Section 5. The paper ends with some conclusions, future work and the bibliography.

2 State of the Art in Web Service Coordination

Nowadays there are two parallel standardization initiatives to propose a Web service coordination framework: the *Web Service Coordination* (WS-C) specification [11] and the *Web Service Composite Application Framework* (WS-CAF) [12]. Both initiatives propose a framework to create *coordination contexts*. A coordination context provides functionality to registry: 1) Web services and applications that require coordination; and 2) coordination protocols to make their coordination requirements possible. According to this proposal, Web services and applications must define the used protocols. This solution is more flexible than ours, because it is possible to create many coordination contexts using different protocols (we only propose one free-context protocol based on Linda to coordinate any Web resource). However, it ignores the complexity the definition of new coordination protocols and promotes the use of ah-hoc protocols.

From another perspective, WS-CAF is different from WS-C because it takes the approach that the coordination context should be a first-class entity in a Web service architecture, that is, context is more fundamental than coordination. This allows contexts to be more easily managed in large-scale environments. However, WS-C's notion of context is tightly tied to the coordinator that creates it, making it less flexible than WS-CAF. Therefore, a separate specification for modelling Web-services context data structures is proposed by WS-CAF (*Web Service Context* specification).

In any case, with these coordination frameworks, more complex and standard protocols have been defined for transactions (*WS-Transaction* is a superset of the WS-C and *Web Service-Transaction Management* is based on WS-CAF), event-based notification (*WS-Notification* and *WS-Eventing*), security (*WS-Security*) and reliable messaging (*WS-Message Delivery* and *WS-Reliable Messaging*). Researchers have also exploited the Linda model for implementing this type of protocols. However they have not defined new and specific protocols, but semantic extensions of the Linda primitives. Some well-known proposals are, for example, TSpaces[1] for persistence and transactions, Bettini et al.'s proposal [13], WSSecSpaces [14] and Ruple[2] for security, and finally WorkSpaces [15] for

---
[1] TSpaces, http://www.almaden.ibm.com/cs/TSpaces/
[2] Ruple Project, http://www.roguewave.com/developer/tac/ruple/

composition. Note also that new computing paradigms use Linda-based solutions to support their communication and coordination requirements such as JXTAS-paces[3] in peer-to-peer environments or L2imbo [16] and LIME [17] in mobile and agent-based environments.

3   A Web Coordination Service Based on the Linda Model

To use Linda as the conceptual basis of our *Web Coordination Service* (WCS), we had to enrich some structural aspects of Linda, as it is explained in the following subsection. After that subsection, the general architecture of our WCS is briefly presented.

3.1   Linda for Open Environments

Linda is a well-known proposal based on Generative Communication [5]. To understand our extensions of Linda, it is enough to recall that Linda is composed of a small set of operators, acting on a blackboard, called tuple space, together with a pattern-matching oriented process to deal with tuples.

A Linda tuple is similar to [``Gelernter'',1989], a list of untyped atomic values. Our main extension implies passing from this atomic organization to lists of attribute/value pairs, like: [(author, ``Gelernter''), (year, 1989)]. Although this is still an untyped setting, this bit of structure allows information recovery in distributed environments. The reason is that this kind of structure corresponds with some simple types of XML documents (the tags playing the role of attributes). In addition, the pattern-matching process associated to Linda can be now considered as a complex procedure in which attributes act as keys for the matching. These features have been experimentally found when developing our Web Coordination Service: they are needed to deal with XML documents encapsulated as Java objects (note that JavaSpaces is the basic technology for our WCS). Thus, this complex matching works in two phases: first guided by attributes, and then by values.

This extended Linda model also allows one to perform some kind of semantic matching (as reported in [18]) and includes *reactive behavior*, by means of a system of *subscriptions*. This last aspect have been borrowed from (and can be implemented with the help of) JavaSpaces.

3.2   Design and Implementation of the Web Coordination Service

In a more detailed description, the designed WCS is composed of three software components (see Figure 1).

The *XML-based Space* component encapsulates the tuple space. Its interface provides a collection of operations to write XML tuples into, read them from the tuple-space, and be notified of the writing of a new XML tuple into the encapsulated space, according to the presented extension of Linda. This component has

---

[3] JXTASpaces Project, http://jxtaspaces.jxta.org/

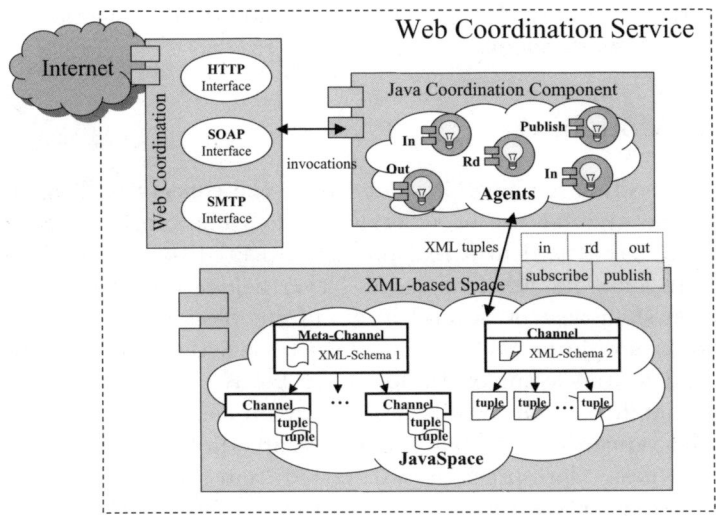

Fig. 1. Software Components of the coordination service

been built from JavaSpaces. The internal communication patterns have been implemented on JavaSpaces to guarantee the semantics of the classical Linda model and later improve the efficiency of the XML-based space (for the details see [4]).

On the other hand, the *Java Coordination* component is the core of the service. It has two different interfaces: the *Basic Coordination Interface* (BCI), which provides the collection of writing and reading operations proposed by Linda and promotes a cooperative style based on blocking readings; and the *Reactive Coordination Interface* (RCI), whose operations allow for a process advertising its interest in generating a specific type of XML tuples, publishing the advertised XML tuples and subscribing its interest in receiving XML tuples of a specific XML schema, encouraging a reactive style of cooperation among processes.

This component is a repository of agents (*agent* in the sense of a computational entity which acts on behalf of other entities in a semi-autonomous way, and performs its tasks in cooperation with other agents; the mobility and learning attributes have been excluded from this context). Every time an external Web service or application invokes a coordination operation, an internal agent is created. These agents are capable of coordinating with other agents by exchanging XML tuples through XML-based spaces. Therefore, the required cooperation by external entities is executed by their respective internal agents. The result of this cooperation is communicated from agents to external entities using an event-based mechanism scalable to the Internet.

Finally, the *Web Coordination* component acts as a Web-accessible interface of the Java Coordination component. It provides the same collection of operations than the Java Coordination component through an interface based on some standard Internet protocols (HTTP, SOAP and SMTP implementations of this component have been developed).

4 Formal Modelling

In the previous section we have presented both the conceptual basis (Subsection 3.1) and the architecture of a Service to coordinate other Web services. This Service has been used in a real project (oriented to location-based Web services [19]) to show its practical value. Nevertheless, the clear semantics of Linda has been shadowed by several layers of complex wrappings from XML documents to Java objects. So, a main concern is whether the final operations for reading and writing complex tuples really behave as Linda operators. To increase the reliability of our WCS, our strategy implied giving a formal model of (a simplified version of) our blackboard architecture, and then we tried to prove its correctness. More precisely, we started from a formal model of Linda (to be interpreted as an abstraction of JavaSpaces), then we defined a set of operations based on the way that the complex matching was implemented in our WCS, and finally we proved that the new defined operations also satisfy Linda equations.

4.1 Linda Coordination Model

As it was explained in the introduction, we chose the mathematical formalism of [8], based on transition systems. Although there are other approaches in the literature, some of them inspired by process calculi [9,10], we chose this formalism for its simplicity and because we were more interested in the kinds of data that could be managed by the coordination medium, through a pattern-matching procedure, than in the behavior of the active components (or processes).

In [8] the tuple-based coordination medium is represented as a software component interacting with coordinated entities by receiving *input events* and sending *output events*. The main elements of this model are: a set of tuples t ranging over T ($\overline{T}$ denotes the set of multisets over T); a set of templates $templ$ ranging over $Templ$; a matching predicate $mtc(templ, t)$ between templates and tuples; and a choice predicate $\mu(templ, \overline{t}, \widehat{t})$, where $\overline{t}$ is a multiset of tuples ranging over $\overline{T}$, and the symbol $\widehat{t}$ means that it can be a tuple t from the tuple multiset $\overline{t}$ which matches the template $templ$, or is an error element $\perp_T$ if no matching is available in $\overline{t}$ (in general for any set T, we will assume the void value $\perp_T$ is contained in it). This is formally defined in the following matching rules, where $t|\overline{t}$ is used to denote the union of the element t and the multiset $\overline{t}$ (see [8] for details):

$$\frac{mtc(templ, t)}{\mu(templ, t|\overline{t}, t)} \qquad \frac{\nexists\; t \in \overline{t}\;\; mtc(templ, t)}{\mu(templ, \overline{t}, \perp_T)} \tag{1}$$

The status of a tuple space at a given time is characterized as a labelled transition system by the couple $\langle \overline{t}, \overline{w} \rangle$ where $\overline{t}$ is a multiset of tuples and $\overline{w}$ is a multiset of *pending queries*. A pending query is an input event *ie* ranging over $IE \subseteq W$, where W is a set of operations such as the reading operation rd, the reading and removing operation in, or the writing operation out, which are possibly waiting to be served by a tuple space. Output events oe range over OE, where OE is a set of notification messages, with the syntax $\underline{o}v$ representing a message v for entity o.

The semantics of a tuple space is defined by the couple $\langle S, E \rangle$. Here, $S \subseteq W \times \overline{T}$ is a satisfaction predicate for queries, so that $\langle w, \overline{t} \rangle \in S$ means w can be servedunder the current space's content $\overline{t}$ and, in addiction, $E : W \mapsto 2^{(\overline{T} \times \overline{W}) \times OE \times (\overline{T} \times \overline{W})}$ is an evaluation function, so $\langle \overline{t}, \overline{w}, \widehat{oe}, \overline{t}', \overline{w}' \rangle \in E(w)$ means that the evaluation of the pending query w causes the tuple space in state $\langle \overline{t}, \overline{w} \rangle$ to move to $\langle \overline{t}', \overline{w}' \rangle$ and produce output event oe (or nothing).

The semantics of this transition system is defined by the rules (see [8] for details):

$$\frac{\nexists w \in \overline{w} : S(w, \overline{t})}{\langle \overline{t}, \overline{w} \rangle \xrightarrow{ie}_I \langle \overline{t}, ie | \overline{w} \rangle} \qquad \frac{S(w, \overline{t}) \quad \langle \overline{t}, w | \overline{w}, oe, \overline{t}', \overline{w}' \rangle \in E(w)}{\langle \overline{t}, w | \overline{w} \rangle \xrightarrow{oe}_O \langle \overline{t}', \overline{w} \rangle}$$

The predicate S is defined as the least relation satisfying the rules:

$$S(rdp(templ)^o, \overline{t})$$

$$S(inp(templ)^o, \overline{t})$$

$$\frac{mtc(templ, t)}{S(rd(templ)^o, t|\overline{t})}$$

$$\frac{mtc(templ, t)}{S(in(templ)^o, t|\overline{t})}$$

$$S(out(t)^o, \overline{t})$$

The evaluation function E, defining the actual operations' semantics, is defined by rules for every pending query corresponding to each primitive operation allowed:

$$\frac{\mu(templ, \overline{t}, \widehat{t})}{\langle \overline{t}, \overline{w}, \underline{o}\widehat{t}, \overline{t}, \overline{w} \rangle \in E(rdp(templ)^o)}$$

$$\frac{\mu(templ, \overline{t}, \widehat{t})}{\langle \overline{t}, \overline{w}, \underline{o}\widehat{t}, \overline{t} \backslash \widehat{t}, \overline{w} \rangle \in E(inp(templ)^o)}$$

$$\frac{mtc(templ, t)}{\langle t|\overline{t}, \overline{w}, \underline{o}t, t|\overline{t}, \overline{w} \rangle \in E(rd(templ)^o)}$$

$$\frac{mtc(templ, t)}{\langle t|\overline{t}, \overline{w}, \underline{o}t, \overline{t}, \overline{w} \rangle \in E(in(templ)^o)}$$

$$\langle \overline{t}, \overline{w}, \bot, t|\overline{t}, \overline{w} \rangle \in E(out(t)^o)$$

Once the abstract framework has been presented, in the next subsection we give a definition of tuples and templates as finite and ordered sequences of data, each one over its own domain. Moreover, templates admits wildcards variables. Then we define the matching and choice predicates. We have named this model as the *core model* and some of its definitions are inspired by [20].

4.2 A Core Model Based on Linda

Let $A = (A_1, \ldots, A_m)$ be a list of attributes with domains $D_i = dom(A_i)$, $i = 1, \ldots, m$.

The notation $D = dom(A) = \bigcup_{i=1}^{m} dom(A_i) = \bigcup_{i=1}^{m} D_i$ will be used as a shorthand.

Definition 1. *A **core-tuple** t_{core} over A is a partial and injective mapping:*

$$\mathbf{t_{core}} : \; X \subseteq A \longrightarrow dom(X) \qquad (2)$$

for which the following holds: $\forall A_i \in X \; t_{core}(A_i) = d_i$ *with* $d_i \in D_i$.

Then, core-tuples are finite and ordered sequences of simple data and have the form $t_{core} = (d_1, \ldots, d_n)$ where each $d_i \in D_i$.

Definition 2. *A **core-template** $templ_{core}$ over A is a partial and injective mapping:*

$$\mathbf{templ_{core}} : \; X \subseteq A \longrightarrow dom_?(X) \qquad (3)$$

for which the following holds: $\forall A_i \in X \; templ_{core}(A_i) = td_i$ *with* $td_i \in dom_?(A_i)$ *where* $dom_?(A_i) = dom(A_i) \cup \{?x_i\}$. *The additional value $?x_i$ denotes a wildcard variable that matches any value, and this value is saved in the variable x_i.*

Then, core-templates are finite and ordered sequences of data or wildcards variables and have the form $templ_{core} = (td_1, \ldots, td_n)$ with $td_i \in dom_?(A_i)$.

Hereinafter, T_{core} will denote the set of core-tuples and $Templ_{core}$ will denote the set of core-templates in the core-Linda model. By definition, it is obvious that $T_{core} \subseteq Templ_{core}$, i.e. in particular, each tuple is a template without any wildcard variable.

Definition 3 (Core-Matching Predicate). *Let $t_{core} = (d_1, \ldots, d_n) \in T_{core}$ be a tuple and, $templ_{core} = (td_1, \ldots, td_m) \in Templ_{core}$ be a template. We say that t_{core} **matches** $templ_{core}$ (denoted by $mtc_{core}(templ_{core}, t_{core})$), if the following conditions hold:*

- t_{core} *and* $templ_{core}$ *have the same arity, i.e.* $m = n$.
- *Each non-wildcard field of* $templ_{core}$ *is equal to the corresponding field of* t_{core}, *i.e.* $td_i = ?x_i$ *or* $td_i = d_i$, $1 \leq i \leq n$.

Definition 4 (Core-Choice Predicate)

Given the matching rule $mtc_{core}(templ_{core}, t_{core})$ we define the **core-choice predicate** as the predicate $\mu_{core}(templ_{core}, \bar{t}_{core}, \hat{t}_{core})$ which satisfies the rules of the Linda matching defined in (1).

$$\frac{mtc_{core}(templ_{core}, t_{core})}{\mu_{core}(templ_{core}, t_{core}|\bar{t}_{core}, t_{core})} \qquad \frac{\not\exists\ t \in \bar{t}_{core}\quad mtc_{core}(templ_{core}, t_{core})}{\mu_{core}(templ_{core}, \bar{t}_{core}, \bot_T)}$$

That is, $\hat{t}_{core}$ is a core-tuple from the tuple multiset $\bar{t}_{core}$ that matches a template $templ_{core}$, or is a void value (denoted by $\bot_T$) when no tuple matching occurs in $\bar{t}_{core}$.

4.3 A Structured Extension of the Core Model

The core model presented in the previous subsection can be seen as a naive formalization of JavaSpaces where JavaObjects are represented by tuples and the JavaSpaces matching process corresponds to Definitions 3 and 4. But in this subsection, in order to provide a formal model closer to our WCS, we will work with a version of Linda where tuples and templates are sequences of attribute/value pairs.

Definition 5. A **structured-tuple** t_{str} over A is a partial and injective mapping:

$$\mathbf{t_{str}}: \quad X \subseteq A \longrightarrow X \times dom(X) \qquad (4)$$

for which the following holds: $\forall\ a_i \in X\ t_{str}(a_i) = (a_i, v_i)$ with $v_i \in dom(a_i)$.

Definition 6. A **structured-template** $templ_{str}$ over A is a partial and injective mapping:

$$\mathbf{templ_{str}}: \quad X \subseteq A \longrightarrow X \times dom_?(X) \qquad (5)$$

for which the following holds: $\forall\ a_i \in X\ templ_{str}(a_i) = (a_i, tv_i)$ with $tv_i \in dom_?(a_i)$ where $dom_?(a_i) = dom(a_i) \cup \{?x_i\}$. The additional value $?x_i$ denotes a wildcard variable, that matches with any value and this value is saved in the variable x_i.

Hereinafter, T_{str} will denote the set of structured-tuples and $Templ_{str}$ will denote the set of structured-templates. It is also obvious that $T_{str} \subseteq Templ_{str}$.

Definition 7 (Structured-Matching Predicate)

Let $t_{str} = ((a_1, v_1)\ldots(a_n, v_n)) \in T_{str}$ be a tuple and,
let $templ_{str} = ((ta_1, tv_1)\ldots(ta_m, tv_m)) \in Templ_{str}$ be a template. We say that t_{str} **matches** $templ_{str}$ (denoted by $mtc_{str}(templ_{str}, t_{str})$) if the following conditions hold:

- t_{str} and $templ_{str}$ have the same attribute structure, i.e. $m = n$ and $ta_i = a_i$, $1 \leq i \leq n$.
- Each non-wildcard field of $templ_{str}$ is equal to the corresponding field of t_{str}, i.e. $tv_i =?x_i$ or $tv_i = v_i$, $1 \leq i \leq n$

Definition 8. *The **scheme-tuple** of a structured-tuple, or of a structured-template, is a core-tuple t_{core} in A defined by the mapping:*

$$\mathbf{scheme}: \quad T_{str} \cup Templ_{str} \longrightarrow T_{core_A} \quad (6)$$

provided that $scheme(((a_1, v_1) \ldots (a_n, v_n))) = (a_1, \ldots, a_n)$

That is, given a structured-tuple t_{str}, or a structured-template $templ_{str}$, the scheme mapping returns a core-tuple t_{core} with only the attributes or tags.

Definition 9. *The $\overline{\mathbf{scheme}}$ of a multiset of structured-tuples $\overline{t}_{str}$ is another multiset of core-tuples in A defined by the mapping:*

$$\overline{\mathbf{scheme}}: \quad \overline{T}_{str} \longrightarrow \overline{T}_{core_A} \quad (7)$$

such that: $\overline{scheme}(\overline{t}_{str}) = \{scheme(t_{str}) \mid t_{str} \in \overline{t}_{str}\}$

Definition 10. *The **value-tuple** of a structured-tuple is a core-tuple in D defined by the mapping:*

$$\mathbf{val}: \quad T_{str} \longrightarrow T_{core_D} \quad (8)$$

provided that $val(((a_1, v_1) \ldots (a_n, v_n))) = (v_1, \ldots, v_n)$

That is, given a structured-tuple t_{str}, the val mapping returns a core-tuple t_{core} with only the values.

Definition 11. *The **value-template** of a structured-tuple t_{str} is a core-tuple in $dom_?(D)$ defined by the mapping:*

$$\mathbf{val}: \quad Templ_{str} \longrightarrow T_{core_{dom_?(D)}} \quad (9)$$

provided that $val(((ta_1, tv_1) \ldots (ta_n, tv_n))) = (tv_1, \ldots, tv_n)$

That is, given a structured-template $templ_{str}$, the val mapping returns a core-template with only the values and wildcards.

In the following definition, the notion of *channel* is introduced. This concept appeared first in the implementation of the WCS as a tool both to increase the efficiency of the system and to recover multiple tuples (with the same scheme) by means of JavaSpaces. Here, we give an interpretation of channels in our formal model.

Definition 12. *Given a multiset of structured-tuples $\overline{t}_{str}$ and a structured-template $templ_{str}$, we define the $\overline{\mathbf{channel}}$ of $templ_{str}$ in $\overline{t}_{str}$ as the mapping:*

$$\overline{\mathbf{channel}}: \quad Templ_{str} \times \overline{T}_{str} \longrightarrow \overline{T}_{core_D} \quad (10)$$

for which the following holds:

$\overline{channel}(templ_{str}, \overline{t}_{str}) =$
$\quad \{val(t_{str}) \in \overline{T}_{core_D} | t_{str} \in \overline{t}_{str} \wedge mtc_{core}(scheme(templ_{str}), scheme(t_{str}))\}$

That is, the *channel* of a structured-template in a tuple multiset is the multiset of value-tuples that share the same attribute structure as the template. Furthermore, given a scheme-tuple and a value-tuple, we can define the inverse mapping to rebuild a structured-tuple.

Definition 13. *The* **rebuild_tuple$_{str}$** *operation, which produces a structured-tuple from a pair of core-tuples, is a mapping:*

$$\text{rebuild\_tuple}_{str}: \quad T_{core_A} \times T_{core_D} \longrightarrow T_{str} \qquad (11)$$

for which the following holds: given a scheme-tuple $a = (a_1, \ldots, a_n)$ and a value-tuple $v = (v_1, \ldots, v_m)$ with $n = m$ and $v_i \in D_i$, then:

$$\text{rebuild\_tuple}_{str}(a, v) = ((a_1, v_1) \ldots (a_n, v_n))$$

that is, an ordered sequence of attribute/value pairs.

If a or v are void ($\perp_T$) then the image of the mapping is also void ($\perp_T$)

Definition 14 (Structured-Choice Predicate). *We define the **structure-choice predicate** as the predicate $\mu_{str}(templ_{str}, \overline{t}_{str}, \widehat{t}_{str})$, where $\widehat{t}_{str}$ is defined as:*

$$\widehat{t}_{str} = \text{rebuild\_tuple}_{str}(\widehat{t.\text{scheme}}, \widehat{t.\text{val}})$$

where, the tuple $\widehat{t.\text{scheme}}$ satisfies:

$$\mu_{core}(\text{scheme}(templ_{str}), \overline{\text{scheme}}(\overline{t}_{str}), \widehat{t.\text{scheme}})$$

and, the tuple $\widehat{t.\text{val}}$ satisfies:

$$\mu_{core}(\text{val}(templ_{str}), \overline{\text{channel}}(templ_{str}, \overline{t}_{str}), \widehat{t.\text{val}})$$

That is, $\widehat{t}_{str}$ is built by a scheme-tuple and a value-tuple, and these tuples satisfy the predicate μ in the core model. The scheme-tuple $\widehat{t.\text{scheme}}$ satisfies the predicate μ_{core} in a multiset of schemes, and the value-tuple $\widehat{t.\text{val}}$ satisfies the predicate μ_{core} in the channel of the template.

4.4 Main Results

In this subsection, we will prove that the structured matching meets Linda's matching rules. First, Theorem 1 proves the relationship between the matching predicate in the structured model and the matching predicate defined in the core model. Then, Theorem 2 will prove that the structured pattern matching process defined in Definition 14, and successfully implemented in our WCS, satisfies Linda's matching rules introduced in (1). The first three results are presented without their proofs (they can be easily inferred from the definitions).

Theorem 1. *The following relationship between mtc_{core} and mtc_{str} holds:*

$$mtc_{str}(templ_{str}, t_{str}) = mtc_{core}(\text{scheme}(templ_{str}), \text{scheme}(t_{str}))$$
$$\wedge\, mtc_{core}(\text{val}(templ_{str}), \text{val}(t_{str}))$$

Proposition 1. *Let $t_{str} \in T_{str}$ be a tuple and $\bar{t}_{str} \in \overline{T}_{str}$ be a tuple multiset then $scheme(t_{str})|\overline{scheme(\bar{t}_{str})} = \overline{scheme(t_{str}|\bar{t}_{str})}$*

Proposition 2. *Let $t_{str} \in T_{str}$ be a tuple, $\bar{t}_{str} \in \overline{T}_{str}$ be a tuple multiset and $templ_{str} \in Templ_{str}$ be a template. If $mtc_{core}(scheme(templ_{str}), scheme(t_{str}))$ then $val(t_{str})|\overline{channel(templ_{str}, \bar{t}_{str})} = \overline{channel(templ_{str}, t_{str}|\bar{t}_{str})}$*

Theorem 2. *The defined predicates μ_{str} and mtc_{str} meets Linda's matching rules:*

(i) $$\frac{mtc_{str}(templ_{str}, t_{str})}{\mu_{str}(templ_{str}, t_{str}|\bar{t}_{str}, t_{str})}$$

(ii) $$\frac{\nexists\ t_{str} \in \bar{t}_{str}\quad mtc_{str}(templ_{str}, t_{str})}{\mu_{str}(templ_{str}, \bar{t}_{str}, \perp_T)}$$

Proof. (i) Let $t_{str} \in T_{str}$ be a tuple such that $mtc_{str}(templ_{str}, t_{str})$. Then, by Theorem 1:

$$mtc_{core}(scheme(templ_{str}), scheme(t_{str}))\ \text{and,} \qquad (12)$$

$$mtc_{core}(val(templ_{str}), val(t_{str})) \qquad (13)$$

by Definition 3 and (12)

$$\mu_{core}(scheme(templ_{str}), scheme(t_{str})|\overline{scheme(\bar{t}_{str})}, scheme(t_{str})) \qquad (14)$$

by Proposition 1 and (14)

$$\mu_{core}(scheme(templ_{str}), \overline{scheme(t_{str}|\bar{t}_{str})}, scheme(t_{str})) \qquad (15)$$

by Definition 3 and (13)

$$\mu_{core}(val(templ_{str}), val(t_{str})|\overline{channel(templ_{str}, \bar{t}_{str})}, val(t_{str})) \qquad (16)$$

by Proposition 2 and (16)

$$\mu_{core}(val(templ_{str}), \overline{channel(templ_{str}, t_{str}|\bar{t}_{str})}, val(t_{str})) \qquad (17)$$

Now, we can build a structured-tuple as:

$$t_{str} = rebuild\_tuple_{str}(scheme(t_{str}), val(t_{str}))$$

then, by Definition 14, (15) and (17)

$$\mu_{str}(templ_{str}, t_{str}|\bar{t}_{str}, t_{str})$$

(ii) If $\nexists\ t_{str} \in \bar{t}_{str}$ such that $mtc_{str}(templ_{str}, t_{str})$ then, by Theorem 1, there might be two reasons for it:

The first reason is that a matching of schemes does not hold. That is,

$$\nexists\ t.scheme \in \overline{scheme}(\bar{t}_{str}) : \quad mtc_{core}(scheme(templ_{str}), t.scheme)$$

by Definition 4

$$\mu_{core}(scheme(templ_{str}), \overline{scheme}(\bar{t}_{str}), \perp_T)$$

and then, by Definition 13

$$\mu_{str}(templ_{str}, \bar{t}_{str}, \perp_T)$$

The second reason is that a matching of schemes holds but a matching of values does not hold into the channel of $templ_{str}$. That is,

$$\exists\ t_{str} \in \bar{t}_{str} : \quad mtc_{core}(scheme(templ_{str}), scheme(t_{str}))$$

but,

$$\nexists\ t_{str} \in \overline{channel}(templ_{str}, \bar{t}_{str}) : \quad mtc_{core}(val(templ_{str}), val(t_{str}))$$

by Definition 4

$$\mu_{core}(val(templ_{str}), \overline{channel}(templ_{str}, \bar{t}_{str}), \perp_T)$$

and then, by Definition 14 and by Definition 13

$$\mu_{str}(templ_{str}, \bar{t}_{str}, \perp_T). \qquad \square$$

5 Applying Formalization and Implementation Experiences to the Coordination of Web Services

The previous theorems have shown that the strategy used to implement our WCS (abstracting from programming or networking technologies) is sound. This is one of the main objectives of using formal methods application to practical Software Engineering, in which an already-in-use software system is reified in a mathematical model. Here, the important point is not the originality of the techniques nor the difficulty of the proofs, but rather the accuracy of the relation between the model and reality.

Another possible application of this re-engineering work is the analysis of the formal model to look for improvements and extensions. The abstract version often is clearer than its real counter-part, and this allows the modeler to have a more comprehensive view of the system. After looking at different formal coordination models (Petri nets [21], rules [22,23], Linda, etc.), we observed that they all used templates to recognize particular states and represent constraints and therefore a pattern-matching algorithm to interpret themselves. These similarities between models helped us to discover different ways to extend the classical Linda model, for example, an extension of its matching functions to support multiple templates. This opens up the possibility to express more complex coordination restrictions and transactional capabilities. This extension is similar to High Level Petri nets, which provide more compact and manageable descriptions than ordinary Petri nets [24]. In [25], the RETE-like algorithm is used for an efficient implementation of a Web Coordination service that supports reading operations with multiple templates. Similar works can be found in the literature, for example, an adaptation of the RETE algorithm to interpret High Level Petri nets with multiple labelled arcs [26,27].

The idea is to extend Linda with a multiple matching predicate (see [25]) in order to provide operations to extract atomically a group of tuples instead of a single item. This predicate has the form (in case of two templates):

$$mtc^2(templ_1, templ_2, t_1, t_2, R)$$

and it is satisfied if the following conditions hold:

1. $mtc_{str}(templ_1, t_1)$
2. $mtc_{str}(templ_2, t_2)$
3. tuples t_1 and t_2 holds relation R, where R is a boolean restriction function defined over the union of the domains of the data and the wildcards of the templates $templ_1$ and $templ_2$

Given this multiple-matching predicate, a multiple-choice predicate μ^2 is also defined and can be implemented by a RETE-like algorithm (see [25] again).

6 Conclusions and Further Work

In this work, we have reported an application of formal methods to practical Software Engineering. More precisely, we have presented a mathematical model for a Web Coordination Service (WCS). The main results prove that the WCS primitives satisfy Linda semantics (at least if the model faithfully reflects the actual WCS). In addition, another kind of application has been illustrated: the formal model can be used to propose new enhancements of the initial systems. In our case, the model inspired a parallelism with ruled-based systems, and then a promising way to incorporate transactional capabilities through a variant of the RETE algorithm.

At present, we are going back to practice from theory, rebuilding our WCS to include the new features designed at the formal level. First, using Linda model as a mediator for integrating applications in service oriented architectures and multi-agent systems. Second, extending the matching functions to support multiple templates which opens up the possibility to express more complex coordination restrictions and transactional capabilities. And finally, extending the matching functions to provide a semantic matching process. This work is in progress, but when finished, it will allow us to evaluate the methodology used, and, more generally, whether it can be fruitfully applied when dealing with real-life coordination of Web services.

References

1. Alonso, G., Casati, F., Kuno, H., Machiraju, V.: Web Services. Concepts, Architectures and Applications. Springer, Heidelberg (2004)
2. Vinoski, S.: Putting the Web into Web Services. Web Services Interaction Models, Part 2. IEEE Internet Computing 6(4), 90–92 (2002)
3. Buschmann, F., Meunier, R., Rohnert, H., Sommerlad, P., Stal, M.: A System of Patterns. Wiley, Chichester (1996)
4. Álvarez, P., Bañares, J.A, Muro-Medrano, P.: An Architectural Pattern to Extend the Interaction Model between Web-Services: The Location-Based Service Context. In: Orlowska, M.E., Weerawarana, S., Papazoglou, M.M.P., Yang, J. (eds.) ICSOC 2003. LNCS, vol. 2910, pp. 271–286. Springer, Heidelberg (2003)
5. Gelernter, D.: Generative communication in Linda. ACM Transactions on Programming Languages and Systems 7(1), 80–112 (1985)
6. Carriero, N., Gelernter, D.: Linda in context. Communications of the ACM 32(4), 444–458 (1989)
7. Freeman, E., Hupfer, S., Arnold, K.: JavaSpaces. Principles, Patterns, and Practice. Addison-Wesley, London, UK (1999)
8. Viroli, M., Ricci, A.: Tuple-Based Coordination Models in Event-Based Scenarios. In: IEEE 22nd International Conference on Distributed Computing Systems (ICDCS 2002 Workshops) - DEBS'02 International Workshop on Distributed Event-Based Sytem, Vienna, Austria (2002)
9. Busi, N., Gorrieri, R., Zavattaro, G.: A Process Algebraic View of Linda Coordination Primitives. Theoretical Computer Science 192(2), 167–199 (1998)
10. Busi, N., Gorrieri, R., Zavattaro, G.: Process Calculi for Coordination: from Linda to JavaSpaces. In: Rus, T. (ed.) AMAST 2000. LNCS, vol. 1816. Springer, Heidelberg (2000)
11. Cabrera, F., Coopeland, G., Freund, T., Klein, J., Langworthy, D., Orchand, D., Schewchuk, J., Storey, T.: Web Service Coordination (WS-Coordination). Technical report, IBM & Microsoft Corporation & BEA System (2002)
12. Bunting, D., Chapman, M., Hurley, O., Little, M., MischKinsky, J., Newcomer, E., Webber, J., Swenson, K.: Web Service Coordination Framework(WS-CF). Technical report, Arjuna Technologies & Fujitsu Limited & IONA Technologies & Sun Microsystems & Oracle Corporation (2004)
13. Bettini, L., Nicola, R.D.: A Java Middleware for Guaranteeing Privacy of Distributed Tuple Spaces. In: Guelfi, N., Astesiano, E., Reggio, G. (eds.) FIDJI 2002. LNCS, vol. 2604, pp. 175–184. Springer, Heidelberg (2003)
14. Lucchi, R., Zavattaro, G.: WSSecSpaces: a Secure Data-Driven Coordination Service for Web Services Applications. In: Proceedings of the 2004 ACM Symposium on Applied Computing (SAC04), pp. 487–491 (2004)
15. Tolksdorf, R.: Workspaces: a Web-based Workflow Management System. IEEE Internet Computing 6(5), 18–26 (2002)
16. Davies, N., Friday, A., Wade, S.P., Blair, G.: L2imbo: A distributed systems platform for mobile computing. Mobile Networks and Applications 3(2), 143–156 (1998)
17. Picco, G., Murphy, A., Roman, G.: LIME: Linda Meets Mobility. In: Garlan, D., Kramer, J. (eds.) Proceedings of the 21st International Conference on Software Engineering (ICSE'99), pp. 368–377. ACM Press, New York (1999)
18. Álvarez, P., Bañares, J.A., Mata, E., Muro-Medrano, P., Rubio, J.: Generative Communication with Semantic Matching in Distributed Heterogeneous Environments. In: Moreno-Díaz Jr., R., Pichler, F. (eds.) EUROCAST 2003. LNCS, vol. 2809, pp. 231–242. Springer, Heidelberg (2003)

19. Álvarez, P., Bañares, J.A., Muro-Medrano, P., Nogueras, J., Zarazaga, F.: A Java Coordination Tool for Web-Service Architectures: The Location-Based Service Context. In: Guelfi, N., Astesiano, E., Reggio, G. (eds.) FIDJI 2002. LNCS, vol. 2604, pp. 1–14. Springer, Heidelberg (2003)
20. Bravetti, M., Gorrieri, R., Lucci, R., Zavattaro, G.: Probabilistic and Prioritized Data Retrieval in the Linda Coordination Model. In: De Nicola, R., Ferrari, G.L., Meredith, G. (eds.) COORDINATION 2004. LNCS, vol. 2949, pp. 55–70. Springer, Heidelberg (2004)
21. Murata, T.: Petri Nets: Properties, Analysis and Applications. Proceedings of the IEEE 16(1), 39–50 (1990)
22. Kappel, G., Lang, P., Rausch-Schott, S., Retschitzegger, W.: Workflow Management Based on Objects, Rules, and Roles. Data Engineering Bulletin 18(1), 11–18 (1995)
23. Bonifati, A., Ceri, S., Paraboschi, S.: Pushing reactive services to XML repositories using active rules. In: The Tenth internatinional World Wide Web conference on World Wide Web, Hong Kong, pp. 633–641. ACM Press, New York (2001)
24. Jensen, K., Rozenberg, G. (eds.): High-level Petri Nets. Springer, Heidelberg (1991)
25. Mata, E., Álvarez, P., Bañares, J.A., Rubio, J.: Towards and Efficient Rule-Based Coordination of Web Services. In: Lemaître, C., Reyes, C.A., González, J.A. (eds.) IBERAMIA 2004. LNCS (LNAI), vol. 3315, pp. 73–82. Springer, Heidelberg (2004)
26. Valette, R., Bako, B.: Software Implementation of Petri Nets and Compilation of Rule-based Systems. In: Rozenberg, G. (ed.) Advances in Petri Nets 1991. LNCS, vol. 524, pp. 296–316. Springer, Heidelberg (1991)
27. Muro-Medrano, P.R., Bañares, J.A., Villarroel, J.L.: Knowledge representation-oriented nets for discrete event system applications. IEEE Trans. on Systems, Man and Cybernetics - Part A 28(2), 183–198 (1998)

Using Constraints and Process Algebra for Specification of First-Class Agent Interaction Protocols

Tim Miller and Peter McBurney

Department of Computer Science,
The University of Liverpool, Liverpool, L69 7ZF, UK
{tim,p.j.mcburney}@csc.liv.ac.uk

Abstract. Current approaches to multi-agent interaction involve specifying protocols as sets of possible interactions, and hard-coding decision mechanisms into agent programs in order to decide which path an interaction will take. This leads to several problems, three of which are particularly notable: hard-coding the decisions about interaction within an agent strongly couples the agent and the protocols it uses, which means a change to a protocol involves a changes in any agent that uses such a protocol; agents can use only the protocols that are coded into them at design time; and protocols cannot be composed at runtime to bring about more complex interactions. To achieve the full potential of multi-agent systems, we believe that it is important that multi-agent interaction protocols exist at runtime in systems as entities that can be inspected, referenced, composed, and shared, rather than as abstractions that emerge from the behaviour of the participants. We propose a framework, called $\mathcal{RASA}$, which regards protocols as first-class entities. In this paper, we present the first step in this framework: a formal language for specification of agent interaction protocols as first-class entities, which, in addition to specifying the order of messages using a process algebra, also allows designers to specify the rules and consequences of protocols using constraints. In addition to allowing agents to reason about protocols at runtime in order to improve their the outcomes to better match their goals, the language allows agents to compose more complex protocols and share these at runtime.

1 Introduction

Research into multi-agent systems aims to promote autonomy and often intelligence into agents. Intelligent agents should be able to interact socially with other agents, and adapt their behaviour to changing conditions. Despite this, research into interaction in multi-agent systems is focused mainly on the documentation of interaction protocols, which specify the set of possible interactions for a protocol in which agents engage. Agent developers use these specifications to hard-code the interactions of agents. We identify three significant disadvantages with this approach: 1) it strongly couples agents with the protocols they use — something which is unanimously discouraged in software engineering — therefore requiring agent code to changed with every change in a protocol; 2) agents can only interact using protocols that are known at design time, a restriction that seems out of place with the goals of agents being intelligent and adaptive; and 3) agents cannot compose protocols at runtime to bring about more complex interactions, therefore restricting them to protocols that have been specified by human designers.

We propose a framework, called $\mathcal{RASA}$, which regards protocols as *first-class* entities. These first-class protocols are documents that exists within a multi-agent system, in contrast to hard-coded protocols, which exist merely as abstractions that emerge from the messages sent by the participants. To promote decoupling of agents from the protocols they use, we propose a formal, executable language for protocol specification. This language combines two well-studied and well-understood fields of computer science: process algebra, which are used to specify the messages that can be sent; and constraints, which are used to specify the *rules* governing under which conditions messages can be sent, and the *effects* that sending messages has on a system. Therefore, rather than a protocol being represented as a sequence of arbitrary tokens, each message contains a meaning represented as a constraint. Instead of hard-coding the decision process of when to send messages, agent designers can implement agents that reason about the effect of the messages they send and receive, and can choose the course of action that best achieves their goals. Agents able to reason about protocols can therefore learn of new protocols at runtime, making them more adaptable, for example, by being able to interact with new agents that insist on using specific protocols. The $\mathcal{RASA}$ language also allows protocols to be composed to bring about more complex interactions.

In this paper, we define a syntax and semantics for the $\mathcal{RASA}$ protocol specification language, which forms part of the $\mathcal{RASA}$ framework, a framework for modelling agent interaction as first-class entities. They key ideas that were taken into account in the design of the $\mathcal{RASA}$ language were the following:

- *protocols as first-class entities*: rather than protocol specifications emerging from an abstraction of the behaviour of participating agents, $\mathcal{RASA}$ protocols are *first-class*, meaning that they exist as entities in multi-agent systems;
- *inspectable*: agents are able to inspect and reason about the set of interactions permitted by a protocol, when they can occur, and what their effects are, so that the agents can devise strategies at runtime, therefore de-coupling them from the protocols they use;
- *layered*: other languages used for inspectable interaction protocols, such as OWL-P [7] and Yolum and Singh's Event Calculus extension [27], use the same language for specifying rules and effects as they do for specifying the sequencing of messages. We take a layered approach, in which the language for specifying the sequence of messages is separated from the language for specifying rules and effects. This allows us to develop the $\mathcal{RASA}$ framework independent of the underlying constraint language, and does not enforce the use of a particular language; and

- *reusable, composable and extendable*: existing protocol specifications can be extended and composed with other protocols to bring about new protocols. In addition, composable protocols permits designers to break up their task into smaller subtasks, simplifying the design process.

1.1 $\mathcal{RASA}$ Outline

The idea of first-class protocols is novel, but not entirely new. Desai et al. [7], and Yolum and Singh [27] present some initial work in using OWL and the Event Calculus respectively to model protocols as *first-class* entities (although they do not use this term). These approaches adapt existing declarative languages by adding definitions, written in that language, that specify the rules and effects of protocols. However, there are two major downsides to taking this approach. Firstly, the effects and rules must be specified in the declarative language (OWL or the Event Calculus), which is too restrictive. Secondly, the message sequencing is also specified in the language itself. For example, to specify that message a is sent before b, one must write a predicate resembling the following:

$$Happens(a, t_1) \land Happens(b, t_2) \land t_1 < t_2$$

This means that event a happens at time t_1, event b happens at time t_2, and event a occurs before event b, specified by the predicate $t_1 < t_2$. This can be specified in a process algebra as $a; b$, which we believe is more intuitive to the human reader, and is no less expressive. Robertson [22] takes a similar approach of using a process algebra and an underlying language to specify first-class protocols. We extend his work by, among other things, providing an additional language construct — local variable declaration — and by formalising the relationship between the process algebra and the underlying language.

Using $\mathcal{RASA}$, protocols can be visually represented as annotated trees outlining the interactions that can occur. As well as being annotated with a transitional message, each arc in the tree is annotated with a precondition and postcondition, in which the precondition must be enabled for the message to be sent, and the postcondition represents the effect of sending a message. The nodes represent the states that result from the corresponding postcondition. We incorporate states into the language to allow designers and agents to calculate the effect of a series of transactions; that is, if there are two messages sent in sequence, the effect of the second depends on the state resulting from the first. The root node of the tree is the initial state of the protocol, and the leaf nodes represent terminating states. Branches in the tree represent choices to be made by one or more agents.

Protocol rules, effects, and states are specified using declarative constraint languages. Using such languages allow agents to reason about which messages to send by calculating the which paths best achieve their goals, with each path from the root node to a leaf node representing a possible sequence of interaction. Agents can also reason about sub-protocols, by taking the root node of a sub-protocol as the starting point. We do not insist on a particular constraint language, but instead assume that it contains a few basic operators and constants common to most constraint languages. Such an approach fulfils our requirement that the language is inspectable, because agents can be equipped with the necessary constraint solvers, while maintaining flexibility by not enforcing a particular language. The constraint language is separate from the language for specifying the possible sequences of interaction in the protocol.

Protocols can be referenced and composed with others to form new protocols. The composition of these interactions provides a precondition and postcondition for an entire protocol. $\mathcal{RASA}$ allows one type of atomic event, comprising a precondition, message, and postcondition. These correspond to an arc in the tree. An atomic event is itself a protocol, meaning that the syntax and semantics of composing existing protocols is the same as composing a single protocol.

We envisage systems in which agents have access to bases of protocol specifications; either locally or centrally. Agents can search through these bases at runtime to find protocols that best suit the goal they are trying to achieve, and can share these protocol specifications with possible future participants. If no single protocol is suitable for the agent, composition of these may offer an alternative.

This paper is structured as follows: Section 2 presents the assumptions we make regarding the constraint language used by the agents engaged in interaction. As will be seen, these assumptions are quite general, allowing wide applicability of the framework. Section 3 then presents the formal syntax of the $\mathcal{RASA}$ modelling language, with an operational semantics for this language presented in Section 4. The paper follows this with a section discussing reuse and composition of protocols. Section 6 then presents a discussion of related work, before Section 7 concludes the paper.

2 Modelling Information

Communication in multi-agent systems is performed across a *universe of discourse*. Agents send messages expressing particular properties about the universe. We assume that these messages refer to *variables*, which represent the parts of the universe that have changing values, and use other *tokens* to represent relations, functions, and constants to specify the properties of these variables and how they relate to each other. We also assume that agents share an *ontology* that provides a shared definition of these relations, functions, and constants.

In this section, we discuss the minimum requirements for a constraint language that can be used in $\mathcal{RASA}$. We do not believe that these requirements are unreasonable — many languages can be used within the framework. For example, there are many description logics [2], constraint programming languages [23], commitment logics [27], or even predicate and modal logics [3] that contain the necessary constructs, although some of these languages may not be executable, and therefore the protocols would not be inspectable. The content languages proposed by FIPA [9] would also be suitable candidates.

Definition 1. *Constraint*

A *constraint* is a piece of information reducing the set of values that are possible for variables in a universe. For example, if an agent wishes to specify that the price of an item, *Item*, is 10 units, it may express this as follows:

$$PriceOf(Item, 10)$$

In this constraint, *Item* is a variable, *PriceOf* a relation, and '10' a constant. Operators used to express constraints over the universe must be defined.

Definition 2. *Constraint System*

We assume agents communicate using a communication language, which has a set of operators used to express constraints over variables. We will refer to such as language as a *constraint language*. Rather than define the syntax and semantics of a new language, or present the details of an existing one, we take the approach that any language can be used as the communication language in our framework, provided it contains a few basic constants and operators with certain properties. As well as using this as a communication language, we assume that this language is used to specify the preconditions and consequences of protocols. We refer to this as the *underlying constraint language* or just *underlying language*. This constraint language is denoted $\mathcal{L}$.

We use the definition of a *constraint system* proposed by De Boer et al. [5]. They define a constraint system as a complete algebraic lattice

$$\langle C, \sqsupseteq, \sqcup, \text{true}, \text{false}\rangle$$

In this structure, C is the set of atomic propositions in the language, for example $1 \leq 2$, $\sqsupseteq$ is an entailment operator, true and false are the least and greatest elements of C respectively, and $\sqcup$ is the least upper bound operator. The shorthand $c = d$ is equivalent to $c \sqsupseteq d$ and $d \sqsupseteq c$.

The entailment operator is a partial order over C, in that, for any atomic propositions, c and d, $c \sqsupseteq d$ means that c contains more information than d. This is read that d is provable from c, which means that any values that satisfy the variables in d also satisfy c. For example, $x \leq 5 \sqsupseteq x \leq 6$ specifies that if x is less than or equal to 5, then it is less than or equal to 6, which is trivially true because there exists no value for x that satisfies $x \leq 5$ that does not also satisfy $x \leq 6$. The $\sqcup$ operator specifies the addition of information. For example, to specify the prices of *ItemA* and *ItemB* are 5 and 10 units respectively, one could write $PriceOf(ItemA, 5) \sqcup PriceOf(ItemB, 10)$. This is is analogous to conjunction in logic, in that $c \sqcup d$ is the joining of information. Therefore, $c \sqcup d \sqsupseteq d$ is true for any c and d.

A *cylindric constraint system* is a constraint system with an operator for hiding variables. De Boer et al. [5] define a cylindric constraint system as a structure, $\langle C, \sqsupseteq, \sqcup, \text{true}, \text{false}, Var, \exists\rangle$, in which Var is a set of variables, and $\exists$ the hiding operator. To hide a variable x in a constraint c, one would write $\exists_x c$. The hiding operator has the following properties:

- $c \sqsupseteq \exists_x c$
- $c \sqsupseteq d$ implies $\exists_x c \sqsupseteq \exists_x d$
- $\exists_x(c \sqcup \exists_x d) = \exists_x c \sqcup \exists_x d$
- $\exists_x \exists_y c = \exists_y \exists_x c$

For example, to specify that the price of *Item* is between 5 and 10 units inclusive, one could write $\exists_{Price}(PriceOf(Item, Price) \sqcup 5 \leq Price \sqcup Price \leq 10)$.

We use shorthand to represent negation in $\mathcal{L}$. For constraints c and d, $c \sqsupseteq \neg d$ is true if and only if $c \sqsupseteq d$ is not. That is, c entails $\neg d$ if and only if c does not entail d. Note that, from this definition, we assume only that negation can occur on the right hand side of the entailment operator. Depending on the underlying constraint language used, this restriction could be relaxed, but this is not necessary to fit into the framework.

Throughout this paper, constraints will adhere to the following grammar, although a suitable language need not adhere to this grammar to be used in the framework:

$$\phi ::= c \mid \phi \sqcup \phi \mid \neg \phi \mid \exists_x \phi$$

In this grammar, c is any atomic constraint in C, and x any variable in Var. We use ψ and ϕ as meta-variables representing constraints that follow this grammar, adding subscripts and superscripts to denote distinct meta-variables. Brackets are used to remove syntactic ambiguity, although to reduce the need for brackets, we specify that $\neg$ and $\exists$ both bind tighter than $\sqcup$, so $\neg \phi \sqcup \psi$ is $(\neg \phi) \sqcup \psi$, and $\exists_x \phi \sqcup \psi$ is $(\exists_x \phi) \sqcup \psi$.

We introduce a renaming operator, which we will write as $[x/y]$, such that $\phi[x/y]$ means 'replace all references of y in ϕ with x'. The reader may have already noted that $\phi[x/y]$ is shorthand for $\exists_y(y = x \sqcup \phi)$. We also introduce the shorthand $\phi \neq \psi$ for $\neg(\phi = \psi)$, and $\exists_{x,y}\phi$ for $\exists_x \exists_y \phi$.

Definition 3. *Free Variables*

The function, $free \in \mathcal{L} \to \wp(Var)$, returns the set of free variables in any constraint; that is, variables referenced in ϕ that are not hidden using $\exists$. For example, $free(x \leq 5) = \{x\}$, and $free(\exists_x(x \leq 5 \sqcup y = x)) = \{y\}$. Calculating the free variables in a constraint can be done at a syntactic level using an inductive definition over the constraints:

$$\begin{aligned}
free(c) &= \ldots \\
free(\text{true}) &= \emptyset \\
free(\text{false}) &= \emptyset \\
free(\phi \sqcup \psi) &= free(\phi) \cup free(\psi) \\
free(\neg \phi) &= free(\phi) \\
free(\exists_x \phi) &= free(\phi) \setminus \{x\}
\end{aligned}$$

We do not define $free(c)$ because that is specific to $\mathcal{L}$. For readability, we use the shorthand $\exists_{\hat{x}} \phi$ to represent $\exists_{free(\phi) \setminus \{x\}} \phi$. That is, $\exists_{\hat{x}} \phi$ means that we quantify over all free variables in ϕ except x.

3 $\mathcal{RASA}$ Protocols

In this section, we present the language for modelling $\mathcal{RASA}$ protocols, and some definitions relevant to this.

The $\mathcal{RASA}$ protocol specification language resembles that of other process algebras, such as CSP [11]. However, we specify the rules and effects of protocols using an underlying constraint language, which would be declarative by definition. This allows agents equipped with the necessary tools to reason about this language to determine when rules are satisfied, to calculate the effect of sending a particular message, and to devise strategies for interaction at runtime.

Definition 4. *Communication Channel*

We assume that a *communication channel* is a one-to-one connection between two agents. The notation $c(i,j)$ denotes the communication channel between the sending agent with identity i, and the receiving agent with identity j, in which identities are represented in the underlying language.

We employ the notation $c(i,j).\phi_m$ to represent the message ϕ_m being sent by agent i to agent j via the channel $c(i,j)$. The event of agent i sending a message to j is the same event as agent j receiving this message. That is, the event is the communication over the channel. Agent identities are omitted when the sending and receiving agents are not relevant; that is, we write $c.\phi_m$.

An alternative way to represent communication between agents is to have many-to-many channels, with the sender and receiver identities as part of the message. However, we choose the first approach so that we can reason about communication in our framework language, rather than mixing this with the underlying constraint language.

Definition 5. $\mathcal{RASA}$ *Protocol*

A $\mathcal{RASA}$ *protocol* is an annotated tree of interactions between entities. An *annotation* is a triplet of constraints, in which the first constraint represents the precondition that must hold for a transition to occur, the second constraint represents the message to be sent, and the third constraint is a postcondition, which must hold after an enabled transition occurs. Branches in the tree represent choices to be made by one or more agents.

Let ϕ represent constraints defined in constraint language, c communication channels, N protocol names, and x a sequence of variables. Protocol definitions adhere to the following grammar.

$$\pi ::= \epsilon \mid \phi \xrightarrow{c.\phi} \phi \mid \pi;\pi \mid \pi \cup \pi \mid N(x) \mid \mathbf{var}_x^\phi \cdot \pi$$

We use π as a meta-variable to refer to protocols; subscripts and superscripts are used to denote distinct meta-variables. ϵ represents the empty protocol, in which no message is sent and there is no change to the protocol state. A protocol of the format $\phi \xrightarrow{c.\phi_m} \phi'$ is an atomic protocol. It represents the value ϕ_m being sent over channel c if ϕ holds in the current state. After the value is sent, the new state of the protocol is updated using ϕ'. We use this to specify rules and effects of protocols: the precondition represents a rule for a protocol because ϕ_m can only be sent if this precondition is true; and the postcondition represents the effect that sending ϕ_m has. For atomic protocols, meta-variables with a prime (') are used to refer to the postconditions; that is, ϕ is the precondition and ϕ' the postcondition. We use ϕ_m (that is, constraints subscripted with m) to denote message constraints.

The protocol $\pi_1; \pi_2$ denotes the sequential composition of two protocols, such that all of protocol π_1 is executed, then protocol π_2. The protocol $\pi_1 \cup \pi_2$ denotes a choice of two protocols. $N(x)$ denotes a reference to a protocol $N(y)$, with variables y renamed to x, such that the referenced protocol is expanded into this protocol. For brevity, we use x and y to represent sequences of variables as well as single variables. Protocols can reference themselves, and can mutually reference each other, which introduces the possibility of non-terminating protocols. The protocol $\mathbf{var}_x^\phi \cdot \pi$ denotes the declaration of a local variable x, with the constraints ϕ on x. The scope of x is limited to the protocol π.

We permit brackets to group together protocols, and to reduce the use of brackets, operators have a strict ordering. The infix operators always bind tighter than variable declaration, with sequential composition binding tighter than choice. Therefore, the protocol $\mathbf{var}_x^\psi \cdot \pi_1; \pi_2 \cup \pi_3$ would be equivalent to $\mathbf{var}_x^\psi \cdot ((\pi_1; \pi_2) \cup \pi_3)$.

Definition 6. *Protocol Specification*

Let N be a name, y be a sequence of variable names, and π be a protocol. A *protocol specification* is defined as a set of definitions of the form

$$N(y) \mathrel{\hat{=}} \pi$$

Definition 7. *Protocol Instance*

Let D be a protocol specification, π a protocol, and ϕ a constraint. A *protocol instance* is a tuple, $\langle D, \pi, \phi \rangle$, in which π can reference the protocol names defined in D, and ϕ is a constraint representing the state of the protocol at that instance. Instances evolve via message sending and the changing of the state.

Example 1. As an example of specifying a protocol, we use a simple negotiation protocol. In this example, a buyer, B, is bidding for an item. If the price that the buyer suggests, $Price$, is greater than the current price in the protocol state, the seller, S, accepts the bid, otherwise rejecting it. In the case that the bid is accepted, $Price$ becomes the new current price. Such a protocol could be a sub-protocol of an English auction protocol, and would be iterated over until no more bids are received, or until some timeout is reached. This protocol is specified as follows:

$Buy(Item, Price, B, S) \cong Bid; AcceptOrReject$

$Bid(Item, PriceB, S) \cong isItem(Item) \xrightarrow{c(B,S).bid(Item,Price)} true$

$AcceptOrReject(Item, Price, B, S) \cong \mathbf{var}_{Curr}^{PriceOf(Item,Curr)} \cdot (Accept \cup Reject)$

$Accept(Item, Price, Curr, B, S) \cong$
$\quad Curr < Price \xrightarrow{c(S,B).accept(PriceOf(Item,Price))} PriceOf(Item, Price)$

$Reject(Item, Price, Curr, B, S) \cong$
$\quad Curr \geq Price \xrightarrow{c(S,B).reject(PriceOf(Item,Price))} true$

So, the bidder sends a bid to the seller. The declaration of the local variable $Curr$ represents the current bid; that is, $Curr$ is equivalent to the value that satisfies $PriceOf(Item, Curr)$ in the state. If the bid is greater than $Curr$, the seller sends an acceptance, and the constraint $PriceOf(Item, Curr)$ is added to the constraint store, overriding any previous constraints on $Item$. If the bid is less than the current price, the bid is rejected and the state remains unchanged. This specification corresponds to the following tree, in which the nodes refer to the consequence of the previous action, and the arcs are of the format $\psi \Longrightarrow c.\phi_m$, interpreted as "if ψ holds, then the transition $c.\phi_m$ can occur." For presentation, we have left out some details that are in the specification.

$$\begin{array}{c} \mid isItem(Item) \Longrightarrow c(B,S).bid \\ \top \\ PriceOf(Item, Curr) \sqcup Curr < Price \diagup \quad \diagdown PriceOf(Item, Curr) \sqcup Curr \geq Price \\ \Longrightarrow c(S,B).accept \qquad \qquad \Longrightarrow c(S,B).reject \\ PriceOf(Item, Price) \qquad \top \end{array}$$

An agent with ID represented by the variable a, wishing to sell an item, i, to another agent b, may propose that this protocol is used to determine a price by proposing the following protocol instance:

$$\langle P, Bid(i, Price, b, a), PriceOf(i, 0) \rangle$$

In which P is the protocol specification above, $Bid(i, Price, b, a)$ is the Bid protocol with renamed variables, and $PriceOf(i, 0)$ is the initial state. The task is now that the agents must find an instantiation for the variable $Price$ on which they both agree.

Clearly, agreeing on a protocol instance from which to begin the negotiation is itself a negotiation problem, which would likely be solvable with another protocol. Such *meta-protocols* are, in the context of a system, at a higher level than other protocols, such as the negotiation protocol above. However, this does not rule out the option of using the same protocol at both levels.

Meta-protocols, their use, and their control are dependent on the system in which they are employed, and on the agents within these systems, so proposing a general solution for this problem is not possible. For example, some agents may adopt a "take it or leave it" approach, in which other participants either use a certain protocol to interact with them, or do not interact with them at all. In other cases, adopting a specific meta-protocol may be a condition of entry into the system. Other systems may leave it up to the agents themselves to decide. In future work, we plan to specify a collection of meta-protocols that can be used to negotiate which $\mathcal{RASA}$ protocol to use, and look at the contexts in which these protocols can be employed.

4 Semantics

In this section, we define and discuss the semantics of the $\mathcal{RASA}$ protocol specification language.

4.1 Structural Operational Semantics

We view the semantics of protocols as commands on a virtual machine, in which states incorporate protocol instances, and the commands are the messages being sent over communication channels. To model these semantics, we make use of *structural operational semantics*, as defined by Plotkin [20].

Using structural operational semantics, a system is defined as a set of transitions, with each transition linking two states. In the case of our protocol semantics, a state is defined as a protocol instance. Recall from Definition 7 that a protocol instance is a tuple $\langle D, \pi, \phi \rangle$, in which D is a protocol specification, π is the protocol that is to be executed, and ϕ a constraint representing the state of the protocol. Thus, a transition takes the form:

$$\langle D, \pi, \phi \rangle \xrightarrow{l} \langle D, \pi', \phi' \rangle$$

This denotes the protocol π being executed in the state ϕ, the transition l occurring at this point. π' denotes the part of the protocol left to execute, and ϕ' denotes the new state of the protocol. l refers to either the communication of a constraint ϕ_m over a channel, written $c.\phi_m$, or the empty transition. We use $\xrightarrow{\varepsilon}$ to represent the empty transition. D is invariant over the course of execution, therefore, we omit it whenever it is not referenced in a transition.

As a shorthand, we use the following to indicate that $\langle D, \pi, \phi \rangle$ evolves to $\langle D, \pi', \phi' \rangle$ over the sequence of transitions $l_1, \ldots, l_n$:

$$\langle D, \pi, \phi \rangle \xrightarrow{l_1, \ldots, l_n} \langle D, \pi', \phi' \rangle$$

We specify the semantics as inference rules of the following form:

$$\frac{antecedents}{conclusion} \quad conditions$$

In which *antecedents* are the assertions about what can occur at the current state, *conditions* are the side conditions under which this rule is enabled, and *conclusion* is the transition that can occur. We use the special protocol, E, to represent the protocol whose execution is complete.

4.2 Renaming

To define the semantics of protocols specified using $\mathcal{RASA}$, we include syntax for renaming over protocols. Specifically, renaming is defined inductively over the structure of the protocols such that the protocol $\pi[x/y]$ represents the protocol π, with every free occurrence of the variable y substituted with the variable x, including variables in the constraints. Formally, renaming is defined as follows, in which I denotes either of the binary infix operators:

$$(\psi \xrightarrow{c.\phi_m} \psi')[x/y] = \psi[x/y] \xrightarrow{c.(\phi_m[x/y])} \psi'[x/y]$$
$$(\pi_1 \ I \ \pi_2)[x/y] = \pi_1[x/y] \ I \ \pi_2[x/y]$$
$$N(z_1, \ldots, y, \ldots, z_n)[x/y] = N(z_1, \ldots, x, \ldots, z_n)$$
$$(\mathbf{var}_y^\psi \cdot \pi)[x/y] = \mathbf{var}_x^{\psi[x/y]} \cdot \pi[x/y]$$
$$(\mathbf{var}_z^\psi \cdot \pi)[x/y] = \mathbf{var}_z^{\psi[x/y]} \cdot \pi[x/y]$$

Informally, this says that renaming y to x in an atomic protocol is equivalent to performing the same rename over the constraints in the protocol. Renaming a infix composition is equivalent to renaming the two sub-protocols of that composition. Renaming y to x in a protocol reference consists of renaming any instances of y in the variable list to x — the name of the protocol is not renamed. Variable declaration has two rules: the first if the declared variable is y, in which case the variable is changed to x; the second if the declared variable is not y, in which case the variable remains the same. In both cases, the constraint on x and the protocol in the scope of the variable are both renamed.

As an example of protocol renaming, we use the negotiation example from Section 3. Suppose that we wish to use this within an auction protocol, with an auctioneer represented by the variable $Auctioneer$, then one could use the renamed protocol $Buy[Auctioneer/S]$ to represent the same protocol, but in which every occurrence of the variable S replaced with $Auctioneer$. Therefore, the message representing a bid on the item would be $c(B, Auctioneer).bid(Item, Price)$.

4.3 Operational Semantics of Protocol Operators

Now we have some basic definitions covered, we define the semantics of the protocol operators in the $\mathcal{RASA}$ language. That is, of executing a protocol within the context of a protocol specification and a constraint representing the protocol state.

Definition 8. *Semantics of the Empty Protocol*

The empty protocol terminates under no transition, and has no effect on the protocol state.

$$\langle \epsilon, \phi \rangle \xrightarrow{\varepsilon} \langle E, \phi \rangle$$

Definition 9. *Semantics of Atomic Protocols*

Firstly, we define the semantics for the atomic protocol. That is, the protocol consisting only of a message being sent over a channel if the precondition is satisfied, resulting in a new protocol state.

$$\frac{}{\langle \psi \xrightarrow{c.\phi_m} \psi', \phi \rangle \xrightarrow{c.\phi'_m} \langle E, \phi' \rangle} \quad \text{if } \phi \sqsupseteq \psi \text{ and } \phi'_m \sqcup \phi' \sqsupseteq \phi_m \sqcup \mathcal{O}(\phi, \psi')$$

This states that, if the precondition ψ is true under the model ϕ, then the transition can occur. This transition can be the constraint, ϕ_m, but can also be a constraint, ϕ'_m, that contains more information that ϕ_m, such that $\phi'_m \sqsupseteq \phi_m$. This allows the message sender to place additional constraints on the message, and, as a consequence, the resulting state. The resulting state is $\mathcal{O}(\phi, \psi')$, in which $\mathcal{O} \in (\mathcal{L} \times \mathcal{L}) \to \mathcal{L}$ is an overriding function defined as $\mathcal{O}(\phi, \psi') = \psi' \sqcup \exists_{free(\psi')}\phi$. Therefore, $\mathcal{O}$ defines a new constraint such that the values of any free variables in ψ' are overridden with the values constrained by ψ', while the free variables in ϕ that are not otherwise in ψ' maintain their pre-state values. However, any additional information in the message, ϕ'_m must also apply to the resulting state. For example, considering the following atomic protocol from the auction example in Section 3:

$$Curr < Price \xrightarrow{c.accept(PriceOf(Item,Price))} PriceOf(Item, Price)$$

The sender confirms that the bid for *Item* at the price *Price* has been accepted, in which *Item* and *Price* are variables. As part of the interaction, the sender would like to instantiate both variables — *Item* with an item, and *Price* with a number. If the sender wants to confirm that the price of the *Item* is 10, then the message will be $c.accept(PriceOf(Item, Price) \sqcup Price = 10)$. The constraint $Price \sqcup 10$ needs to be shared with the postcondition. The semantics enforces this: $\phi'_m \sqcup \phi' \sqsupseteq \phi_m \sqcup \mathcal{O}(\phi, \psi')$. In this example, the only solution for *Price* in this constraint would be $Price = 10$, therefore, the post-state is $PriceOf(Item, Price) \sqcup Price = 10$, which simplifies to $PriceOf(Item, 10)$. Such an approach allows protocol specifications to be general, and then instantiated at runtime.

Definition 10. *Semantics of Sequential composition*

Sequential composition is defined as executing the left-hand protocol until it terminates, and then executing the right-hand protocol until it terminates. The semantics of this is given by two rules.

$$\frac{\langle \pi_1, \phi \rangle \xrightarrow{l_1,\ldots,l_n} \langle E, \phi' \rangle}{\langle \pi_1; \pi_2, \phi \rangle \xrightarrow{l_1,\ldots,l_n} \langle \pi_2, \phi' \rangle} \qquad \frac{}{\langle E; \pi_2, \phi \rangle \xrightarrow{\varepsilon} \langle \pi_2, \phi \rangle}$$

The first of these rules specifies that if sequence of transitions, $l_1, \ldots, l_n$, can be made from $\langle \pi_1, \phi \rangle$ taking the system to the state $\langle E, \phi' \rangle$, then we can perform this transition under the state $\langle \pi_1; \pi_2, \phi \rangle$, leaving us to execute π_2 under the protocol state ϕ'. The second rule specifies that at all times, $E; \pi$ is equivalent to π.

Definition 11. *Semantics of Non-Deterministic Choice*

Non-deterministic choice is defined using two rules.

$$\frac{\langle \pi_1, \phi \rangle \xrightarrow{l_1,\ldots,l_n} \langle E, \phi' \rangle}{\langle \pi_1 \cup \pi_2, \phi \rangle \xrightarrow{l_1,\ldots,l_n} \langle E, \phi' \rangle} \qquad \frac{\langle \pi_2, \phi \rangle \xrightarrow{l_1,\ldots,l_n} \langle E, \phi' \rangle}{\langle \pi_1 \cup \pi_2, \phi \rangle \xrightarrow{l_1,\ldots,l_n} \langle E, \phi' \rangle}$$

These two rules state that, for a protocol $\pi_1 \cup \pi_2$, if one of the arguments can progress, then that argument progresses, and the other is discarded. If both can progress, then a non-deterministic choice is made between the two of them, and the other is discarded. The new protocol state of the argument that is chosen is the new protocol state of the entire transition. The entire protocol terminates when the chosen protocol terminates. A choice between a protocol and the terminated protocol, E, cannot occur because the choice is made before progressing, and because E is not a part of the language syntax.

Definition 12. *Semantics of Protocol References*

$$\frac{}{\langle D, N(x), \phi \rangle \xrightarrow{\varepsilon} \langle D, \pi[x/y], \phi \rangle} \quad \text{if } N(y) \mathrel{\hat=} \pi \in D$$

This rule specifies that, if there is a protocol named N with variables y and protocol π in the set D, then the reference $N(x)$ is equivalent to the protocol π, with the variables y renamed to x.

For example, the $Bid(Item, Price, B, S)$ protocol, from Section 3, can be referenced as with $Bid(i, p, a, b)$, in which $i, p, a,$ and b are variables. This is equivalent to the following:

$$isItem(i) \xrightarrow{c(a,b).bid(i,p)} true$$

In which $Item, Price, B,$ and S are renamed to $i, p, a,$ and b respectively.

Definition 13. *Semantics for Variable Declaration*

A naïve attempt to specify the semantics for variable declaration would give the following.

$$\frac{\langle \pi, \psi \wedge \phi \rangle \xrightarrow{l_1,\ldots,l_n} \langle E, \phi' \rangle}{\langle \mathbf{var}_x^\psi \cdot \pi, \phi \rangle \xrightarrow{l_1,\ldots,l_n} \langle E, \phi' \rangle}$$

The protocol $\mathbf{var}_x^\psi \cdot \pi$ specifies that a new variable x is declared with constraints ψ, and then the protocol π, which may refer to x, is executed. Thus, if the protocol π can progress under the protocol state $\psi \wedge \phi$ to the protocol π' and state ϕ', then make this transition.

We have labelled this definition "naïve" because it does not consider three cases. Firstly, it does not consider that case that x is already a free variable in the state. This will cause problems because the behaviour would be to evaluate π in the protocol state $\psi \wedge \phi$, which could be inconsistent. A designer writing the protocol $\mathbf{var}_x^\psi \cdot \pi$ would surely want any references of x in π to refer to the most recently declared x, therefore, in the antecedent of the rule, x is hidden in ϕ using the cylindric operator, so the constraints of the x in the state are hidden. Secondly, it does not remove the local variable x from the state after execution, meaning that the scope of x is not restricted to π. Finally, it does not maintain the constraints on x over the protocol. A second attempt to specify this rule leads to the following:

$$\frac{\langle \pi, \psi \wedge \exists_x \phi \rangle \xrightarrow{l_1,\ldots,l_n} \langle E, \phi' \rangle}{\langle \mathbf{var}_x^\psi \cdot \pi, \phi \rangle \xrightarrow{l_1,\ldots,l_n} \langle E, \exists_x \phi' \sqcup \exists_{\hat{x}} \phi \rangle} \quad \text{where } \exists_{\hat{x}} \phi = \exists_{\hat{x}} \phi'$$

In this definition, the state $\psi \wedge \exists_x \phi$ has all references to the previously declared x hidden. The side condition, $\exists_{\hat{x}} \phi = \exists_{\hat{x}} \phi'$, says that hiding all variables except x in the pre-state and the post-state will result in the same constraint, therefore, the constraints on x are the same in the post-state and pre-state. Finally, the post-state, $\exists_x \phi' \sqcup \exists_{\hat{x}} \phi$ hides all references to the local variable x in ϕ', and reinstates the global reference by hiding all variables except the global x in the pre-state, ϕ.

Example 2. We turn to an example to help with the understanding of variable declaration, as its definition is not straightforward. Take the following protocol, which is an expanded version of accepting a bid from the example in Section 3:

$$\mathbf{var}_{Curr}^{PriceOf(Item,Curr)} \cdot Curr < Price \xrightarrow{c(S,B).accept(PriceOf(Item,Curr))} PriceOf(Item, Price)$$

If this is executed with the protocol state $PriceOf(Item, 2) \sqcup Curr = 10$, in which $Curr$ is a variable unrelated to the current bid, then the state value of $Curr$ is inconsistent with the constraints on the local variable $Curr$. So, it is executed under the following state (corresponding to $\psi \wedge \exists_x \phi$ in the definition):

$$PriceOf(Item, Curr) \sqcup \exists_{Curr}(PriceOf(Item, 2) \sqcup Curr = 10)$$

This ensures that any reference to $Curr$ will be referring to the local variable name instead of the global name, and will be constrained to the price of *Item*, which is 2. Consider a message in which the buying agent constrains *Price* to be 3; that is, sends the message $PriceOf(Item, Price) \sqcup Price = 3$. After this message, the state would be (corresponding to ϕ' in the definition):

(1) $\quad PriceOf(Item, Price) \sqcup Price = 3 \sqcup$
(2) $\quad \exists_{Item, Price}(PriceOf(Item, Curr) \sqcup PriceOf(Item, 2))$

From the atomic protocol semantics, we keep the constraints for variables not referenced in the post-state by hiding these variables in the pre-state constraint (line 2), and conjoining this with the postcondition (line 1). This constraint can be simplified to the following:

$$PriceOf(Item, 3) \sqcup Price = 3 \sqcup Curr = 2$$

However, the scope of the local declaration $Curr$ ends there, so we want to remove all references to this $Curr$, and reinstate the global $Curr$ with the same constraints it had prior to the local declaration. Therefore, the local $Curr$ in this constraint is hidden, and the resulting constraint conjoined with the pre-state with all variables except $Curr$ hidden (corresponding to $\exists_x \phi' \sqcup \exists_{\hat{x}} \phi$ in the definition):

(1) $\quad\quad\quad \exists_{Item, Price}(PriceOf(Item, 2) \sqcup Curr = 10) \sqcup$
(2) $\quad\quad\quad \exists_{Curr}(PriceOf(Item, 3) \sqcup Price = 3 \sqcup Curr = 2)$

Line 1 is the constraint that reinstates the global $Curr$ with the constraints it had prior to the local declaration. The constraint in Line 2 hides the local variable of $Curr$. This can be simplified to the following:

$$Curr = 10 \sqcup PriceOf(Item, 3) \sqcup Price = 3$$

Which is the expected end state of the protocol. With this simplification, one may wonder why we hide the local reference of $Curr$ rather than just removing all references to it. This is because the postcondition may refer to the local variable, in which case $\exists_{Curr}$ could not be removed, because the constraints on $Curr$ may also constrain other variables, such as $Item$ or $Price$.

5 Reusing, Composing, and Reasoning About Protocols

So far, we have outlined how protocol designers can specify a protocol, and have defined the semantics for a protocol given a protocol specification and an initial state. However, crucial to the goals of agent-oriented software engineering is the fact that first-class interaction protocols should be *reusable* and *composable*, and *inspectable* such that agents can reason about protocols to decide their course of action. How designers and agents reuse, compose, and reason about $\mathcal{RASA}$ protocols is not the topic of this paper, however, in this section, we briefly outline how the $\mathcal{RASA}$ protocol specification language supports these requirements. In future work, we plan to investigate these areas in more detail.

Protocols can be referenced via their name, which allows protocols and protocol specifications to be reused to create larger, compound protocols. For example, the simple negotiation protocol specified in Section 3 could be embedded within an auction protocol. Assuming the existence of protocols called *Start*, *DeclareWinner*, and *NoBids*, which represent the auction starting, a winner being declared, and no bids received before a certain condition is met, such as a timeout, one could specify an English auction as follows, in which Bs is a set of bidders:

$Auction(Item, Price, Bs, S) \triangleq Start; Bids; (DeclareWinner \cup NoWinner)$

$Bids(Item, Price, Bs, S) \triangleq \epsilon \cup ((\mathbf{var}_B^{B \in Bs} \cdot Buy); Bids)$

The protocol *Bids* is zero of more iterations of the *Buy* protocol, in which one bidder from the set of bidders, Bs, submits a bid, and it is either accepted or rejected.

$\mathcal{RASA}$ is well-suited for composition. The syntax and semantics of protocol composition operators treat all protocols the same; that is, atomic protocols are complete protocols themselves. Therefore, the syntax and semantics for constructing protocols from atomic protocols can be used to create compound protocols from other compound protocols. The auction example above is an example of composing new protocols from existing protocols.

Protocol composition need not be restricted to protocol designers. Agents able to reason about protocols specified using $\mathcal{RASA}$ could be equipped, in a straightforward manner, with the ability to compose new protocols from existing protocols using planning techniques.

As an example, consider an intelligent agent, I, that believes it can buy a particular item of a rather unintelligent agent, U, and then sell it back to the same agent at a profit. U may propose that itself and I agent engage in two rounds of the *Buy* protocol from Section 3, but with the buyer and seller swapped in each case:

$$Buy[U/S, I/B]; Buy[I/S, U/B]$$

This composed protocol represents the unintelligent agent acting as the seller, S, and the intelligent agent representing the buyer, B, followed by the same protocol, but with the buyer and seller swapped. Depending on the initial state of the protocol, if I can convince U to engage in this protocol, then I may be able to make a profit. Note that this is different to I proposing $Buy[U/S, I/B]$, and then after this interaction has taken place, proposing $Buy[I/S, U/B]$, because U may not agree to the second protocol, leaving I stuck with the item it does not want. If, however, U agrees to the composed protocol, it is forced to put in a bid for the item if I buys it. This example is fabricated, and would require a highly intelligent agent to devise such a strategy, but should an agent be intelligent enough to exploit this, this example demonstrates that deriving the composite protocol would be straightforward.

There are cases in which protocol composition can lead to problems. For example, take the sequential composition of two protocols: $\pi_1; \pi_2$. For all possible states resulting from the protocol π_1, the precondition from at least one of the paths in π_2 must be enabled, otherwise the execution of the protocol can become *stuck*, in which there are no possible messages that can be sent. To compose this protocol, one must prove that the protocol can never become stuck. Determining such proof obligations, and defining a proof system to help discharge such proof obligations, are part of our ongoing work on the $\mathcal{RASA}$ framework.

Agents can reason about which messages to send by calculating the best course of action at each point in which they can send a message. Each path from the root node to a leaf node represents a possible sequence of interaction. To choose a course of action, classical planning or reactive planning techniques and algorithms can be adapted and used; or more likely, a combination of both. For example, an agent can calculate the end state of all possible interaction sequences of a protocol to decide the next message they send. However, an agent would have to react to changes when it receives a message from another agent, which would likely reduce the choices for its next move.

6 Related Work

There are many languages that have been designed to model agent interactions, such as Social Integrity Constraints [1], FIPA [9], and Agentis [8]. AgentUML [17] has been given a formal semantics for modelling agent interactions [4]. In this section, we discuss and contrast some of the approaches most relevant to the $\mathcal{RASA}$ framework.

Process algebras, such as CSP [11], CCS [15], and the π-calculus [16] are used to model processes and their interactions. While the combination of processes can form the basis of a protocol specification, these languages cannot be used to specify rules and effects. Languages such as Object-Z/CSP [25], which mixes process algebras with state-based languages, are often not inspectable.

Viroli and Ricci [26] propose a method for formalising *operating instructions* for use on mediating coordination artifacts. Sequences of operation instructions resemble our first-class protocols; however their language does not provide the necessary constructs to document the rules or outcomes of protocols. In addition, Viroli and Ricci explicitly comment that their goals are to provide a methodology for environment-based coordination, rather than a general approach to agent interaction semantics.

De Boer *et al.* [6] present a language that uses constraints and process algebra to model agent interactions. However, like many interaction modelling languages, the interaction is emergent from the model of the participants, rather than being first class.

Propositional dynamic logic (PDL) [10] resembles our notion of protocols. For example, PDL allows one to define a collection of sequences of actions, each with an outcome specified as a predicate. In fact, PDL has been extended [19] with belief and intention modal operators to define a language, PDL-BI, for modelling agent interaction. The main differences between these approaches and our approach is that PDL and PDL-BI are declarative languages, while the $\mathcal{RASA}$ language is algebraic; and our language does not require the use of a specific language to model protocol rules and effects. In addition, the class of protocols describable using PDL(-BI) is *regular*, while $\mathcal{RASA}$ allows a larger class; for example, two named protocols with mutually recursive references.

Related work on inspectable protocol specifications also exists in the literature. OWL-P [7] is a language and ontology for modelling protocols, which is coded in the OWL web ontology language [18]. While the approach and goals are different to ours, OWL-P protocols can be used as first-class protocols, and agents would be able to successfully reason about these using OWL tools, and is therefore of interest to us. The syntax and semantics of OWL-P are significantly different to ours, but, like $\mathcal{RASA}$ protocols, OWL-P protocols can be composed. However, unlike $\mathcal{RASA}$, in which composing two protocols has the same syntax and semantics at all levels, OWL-P protocol composition uses a new process with new syntax for composition. We believe this to be a significant advantage of our approach. In addition, OWL-P is not layered; that is, the message sequencing is specified in the same language as the protocol rules and effects — OWL.

This restricts designers to using OWL for specification. We also believe that using a declarative language to specify message sequencing is less intuitive for human readers.

Yolum and Singh [27] present an extension to the Event Calculus [12] that is tailored to first-class protocol specification. The language is declarative, and the authors discuss the use of an abductive planner for agents to plan their execution paths. Although not explicitly discussed by Yolum and Singh, it appears that protocol composition would be possible using this language. This approach is different to ours in the same way that OWL-P is: the language is not layered, and message sequencing is declaratively specified.

Robertson [22] presents the Lightweight Coordination Calculus (LCC). The goals of Robertson are similar to ours, and indeed, one could view the $\mathcal{RASA}$ language as an extended version of LCC, in which we have taken more consideration of the relationship between the protocol specification language and the underlying constraint language by formalising the behaviour of atomic protocols. However, our language differs from LCC in several ways. Firstly, we take a global view of protocols, whereas LCC takes a local view; that is, two interacting agents will each have a specification of their view of the protocol. We believe it would be straightforward to switch between such views in either language. Secondly, our formalisation of atomic protocols treats protocol state differently. Thirdly, we provide a local variable construct, which is useful for defining constraints over sub-protocols, as demonstrated by the example in Section 3. McGinnis [14] has successfully composed LCC protocols at runtime (although McGinnis refers to this as *synthesis*) — a goal clearly in line with our idea of protocol composition.

Serrano *et al.* [24] describe a multi-agent programming framework in which interactions are represented by first-class objects. These objects assert some control over message passing at runtime to guide the interaction. However, this requires the identification of roles at design time, and appears to force participating agents to implement certain interfaces, which we explicitly aim to prevent.

There is also work related to protocol composition in the agent communications literature. McBurney and Parsons [13] propose a formalism for composing dialogue game protocols, which enables similar types of composition, but over a more restricted class of protocols. Reed *et al.* [21] present a framework which allows agents to assign meanings to messages at run-time, and thus, indirectly, to create new interaction protocols.

7 Conclusions and Future Work

In this paper we have presented a novel language for the specification of agent interaction protocols, defining both the syntax and the semantics formally. This language forms part of the $\mathcal{RASA}$ framework, a framework for creating multi-agent interaction protocols as first-class entities. $\mathcal{RASA}$ is general across both the type of interaction protocol and in the language or ontology used by the agents engaged in interaction. By treating interaction protocols as first-class entities, $\mathcal{RASA}$ permits protocols to be inspected, referenced, composed, and shared, by ever-changing collections of agents engaged in interaction. The task of protocol selection and invocation may thus be undertaken by agents rather than agent-designers, acting at run-time rather than at design-time. Frameworks such as this will be necessary to achieve the full vision of multi-agent systems.

Before such visions are realised, significant further work is required. We aim to develop a proof system for the $\mathcal{RASA}$ framework, which, as well as providing a system for designers to verify properties about their protocols, will provide agents with a way to make decisions about their actions, and to verify protocols composed at runtime. Also, further work is needed on the verification of protocols using this proof system, and the development of verifiable semantics for them within this framework. In addition, we plan to specify a collection of meta-protocols for negotiating which protocols to use, and identify in which contexts each meta-protocol would be useful. To develop and test these ideas, we plan a prototype implementation in which agents negotiate the exchange of information using protocols specified using the $\mathcal{RASA}$ framework.

Acknowledgements. We are grateful for financial support from the EC PIPS project (EC-FP6-IST-507019) and the EPSRC Market-Based Control project (GR/T10664/01). We also thank Jarred McGinnis, the anonymous referees and participants at the ESAW 2006 meeting for their comments.

References

1. Alberti, M., Daolio, D., Torroni, P., Gavanelli, M., Lamma, E., Mello, P.: Specification and verification of agent interaction protocols in a logic-based system. In: Haddad, H.M., Omicini, A., Wainwright, R.L. (eds.) Proceedings of the 2004 ACM Symposium on Applied Computing, pp. 72–78. ACM Press, New York (2004)
2. Baader, F., Calvanese, D., McGuinness, D.L., Nardiand, D., Patel-Schneider, P.F.: The Description Logic Handbook: Theory, Implementation, Applications. Cambridge University Press, Cambridge (2003)
3. Blackburn, P., van Benthem, J., Wolter, F.: Handbook of Modal Logic. Elsevier, North Holland (2006)
4. Cabac, L., Moldt, D.: Formal semantics for AUML agent interaction protocol diagrams. In: Odell, J.J., Giorgini, P., Müller, J.P. (eds.) AOSE 2004. LNCS, vol. 3382, pp. 47–61. Springer, Heidelberg (2005)
5. De Boer, F.S., Gabbrielli, M., Marchiori, E., Palamidessi, C.: Proving concurrent constraint programs correct. ACM Transactions on Programming Languages and Systems 19(5), 685–725 (1997)
6. de Boer, F.S., de Vries, W., Meyer, Ch. J.-J., van Eijk, R.M., van der Hoek, W.: Process algebra and constraint programming for modelling interactions in MAS. Applicable Algebra in Engineering, Communication and Computing 16, 113–150 (2005)
7. Desai, N., Mallya, A.U., Chopra, A.K., Singh, M.P.: OWL-P: A methodology for business process modeling and enactment. In: Kolp, M., Bresciani, P., Henderson-Sellers, B., Winikoff, M. (eds.) AOIS 2005. LNCS (LNAI), vol. 3529, pp. 79–94. Springer, Heidelberg (2006)
8. d'Inverno, M., Kinny, D., Luck, M.: Interaction protocols in Agentis. In: Proceedings of the 3rd International Conference on Multi Agent Systems, pp. 112–119. IEEE Press, New York (1998)

9. FIPA: FIPA communicative act library specification (2001)
10. Harel, D., Kozen, D., Tiuryn, J.: Dynamic Logic. MIT Press, Cambridge (2000)
11. Hoare, C.A.R.: Communicating Sequential Processes. Prentice-Hall International (1985)
12. Kowalski, R., Sergot, M.: A logic-based calculus of events. New Generation Computing 4, 67–95 (1986)
13. McBurney, P., Parsons, S.: Games that agents play: A formal framework for dialogues between autonomous agents. Journal of Logic, Language and Information 11(3), 315–334 (2002)
14. McGinnis, J., Robertson, D., Walton, C.: Protocol synthesis with dialogue structure theory. In: Parsons, S., Maudet, N., Moraitis, P., Rahwan, I. (eds.) ArgMAS 2005. LNCS (LNAI), vol. 4049, pp. 199–216. Springer, Heidelberg (2006)
15. Milner, R.: A Calculus of Communicating Systems. Springer, Heidelberg (1980)
16. Milner, R.: Communicating and Mobile Systems: the Pi-Calculus. Cambridge University Press, Cambridge (1999)
17. Odell, J., Van Dyke Parunak, H., Bauer, B.: Extending UML for agents. In: Wagner, G., Lesperance, Y., Yu, E. (eds.) Proceedings of the Agent-Oriented Information Systems Workshop at the 17th National conference on Artificial Intelligence, pp. 3–17 (2000)
18. OWL: Web Ontology Language (February 2004), http://www.w3.org/TR/2004/REC-owl-features-20040210
19. Paurobally, S., Cunningham, J., Jennings, N.R.: A formal framework for agent interaction semantics. In: Proceedings of the 4th International Joint Conference on Autonomous Agents and Multiagent Systems, pp. 91–98. ACM Press, New York (2005)
20. Plotkin, G.: A structural approach to operational semantics. Technical Report DAIMI FN-19, Aarhus University, Computer Science Department (September 1981)
21. Reed, C., Norman, T.J., Jennings, N.R.: Negotiating the semantics of agent communications languages. Computational Intelligence 18(2), 229–252 (2002)
22. Robertson, D.: Multi-agent coordination as distributed logic programming. In: Demoen, B., Lifschitz, V. (eds.) ICLP 2004. LNCS, vol. 3132, pp. 416–430. Springer, Heidelberg (2004)
23. Saraswat, V.A., Rinard, M., Panangaden, P.: Semantic foundations of concurrent constraint programming. In: Proceedings of the 18th ACM Symposium on Principles of Programming Languages, pp. 333–352. ACM Press, New York (1991)
24. Serrano, J.M., Ossowski, S., Saugar, S.: Reusable components for implementing agent interactions. In: Bordini, R.H., Dastani, M., Dix, J., Seghrouchni, A.E.F. (eds.) Programming Multi-Agent Systems. LNCS (LNAI), vol. 3862, pp. 101–119. Springer, Heidelberg (2006)
25. Smith, G., Derrick, J.: Abstract specification in Object-Z and CSP. In: George, C.W., Miao, H. (eds.) ICFEM 2002. LNCS, vol. 2495, pp. 108–119. Springer, Heidelberg (2002)
26. Viroli, M., Ricci, A.: Instructions-based semantics of agent mediated interaction. In: Jennings, N.R., Sierra, C., Sonenberg, L., Tambe, M. (eds.) Third International Joint Conference on Autonomous Agents and Multiagent Systems, pp. 102–109. IEEE Computer Society Press, Los Alamitos (2004)
27. Yolum, P., Singh, M.P.: Reasoning about commitments in the event calculus: An approach for specifying and executing protocols. Annals of Mathematics and AI 42(1–3), 227–253 (2004)

Dynamic Specifications in Norm-Governed Open Computational Societies

Dimosthenis Kaponis and Jeremy Pitt

Intelligent Systems and Networks Group,
Electrical and Electronic Engineering Dept.,
Imperial College London, SW72BT, London, UK
dkaponis@acm.org, j.pitt@imperial.ac.uk

Abstract. A defining characteristic of Open Computational Societies is the unpredictable behaviour of their participants, resulting from their operational and architectural heterogeneity. This has led to the development of computational frameworks that facilitate the declaration of agent specifications in terms of normative relations. The frameworks offer modelling, simulation and validation, but typically have not supported dynamic modification of the specification at runtime by the agents themselves. This omission can be a limitation in certain scenarios, where agents might be capable of adaptation when faced with unexpected stimuli, but the specifications under which they operate did not allow for it. In this paper we extend an existing normative computational framework to facilitate well-defined dynamic normative modification of a specification by the agents themselves, given a well-defined meta-specification. We complement the framework with a mathematical model of the 'specification space'. We argue that the introduced dynamism preserves several of the advantages of static normative frameworks while allowing for more flexible, highly autonomous systems, simpler specification authoring and generic protocol reuse.

1 Introduction

Artikis *et al.* define Open Agent Societies[1][2], as Open Systems[3][4] characterised by the lack of a common goal among the peers and the possibility for architectural heterogeneity, with reasoning capability and processes that are unknown to the designer and the society as a whole. Artikis' work is largely based on prior theoretical work by Santos, Sergot and Jones[5][6], where they present a formal account of the notions of *normative position* and *institutional power* (introduced by Searle in [7]) between two agents.

Artikis *et al.* show that treating Open Agent Societies as instances of norm-governed systems has significant advantages in simulating, modelling and evaluating the performance of such societies. That body of work has been the basis for a range of protocol specifications, including auctions[2], resource sharing[8] and voting[9].

In some types of societies, where the environmental and social charactcristics are particularly volatile and/or unexpected there might be a requirement for very complex specifications in order to cater for a wide range of possible outcomes. Providing a means for well-defined, bounded dynamic modification of the specification at run-time shifts a larger proportion of the decision-making process to the agents themselves.

In this paper we present an extended computational framework for specifying dynamic specifications (meta-specifications) for norm-governed systems, based on the prior work by Artikis et al. The extended framework could be used to:

- Allow for meta-specification authoring containing variable components (numeric parameters, enumerations and composite protocols) modifiable by the agents at runtime within well-defined limits.
- Enable simplified protocol and, subsequently, specification authoring and higher reusability of existing resources through parameterisation.

This work is a first step towards achieving this: the social requirements of Open Computational Societies (OCS) already place considerable weight upon the sophistication of the participating agent architectures. The introduction of social dynamics affecting the operational characteristics of the system significantly enhances the role and importance of the individual agent architecture and capabilities. This is an important aspect of dynamically specified OCS, that while we are aware of, we do not explicitly cover in this paper. We assume the existence of suitable agent architectures, able to socially interact for the purposes of achieving individual and social goals, whether private or social, e.g. those involving the modification of the specification during execution. We assume that those agents are capable of autonomous reasoning and consideration of the changes to the norms their operation might require during execution.

We complement our framework with a brief presentation and discussion of a novel mathematical model based on the theory of Metric Spaces, used in representing the multitude of specifications possible through the use of a dynamic specification for norm-governed systems and encoding designer/evaluator preferences beyond the performance of the specified object protocols. In turn, the model alone, or in conjunction with additional protocol and social metrics, allows for the evaluation and study of a meta-specification as well as the evolution of a society over time and can be used in the design, evaluation and simulation of such societies.

Finally, we briefly discuss the design and implementation of software tools aimed to assist with the evaluation and experimentation of the framework. The methodology and function of those tools can also aid in the design of OCS, although this paper will not focus on this aspect as such. Finally, throughout the paper, we exemplify our work through an existing voting protocol for online deliberative assemblies[9].

The rest of the paper is organised as follows: Section 2, continues with a presentation of the foundations of this work, while Section 3 presents the framework for dynamic specifications. Section 4 discusses a modelling approach to evaluating and designing dynamic norm-governed systems. Section 5 briefly presents preliminary experimentation and implementation aspects. Sections 6 and 7 discuss future work, related work, summarise and present the conclusions drawn from this work.

2 Concepts

2.1 The Event Calculus

The Event Calculus (EC) is an action description language introduced by Kowalski and Sergot[10]. It is based on a many-sorted first-order predicate calculus. *Fluents* in the EC are propositions whose value can change over time. For example, $F = V$ denotes that fluent F has value V. A fluent F takes value V for the period of time after $F = V$ is *initiated* by an action and before it is *terminated* by another. To describe an action in Event Calculus one makes use of the Event Calculus predicates (see Table 1) in axioms that define action occurrences (`happens`) and effects of actions (`initiates` and `terminates`).

Table 1. Main predicates of the Event Calculus

Name	Meaning
holdsAt(F = V, T)	Fluent F is V at time T
happens(Act, T)	Action *Act* happens at time T.
initially(F = V)	Fluent F is V at time 0 (zero)
initiates(Act, F = V, T)	Action *Act* at time T initiates a period of time where the value of fluent F is V
terminates(Act, F = V, T)	Action *Act* at time T terminates a period of time where the value of fluent F is V

We employ the Event Calculus in the formalisation of specifications under our framework. We have chosen this action description formalism as it is fairly well studied, computationally grounded and highly efficient, by virtue of its proximity to first-order logic under a software implementation platform such as Prolog.

2.2 Executable Specifications for Open Computational Societies

Artikis *et al.* specify OCS based on: *social constraints*, expressed in terms of *institutional power*[7], *permission and obligation* and an enforcement strategy (e.g. *sanctions*). The normative positions are associated with agents and social states through *social roles*, whereby each agent observes a role-attribution protocol upon joining the society. Central to the framework is the concept of *objective reasoning*: the process of computing normative positions with respect to social constraints, actions and events. Given the ability to perform objective reasoning and appropriate representations of the communications of a society's members, an *external observer* (or, indeed, a society member) can validate member actions relative to a *predefined* declarative specification of the society, expressed in a suitable action description language, such as the Event Calculus.

For the sake of example, consider a society where a) the population can occasionally be unreachable for extended, yet unknown, periods of time and b) the time it takes for messages to reach agents in the society varies considerably. Let us also assume that those traits, although permeating the society throughout its lifetime, are not necessarily known at design time.

In a 'static' (i.e. invariant during execution) specification of the society, all specified protocols involve hard-wired logic and parameters of the protocol itself. This can, in certain cases, lead to either very low performance or a breakdown of various operational protocols depending on specific agent instances or time-bound responses, and by extension possibly the system as a whole. For example, consider a 'statically formulated' voting protocol, such as the one presented in [9], in a society where, for unknown reasons (e.g. technical difficulties, unexpected environmental conditions), a large part of the population fails to vote within a hard-coded pre-determined *timeout* period. Such a protocol would most probably fail if e.g. it explicitly specified the number of votes cast with respect to the population. Alternatively, it might prove inefficient as would be the case in a society where the population is very high and the hard-coded majority ratio is unattainable. In such cases, the chair of the protocol would be faced with two options, being unable to suggest or perform any modification to the specification itself: either ignore the 'static' specification and override the hard-coded parameters (such as timeout, majority ratio etc.) or comply and repeat the election process repeatedly. In the first case they would be in violation of the protocol and possibly sanctioned or penalised depending on the enforcement strategy. In the second, they would prolong a situation where the election process would probably take a very long time until it produced any effect.

In either case, the partial failure of the specified operational protocols or the — repeated — sanctioning of agents for not following the protocols, would signify the inability of the system specification to cater for the agents' operating environment, despite the agents' capability to adapt and evolve. In the following section we introduce a framework for the specification of dynamic specifications for norm-governed systems that enables well-defined modifications to object-protocols at runtime by the agents themselves.

3 Dynamic Specifications for Norm Governed Systems

We define Dynamic specifications for Norm-Governed systems as a type of specification that, in addition to the operational characteristics of the society, contains rules, procedures and constraints that enable well-defined and constrained runtime modification of its object protocols by agents. In order to enable the authoring of such specifications, we extend Artikis' framework for the specification of OCS. We employ the Event Calculus to formalise the framework and introduce additional fluents and actions at the framework level. This work is inspired by Marín and Sartor's formalisation of External Time of norms presented in [11]. In our framework the modification of specifications is supported through two distinct constructs:

- **Parameters** are numerical variables in the specification that specify some operational characteristics of a norm creation rule, i.e. the maximum time delay between valid agent responses or actions, the number of maximum or minimum participants in a given protocol, the number of seconds required before a motion is considered in a voting protocol and so on.

– **Replaceable Components** are sets of rules that represent the semantics of part or the whole of a protocol and can be replaced at runtime by the society's agents. The existence of Replaceable Components allows for *procedural* modification of a protocol (cf. parametric modification). For example, a voting protocol might contain two separate well-known formulations for determining the winner; one that specifies a Simple Transferrable Vote (STV) and another that specifies a First Past the Post (FPTP). Both could — at different times or for different types of elections — be employed by a society depending on its needs, as is indeed the case in many human societies today.

Meta-specifications can, but are not required to, provide options for both ranges for numeric parameters and all possible alternatives for replaceable components. Indeed, in Section 3.3 we discuss the possibility where more sophisticated societies might employ argumentation protocols and multi-level segmented authority groups to modify these options at runtime.

The extended framework, a superset of the Artikis' specification framework, is presented in Table 2. The following sections present Parameters and Replaceable Components, illustrating the concepts under an Event Calculus formalisation of the voting protocol presented in [9].

Table 2. Fluents and actions of the framework

Name	Type	Description
$valid(ag, act)$	fluent	action act performed by agent ag is valid
$pow(ag, act)$	fluent	agent ag has the power to perform action ag
$per(ag, act)$	fluent	agent ag has the permission to perform action ag
$obl(ag, act)$	fluent	agent ag has the obligation to perform action ag
$role\_of(r, ag)$	fluent	agent ag has the role r
$preconditions(r, ag)$	fluent	agent ag satisfies the preconditions for role r
$assign(r, ag)$	action	assign the role r to agent ag
$parameter(p, val)$	fluent	parameter p has the value val
$rangeOf(p, min, max)$	fluent	parameter p has a range of $[min, max]$.
$setValue(p, val)$	action	value val is assigned to parameter p
$selectComponent(pr, c)$	action	component c is selected in protocol pr
$active(pr, c)$	fluent	component c is active for protocol pr

3.1 Parameters

Numeric parameters are real-valued variables that control aspects of a protocol's procedural characteristics, e.g. the maximum time an agent has to respond to a specific call/request, the number of agents that constitute a majority in a voting protocol, the minimum transmission capability an agent needs to have to be considered for selection as a router in an ad-hoc network etc. They are defined as a 3-tuple specifying the name of the variable and its range: the minimum and maximum values it can take, *parameter(p, min,max)*. In addition to the definition of the parameters, the meta-specification contains:

- Initial values for each parameter
- Rules that make use of the parameter value.
- Rules that modify the parameter value

Empowered agents (e.g. the *chair* of a voting protocol) can set the values of numeric parameters by performing the *SetValue(p,val)* action:

$$\begin{aligned}
&\textbf{pow}(C, setValue(p, V)) = true \text{ holdsat } T \leftarrow \\
&\quad role\_Of(C) = chair \text{ holdsat } T \;\wedge \\
&\quad rangeOf(p) = (Min, Max) \;\wedge \\
&\quad V < Max \;\wedge \\
&\quad V > Min
\end{aligned}$$

Action validation rules, i.e. rules that interpret an action depending on the social state and the acting agent's normative position, employ the parameter values in performing some constraint evaluation.

In the case of an Open Computational Society with a volatile population, such as the one discussed earlier, giving the agents some control over the procedural characteristics of their voting protocol might be desired. For example, agents could be enabled to control (within bounds) the *majority* threshold, i.e. the parameter used in determining whether or not the votes in favour of a motion constitute a majority or not.

The existing formulation of the voting protocol by Pitt et al.[9] embeds this information inside the protocol. In that formulation, the voting protocol's chair has an obligation to announce whether a motion has been carried or not depending on the vote count. This is formalised as follows:

$$\begin{aligned}
&\textbf{obl}(C, declare(C, M, not\_carried)) = true \text{ holdsat } T \leftarrow \\
&\quad role\_of(C, chair) = true \text{ holdsat } T \;\wedge \\
&\quad status(M) = voted \text{ holdsat } T \;\wedge \\
&\quad votes(M) = (F, A) \text{ holdsat } T \;\wedge \\
&\quad F < A
\end{aligned}$$

where F represents votes in favour and A votes against motion M.

While the protocol itself is procedurally *generic*, i.e. it does not describe an application specific process, but could be applied to a number of applications unchanged, the actual formulation is not: the `majority` parameter is hard-coded into the specification. Any modification would require the authoring of a new specification and the recreation of the computational society anew after the specification has been appropriately modified.

In contrast, a dynamic formulation of the rule would be as follows:

$$\mathbf{obl}(C, declare(C, M, not\_carried)) = true \text{ holdsat } T \leftarrow$$
$$role\_of(C, chair) = true \text{ holdsat } T \wedge$$
$$status(M) = voted \text{ holdsat } T \wedge$$
$$votes(M) = (F, A) \text{ holdsat } T \wedge$$
$$parameter(majority, V) = true \text{ holdsat } T \wedge$$
$$(F/(F+A)) < V$$

The addition of the numeric parameter simplifies the reuse of this rule (and the associated rules forming the voting protocol) as it disassociates the *procedure* with the *application-specific* details. In general, numeric parameters modularise protocols by removing application specific information from the protocol itself. While the protocol still contains hard-coded procedural characteristics, requirements and features, it is less likely to contain application-specific details.

3.2 Replaceable Components

Numeric parameters affect some application-specific characteristics of a protocol, but do not affect the structure of the protocol itself. *Replaceable Components*, allow for coarse-grained modifications to the *procedural* characteristics of a protocol through the replacement of its parts by well-defined alternative formulations. Protocols can consist of Replaceable Components (sets of rules) offering different formulations for part (or the whole) of a protocol. The meta-specification might provide a number of possible replaceable components per protocol. Parts of the specification can then be activated or deactivated, dynamically at runtime, through the same norm modification mechanisms employed in modifying numeric parameters.

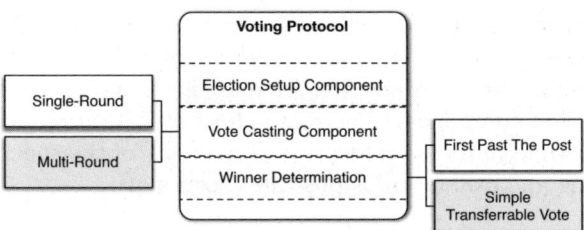

Fig. 1. A Composite Voting Protocol

The voting protocol in our example can be treated as a group of replaceable components, instead of a single monolithic protocol. Figure 1, illustrates how it can be conceptually divided into three components: the *Election Setup Component*, where interaction in announcing an election and announcing a candidacy is specified, the *Vote Casting Component*, where the actual vote casting takes place (e.g. whether a centralised or distributed way for vote gathering is used), and the *Winner Determination Component*, where the rules that determine the winner are specified. A meta-specification can, in addition to numeric parameters, contain one or multiple formulations of each component, thus enabling substitution of one component with an alternative at some point during the lifetime of the society, replacing part of the specification with another. For example, the chair's obligation to declare a motion as *not_carried* in the Simple Transferrable Vote (STV) component would be updated as follows to include a check of the state of the rule:

$$\begin{aligned}
\mathbf{obl}([[voting, stv]], C, declare(C, M, not\_carried)) &= true \text{ holdsat } T \leftarrow \\
active(voting, stv) &= true \text{ holdsat } T \land \\
role\_of(C, chair) &= true \text{ holdsat } T \land \\
status(M) &= voted \text{ holdsat } T \land \\
votes(M) &= (F, A) \text{ holdsat } T \land \\
parameter(majority, V) &= true \text{ holdsat } T \land \\
(F/(F+A)) &< V
\end{aligned}$$

The selection of the active components takes place through any norm modification protocols defined in a meta-specification and completed through the *selectComponent(pr, c)* action. The state of a component, i.e. whether it is active or not at any given moment, is retrieved through the *active(pr,c)* fluent.

Rule Tags. In a software implementation of Replaceable Components in a language such as Prolog, rules specifying a replaceable component would be specified by 'tagging' them with a component and protocol identifier to which they belong. Multiple identifiers for one rule might also be allowed, raising the possibility that a rule is shared between multiple components and/or protocols, although this might significantly affect the readability and clarity of the specification. In this version of the framework, we require that numeric parameters are shared between the replaceable components that refer to them. For example a *timeout* parameter in a vote casting component would need to be referenced and used in all alternative formulations of that component. The reason for this limitation is that we aim for a constant (over time) number of 'degrees of freedom' of the specification so that further modelling and evaluation is possible (Section 4).

3.3 Mechanisms

The actual modification of the specification at runtime is achieved through the following types of mechanisms:

- Role-dependent norms
- Application/State specific norms and
- Generic decision-making protocols, such as argumentation.

In all cases it is assumed that agents are capable of autonomous reasoning based on sensory and social input, and there exists a communications framework capable of secure, error-free transmission. As this paper is not concerned with the implementation details of either agent architectures of application-specific (object) protocols; it is assumed that there exist appropriate specifications defining application specific aspects of the following mechanisms.

Role-dependent norms. These norms are specified through rules taking into account only an agent's role. For example, in an open producer/consumer system, a producer might be *de jure* permitted (or obliged) to initiate a process for re-evaluation of the cost of a service along with the other producers periodically or when it detects an abnormality. This is the typical type of specification found in most statically specified norm-governed systems. Additional dynamics for this class of norms might be introduced through Organisational Self-Design[12] and coordination protocols.

State specific norms. State specific norms refer to cases where all agents might be empowered to perform some modification to the specification in case of exceptional circumstances, as a last resort to avoid system failure. This is tantamount to a passenger being empowered and permitted to open an aircraft door if it has performed an emergency landing and the crew is unable to perform its duty, following an announcement to that effect by the empowered agent (in this case a member of the senior crew on the aircraft); the actual process through which the society is informed about the state is application and implementation specific. Note that there is no role attribution or requirement for the existence of these norms.

Generic Decision-making protocols. Finally, generic decision-making protocols refer to *reusable* decision making, argumentation or negotiation protocols that are further grounded to an application specific context through parameterisation and default values. For example, in more complex societies, an argumentation protocol could be used as the basic meta-protocol employed by the society to modify and adapt object-protocols throughout the system's lifetime.

For example, a multi-role heterogeneous society where the authority with regards to the modification of the object-protocols in place is distributed according to social roles and governed by role-dependent norms, might employ argumentation as a meta-protocol to define the rules for a number of object-protocols employed by the whole population. The object protocols could be application specific, or be in turn lower-level decision making protocols, such as the voting protocol presented earlier.

4 Modelling Dynamic Specifications

In 'static' specifications of norm-governed OCS, system designers and/or evaluators define their expectations of the participating *agents* and complement those expectations with a suitable sanction or enforcement policy. The focus, in these cases, is on the agents conforming to a range of rules and their attributed roles.

The introduction of 'dynamic' specifications, presents the requirement of evaluating how *meta-specifications* perform as a result of the social processes aimed at defining them. In order to be able to model and evaluate how *meta-specifications* perform as a whole, under identical conditions, against each other and against an optimal expectation of the system at or near those conditions, we require a way to measure the performance of a specification over time, or, at a given point in time based on our expectations of the system. Such an evaluation method should ideally provide quantifiable results and be computationally grounded, so that a) comparisons between meta-specifications and protocols would be possible and b) designer performance and operational criteria could be input into a software program, automating the task of evaluating different systems.

In this section we present a mathematical model for dynamic specifications, by considering the *degrees of freedom* of a specification. The model allows semi-automated evaluation of dynamic specifications through the quantification of the operational and performance criteria of the system designer beyond the metrics provided by the individual object-protocols. The model is based on the mathematical concept of Metric spaces[13, p.12].

4.1 The Specification Space

During execution, a meta-specification gives rise to a number of successive 'static' specifications as agents employ meta-protocols in modifying the object-protocols in the meta-specification. In the case of a system employing the framework presented above, these 'static runtime specifications' would differ in their *Numeric Parameters* as well as the procedural characteristics defined by the agents in the form of *Replaceable Components*. We quantify the parameters and enumerate the protocol components so as to create an n-dimensional space consisting of all the *degrees of freedom* of the specification. For example, a specification providing three unbounded real-valued parameters would create the $\mathbb{R}^3$ point set, whereas one with four, the $\mathbb{R}^4$ space and so on. Typically, parameters are bounded by the specification to a suitable, for the application, range. We treat bounded numeric parameters as subsets of $\mathbb{R}$. In contrast, Replaceable Components are treated as enumerations: sets formed by the series of natural numbers starting from 0. We call the point set created by the parameter ranges, and component enumerations, the *Specification Space*.

In general, a Specification Space X, based on $p + c$ degrees of freedom with c Replaceable Component components and p numeric parameters would be defined as follows:

$$X = \underbrace{\{0, 1, \cdots, n_0\} \times \cdots \times \{0, 1, \cdots, n_{c-1}\}}_{c \text{ terms}} \times \quad (1)$$

$$\underbrace{[a_0, b_0] \times \cdots \times [a_{p-1}, b_{p-1}]}_{p \text{ terms}} \quad (2)$$

where $[a_i, b_i] \subset \mathbb{R}, \forall i : i \in \{0, 1, \cdots, p-1\}$.

Each set S specifies the range of values for one dimension (parameter) in the space. A point v in the specification space represents the complete specification at some point in time. It is formally defined as an n-tuple where each 'dimension' variable belongs to the set S_i of each parameter i:

$$v = (x, y, z, ..., n) : x \in S_x, y \in S_y \text{ etc.} \tag{3}$$

The formal definition of a metric would then be:

$$M(k,l) = \sum_{i=0}^{c-1} \frac{|l_i - k_i|}{n_i} + \sum_{i=0}^{p-1} w_i \frac{|l_i - k_i|}{|lmax_i - kmax_i|} \tag{4}$$

The resulting metric is a normalised, weighted Manhattan metric, as the given distance is equivalent to the Manhattan distance between the points in a Euclidean space. The proof for the above generalised metric is omitted due to space limitations, but is immediately derived from the fact that the unidimensional metrics are Euclidean and multiplied by factors in $\mathbb{R}$.

The user then has to specify the weights, w_i for each dimension, where $w_i \in [0, 1]$ and $\sum_i w_i = 1$.

In the context of our example, we will apply the model on the modified voting protocol discussed in the previous section. We will assume three degrees of freedom, two numeric parameters, *majority* (a parameter that specifies what constitutes a majority), that ranges between $\frac{1}{2}$ (50% majority) and $\frac{2}{3}$ (two third's majority) and *timeout*, a parameter that specifies the maximum amount of time agents are allowed before voting and that ranges between 3 and 100 time units. The third degree of freedom comes from two alternative components for the Winner determination: First Past The Post (FPTP) and Single Transferrable Vote (STV). The generated space for these parameters would then be:

$$Sv = [3, 100] \times [\frac{1}{2}, \frac{2}{3}] \times \{0, 1\} \tag{5}$$

At each time instance of this society, the active specification will occupy one point v in the Specification Space Sv where v is defined as:

$$v = (x, y, z) : x \in [3, 100], y \in [\frac{1}{2}, \frac{2}{3}], z \in \{0, 1\} \tag{6}$$

The metric would be as follows:

$$d(a,b) = w_x \frac{|x_b - x_a|}{|100 - 3|} + w_y \frac{|y_b - y_a|}{|\frac{2}{3} - \frac{1}{2}|} + w_z \frac{|z_b - z_a|}{1} \tag{7}$$

4.2 The Metric

The choice of a *metric* is a fundamental part of the model as it encodes part of the designer criteria. In the example above, a weighted normalised Manhattan metric was used for simplicity. The 'importance' of each dimension in this metric is specified through weights for each dimension w_x, w_y, w_z.

In the general case, the selection of the metric can be specified as part of the evaluation process, the specific designer preferences or other organisational guidelines. While the selection of a metric is left to the designer/evaluator of a system, identical metrics are required when comparing the performance between multiple system-specification pairs. For this reason, metrics may be defined at an organisational or standard authoring level and not at the individual designer level. Any function would need to satisfy the four conditions defining a metric (non-negativity, identity, symmetry and the triangle inequality).

4.3 Model-Based Specification Evaluation and Design

Evaluation. Having applied the model to a single, or a number of scenarios, we can perform a series of evaluation tests that can help extract behavioural characteristics related to the meta-specification, both within the context of the application and in terms of the subjective designer criteria as defined by the metric and preferred points (or regions) in the specification space.

The methodology involved in acquiring an evaluating the performance of a society involves the creation of one or more scenarios that expose it to a social or environmental stimulus that exceeds the normal operating stimuli expected by the system specification.

The evaluation methodology involves the acquisition of statistical information about the normative evolution of the specification near or at that point in time, the proximity of the specification in this period . This can be compared with the *a priori* designer expectations of the system, in a way so that specifications that do not conform can be isolated and discarded.

These expectations are encoded in both the metric and a number of points (or regions) in the specification space of significance to the designer. The performance evaluation involves the comparison of the point of the specification at each time point in the specification space with the point of a predefined point or subspace representing the designer's expectations. In the case of the voting example, the unexpected environmental stimulus was the difficulty in communicating due to technical error and the numeric parameter was the *timeout* period between votes.

Statistical analysis through the use of the model provides an evaluation of how the system caters to the incident. In addition to determining whether a specification converges to a desired subspace or point, the metric can provide a series of other statistical information such as the average (distance) for a time-series of points, whether the specification converges to given points in the specification space (and if so, the time it takes to achieve convergence), or the time it takes for the specification to reach a minimum (threshold) distance to a point. In more complex dynamic specifications (i.e. specifications with higher degrees of freedom), the iterative application of the model can correlate aspects of specifications with externally apparent behaviour/phenomena.

In addition to time-series statistics, the overall characteristics of a meta-specification can be extracted through a scatter of points in the specification space under varying operating conditions.

Design. The use of the model, through the definition of a metric and points in the specification space, provides a quantifiable measure of designer criteria not necessarily directly related to the performance or characteristics of one or more object protocols in a society. Such criteria might be related to performance, security or other social traits that an evaluator of a system desires (or wishes to avoid) the society to possess. As such, the model could provide additional information on complex reusable protocols during the design phase. Specifically, a series of patterns between agent and social characteristics and points in their respective specification spaces could be identified and required by designers automatically discarding a number of possible protocols that do not match their requirements. Basic characteristics of an agent architecture might be indicated by the size and 'shape' of the Specification Space in each instance or by the position of the system specification in the space under specific and known social and environmental conditions. The storage and further use of such meta-data about specifications (or protocols) could result in a specification characteristics library that would provide additional insight into the traits of a protocol or specification to designers of such systems beyond its obvious operational characteristics, through its spatial characteristics in the specification space.

5 Experimentation and Implementation

In the context of evaluating and grounding the theoretical work presented earlier, we are developing a suite of software tools (see Figure 2) that assists in the simulation and evaluation of the framework. These are broadly divided into:

- A Simulation Platform
- A Specification Visualiser

5.1 The Simulation Platform

The simulation platform consists of a number of software programs aimed at executing and recording the norm modification events (the 'narrative') of a simulation of executing agents under a well-defined meta-specification. The simulation platform can accept input either from a manually provided action list (script) or through the use of one or more agent architectures. The operator can pause and step the simulation as well as create social or environmental events (stimuli) at specific time points. These are aimed at creating the conditions for the observation of specific social traits. Observed events are recorded in the narrative and include both the norm modification events and — possibly — centralised or distributed performance metrics. For example, in the case of a voting protocol, the election chair agent would be in a unique position to provide statistics

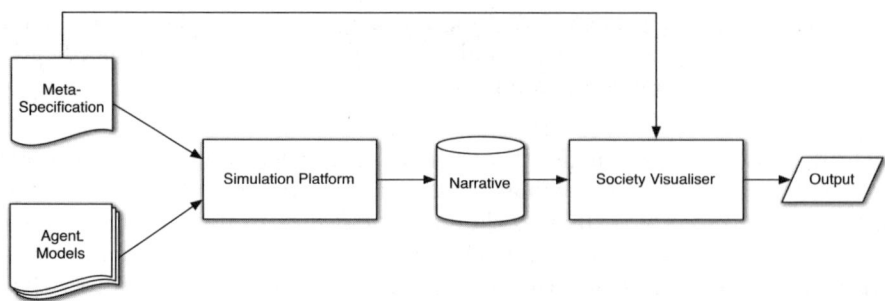

Fig. 2. Illustration of the main flow of the processing by the simulation platform

about the number of voters that abstained or missed the timeout; this would be an instance of a centralised object-protocol performance metric. Both the object-protocol narrative and the meta-protocol narrative are combined with object-protocol data and recorded for later analysis.

5.2 Specification Visualiser

The Specification Visualiser is the main evaluation application implemented in the scope of this work. Its inputs are the meta-specification expressed in Event Calculus, the combined narrative produced by a simulator run, the scenario script containing the time points of interest and a list of performance criteria as well as the designer metric. It subsequently produces a time-series analysis of the social state as well as the specification space defined by the meta-specification along with the trail generated by consecutive specification changes.

By applying the metric space model to the information pool at each time point, a correlation of the specification point and performance metrics of the object-protocols is possible. In addition comparison of the actual specification to designer provided ideal/desired points or regions in that space is also achieved. The specification space can be visualised in 2D or 3D space. The user can inspect the generated specification space, the point scatter (an indication of the regions of the specification where the system resided during the run) and the trail of points over time and visually define reference points in the space for every 2 or 3 degrees of freedom (2D or 3D projection).

The combination of the specification space along with observable (or unobservable in the case of simulation) protocol or system-wide performance metrics allows for the automated evaluation and correlation of specifications at given points in the simulation time. For example, the operator might specify evaluation criteria based on the specification space model — such as the distance from a point in that space or sub-space in the specification space, or the position on a number of dimensions — and allow the computer to automatically evaluate narratives from a number of meta-specification experiments and highlight those that were within the desired parameters. That might translate to convergence to a point or sub-space (or the time-to-converge to a point or sub-space in the specification space), a minimum distance from a point or subspace, etc.

The information returned can then be correlated with the object-protocol performance of the system (such as the number of bytes transferred in a wireless network or the number of votes actually cast in an election). The operator can set the importance for each performance criterion through weighting according to preferences.

The Specification Visualiser provides visual as well as numerical feedback on the performance of the system based on the application of the model (metric, regions and/or points in space). It also presents the object-protocol performance metrics as collected from the population during the simulation allowing for correlation with the position of the system in the specification space and/or individual specification dimensions.

6 Related and Future Work

6.1 Related Work

There is extensive work on the dynamics in Open Computational Societies that bears some similarity to the work presented in this paper. There is also related work in other subfields of Distributed Artificial Intelligence, such as Coordination, Emergence and Learning that shares some characteristics of this work.

Focusing on the context of norm-governed systems, Grossi and Dignum in [14] presenta related approach whereby norms are classified into Abstract and Concrete and the transition between the two is called the *operationalisation of norms*. One of the main examples provided by Grossi concerns a part of the Dutch regulation with regards to the treatment of personal data. The point of concern between abstract and concrete lies in the definition of 'personal data'. In their example, Grossi and Dignum focus on the variability of the meaning of 'personal data' in varying circumstances and highlight the significance of the ontological realm in each circumstance. They proceed to define translation rules between the two ontologies. The work presented here does not observe this distinction: the division of norms into two, largely abstract categories seems artificial and unjustified as the bounds of abstraction are not clearly defined and largely depend on the ontology used. However, this work provides a similar translation mechanism between the 'generic' and the 'specific', through parameterisation and composite protocol modification rules. The generic protocols ('abstract') are translated into application specific protocols ('concrete') by virtue of their operational and functional parameterisation. As such, in the case of the 'personal data' example, this framework would support both the 'abstract norm' (i.e. the obligation to protect personal data) and through parameterisation it would support application specific definitions thereof. Grossi *et al.*[14] also discusses the need for support for exceptional interpretations, under specific circumstances. Again, in the context of this work, this could be achieved through the group-segmentation of authority, as described earlier, whereby a group of agents would be empowered (and potentially obliged) to negotiate changes to the definition of 'personal data' in well-defined circumstances. Finally, the translation between the 'generic' and 'application-specific' norm takes place through the parameterisation defined in the meta-specification. The runtime modification of norms is also discussed in [15]. Excelente-Toledo and Jennings' dynamic selection of coordination mechanisms presented in [16] is similar to 'replaceable components' in this work.

Boella *et al.* present the concept of meta-norms which is similar to the group-segmentation of specification modification powers and permissions discussed in this paper. Specifically, in [17], the notion of *meta-norms* is introduced whereby '[t]he meta-norms of the system ascribe to each level of authority an area of competence (a set of propositions they can permit or forbid and prescribe that the system must respect normative principles like "lex superior derogat inferiori" [)]'. This is similar to the structure presented in Section 3.3 where groups of agents defined by specific roles have the authority to modify social norms in part of the specification, based on their role and through the invocation of an accepted protocol.

There are some similarities in the work on dynamic reconfiguration. Reconfiguration entails the assumption of different roles by agents at runtime, although the organisational structure remains constant[18]. Dynamic reconfiguration might be within the provisions of a 'static' specification for norm-governed systems as it does not entail modification of the specification during run-time. Indeed, most 'static' specification approaches for norm-governed systems include role attribution protocols to agents depending on performance, behaviour and/or conformance to the specification etc. While similar in that both approaches provide a certain level of dynamics in the operation of the system, dynamic reconfiguration does not entail modification of the structure or content of the specification, but attempts to maintain the operation of the system through reconfiguration of the resources available to it to match that pre-defined specification.

Martin and Barber present a somewhat similar premise to the one adopted by the authors of this work and continue to define an adaptive approach for (homogeneous) agents that focuses on authority relationships between agents and adaptation of a decision making framework (DMF) by the agents to suit their needs in [19]. In that work, the society is segmented by virtue of *decision makers*, agents that make decisions for sub-goals taken up by lower-level agents incapable to adapt. Martin and Barber employ voting as their meta-protocol.

Finally, there are considerable similarities between the dynamic norm-governed approach presented here with policy-based agent oriented software frameworks, such as KAoS[20][21]. Conversation policies in the KAoS architecture resemble 'protocols' in terms of actions and responses (messages) between agents. For example, both approaches employ roles as the first-order mechanism for control, and allow for sequences of complex interactions, as opposed to single actions. The KAoS-type conversation policy lies close to the static norm-governed (off-line) specification design, such as that this work is based on, rather than the, more complex, decoupled (generic vs. application specific) dynamic version described here that introduced variability in the 'conversations', given enough time and engaging stimuli. While more recent KAoS developments support dynamic runtime policy changes, those are performed by entities external to the agent society — i.e. performed by a human system administrator/system designer[22].

6.2 Future Work

The preliminary work presented in this paper is further developed in several ways:

- First, through the application of the framework and model to grounded examples that will provide a basis for comparison between different formalisations as well as demonstrate the capabilities of the model. The application of both framework and model will provide sufficient data for further development of both the model and the evaluation methodology.
- Second, through the study of negotiation/argumentation protocols in more dynamic (multi-level) specifications, such as those discussed by Brewka in [23].
- Third, through the further development of software tools for the simulation and analysis of systems specified under this (or a related) framework and the presented model.

7 Discussion

Specifications for norm-governed open agent societies describe how agents ought to behave, what their rights and permissions are and, in some cases, how the society deals with deviations from the norms. Agent societies can be treated as instances of norm-governed systems and it has been shown that such an approach has several distinct advantages: it allows for the application of complex social and legal theories to multiple systems without the requirement for translation to low-level descriptions. It can also be computationally grounded, assuming a suitable formal representation is used; this in turn allows for society-wide simulations and validation of the system output.

This paper presented a framework for specifying OCS with basic support for constrained dynamic specifications (meta-specifications). The framework provides a basis for the dynamic modification of the system specification by the agents themselves at runtime, in response to social and environmental stimuli. The realisation of such a system encompasses several additional unanswered research and implementation questions that we have not covered. Those, largely pertain to the additional requirements placed upon a) the agent architectures, b) host and technical environment and c) agent coordination and interaction. Throughout this paper we assumed the existence of such advanced agent architectures operating in a suitable technical environment that translates the communication, social and computational characteristics defined in and required by the framework.

We argued that the ability to modify the specification at runtime is advantageous at two levels: at the design level, as it simplifies the authoring and evaluation of specifications by separating operational characteristics and generic protocols through numeric parameters and modifiable Replaceable Components and at the operational level, as it allows the participating agents to fully leverage their reasoning, communication and sensory facilities to determine how the system should perform optimally under changing environmental and social conditions through well-defined specification changes. We complemented our framework with a mathematical model of the 'specification space', mapping the degrees of freedom defined in the specification and allowing for statistical and quantifiable evaluation of dynamic specification performance. We presented how the model can be used in evaluating and designing such systems, through simulation, observation and benchmarking of known operating parameters/metrics. Finally, we provided a brief overview of the software tools being designed and implemented towards simulating and evaluating computational societies of this kind.

References

1. Artikis, A., Sergot, M., Pitt, J.: An executable specification of an argumentation protocol. In: Proceedings of International Conference on Artificial Intelligence and Law (ICAIL), pp. 1–11. ACM Press, New York (2003)
2. Artikis, A.: Executable Specification of Open Norm-Governed Computational Systems. PhD thesis, University of London (2003)
3. Hewitt, C., de Jong, P.: Open systems. In: Brodie, M.L., Mylopoulos, J., Schmidt, J.W. (eds.) On Conceptual Modelling: Perspectives from Artificial Intelligence, Databases, and Programming Languages, pp. 147–164. Springer, Heidelberg (1984)
4. Hewitt, C., de Jong, P.: Open systems. Technical report, Massachusetts Institute of Technology (1982)
5. Jones, A., Sergot, M.: On the characterization of law and computer systems: The normative systems perspective. In: Meyer, J.J., Wieringa, R. (eds.) Deontic Logic in Computer Science. John Wiley and Sons, West Sussex, England (1993)
6. Santos, F., Carmo, J.: A deontic logic representation of contractual obligations. In: Meyer, J.J., Wieringa, R. (eds.) Deontic Logic in Computer Science. John Wiley and Sons, West Sussex, England (1993)
7. Searle, J.: Speech Acts: An Essay in the Philosophy of Language. Cambridge University Press, Cambridge (1969)
8. Artikis, A., Kamara, L., Pitt, J., Sergot, M.: A protocol for resource sharing in norm-governed ad hoc networks. In: Leite, J.A., Omicini, A., Torroni, P., Yolum, p. (eds.) DALT 2004. LNCS (LNAI), vol. 3476, pp. 221–238. Springer, Heidelberg (2005)
9. Pitt, J., Kamara, L., Sergot, M., Artikis, A.: Voting in online deliberative assemblies. In: Gardner, A., Sartor, G. (eds.) Proceedings 10th ICAIL, pp. 195–204. ACM Press, New York (2005)
10. Kowalski, R., Sergot, M.: A logic-based calculus of events. New Generation Computing 4, 67–96 (1986)
11. Marín, R., Sartor, G.: Time and norms: a formalisation in the event calculus. In: Proceedings of Conference on Artificial Intelligence and Law (ICAIL), pp. 90–100. ACM Press, New York (1999)

12. Kamboj, S., Decker, K.S.: Organizational self-design in semi-dynamic environments. In: Proceedings of the Fifth International Joint Conference on Autonomous Agents and Multiagent Systems (2006)
13. Bryant, V.: Metric Spaces. Cambridge University Press, Cambridge (1985)
14. Grossi, D., Dignum, F.: From abstract to concrete norms in agent institutions. In: Hinchey, M.G., Rash, J.L., Truszkowski, W.F., Rouff, C.A. (eds.) FAABS 2004. LNCS (LNAI), vol. 3228, pp. 12–29. Springer, Heidelberg (2004)
15. Lopez, F., Lopez, y., Luck, M.: Towards a model of the dynamics of normative multi-agent systems (2002)
16. Excelente-toledo, C.B., Jennings, N.R.: The dynamic selection of coordination mechanisms. Autonomous Agents and Multi-Agent Systems 9, 55–85 (2004)
17. Boella, G., van der Torre, L.: Permissions and obligations in hierarchical normative systems. In: Procs. of ICAIL 03, Edinburgh, pp. 109–118. ACM Press, New York (2003)
18. pa So, Y., Durfee, E.H.: Designing organizations for computational agents. MIT Press, Cambridge, MA, USA (1998)
19. Martin, C., Barber, K.S.: Adaptive decision-making frameworks for dynamic multi-agent organizational change. Autonomous Agents and Multi-Agent Systems 13, 391–428 (2006)
20. Bradshaw, J.M., Dutfield, S., Carpenter, B., Jeffers, R., Robinson, T.: KAoS: A Generic Agent Architecture for Aerospace Applications. In: Finin, T., Mayfield, J. (eds.) Proceedings of the CIKM '95 Workshop on Intelligent Information Agents, Baltimore, Maryland (1995)
21. Bradshaw, J., Dutfield, S., Benoit, P., Woolley, J.: Kaos: Toward an industrial-strength open agent architecture. In: Bradshaw, J. (ed.) Software Agents, pp. 375–418. AAAI Press, Stanford, California, USA (1997)
22. Bradshaw, J., Uszok, A., Jeffers, R., Suri, N., Hayes, P., Burstein, M., Acquisti, A., Benyo, B., Breedy, M., Carvalho, M., Diller, D., Johnson, M., Kulkarni, S., Lott, J., Sierhuis, M., Hoof, R.V.: Representation and reasoning for daml-based policy and domain services in kaos and nomads. In: AAMAS '03: Proceedings of the second international joint conference on Autonomous agents and multiagent systems, pp. 835–842. ACM Press, New York (2003)
23. Brewka, G.: Dynamic Argument Systems: A Formal Model of Argumentation Processes Based on Situation Calculus. J. Logic Computation 11, 257–282 (2001)

Enhancing Self-organising Emergent Systems Design with Simulation

Carole Bernon, Marie-Pierre Gleizes, and Gauthier Picard

IRIT – Paul Sabatier University – 118, Route de Narbonne
31062 Toulouse, Cedex 9, France
Tel.: +33 561 558 294
{bernon,gleizes,picard}@irit.fr

Abstract. Nowadays, challenge is to design complex systems that evolve in changing environments. Multi-agent systems (MAS) are an answer to implement them and many agent-oriented methodologies are proposed to guide designers. Self-organisation is a promising paradigm to make these systems adaptive: the collective function arises from the local interactions and the system design becomes thus bottom-up. The difficulty rests then in finding the right behaviours at the agent-level to make the adequate global function emerge. The aim of this paper is to show how simulation can help designers to find these correct behaviours during the design stage: by simulating a simplified system and observing it during execution, a designer can modify and improve the behaviour of agents. A model of cooperative agents was implemented under the SeSAm platform in order to be integrated into ADELFE, an agent-oriented methodology dedicated to adaptive MAS (AMAS). This model is described here and applied to show how the behaviour of a simple ecosystem can be improved.

1 Introduction

In the last few years, use of computers has spectacularly grown and classical software development methods run into numerous difficulties. The classical approach, by decomposition into modules and total control, cannot guarantee the functionality of software given the complexity of interactions between the increasing and varying number of modules, and the huge size of possibilities. In addition to this, the now massive and inevitable use of network resources and distribution increases the difficulties of design, stability and maintenance. Moreover, we use applications and systems that are plunged into evolving environments and that are more and more complex to build. In such a situation, having an *a priori* known algorithm is not always possible and classical algorithmic approaches are no longer valid. The challenge is to find new approaches to design these new systems by taking into account the increasing complexity and the fact that reliable and robust systems are wanted.

For this, and because of the similarities, it seems opportune to look at natural systems – biological, physical or sociological – from an artificial system builder's point of view, so as to understand the mechanisms and processes which enable their functioning. In Biology for example, a lot of natural systems composed of autonomous individuals exhibit aptitudes to carry out complex tasks without any global control. Moreover, they can adapt to their surroundings either for survival needs or for improving the functioning of the collectivity. This is the case, for example, in colonies of social insects [4] such as termites and ants [6]. The study of swarm behaviours by migratory birds or fish shoals also shows that the collective task comes out of the interactions between autonomous individuals. Non-supervised phenomena resulting from the activity of a huge number of individuals can also be observed in human activities such as the synchronisation of clapping in a crowd or traffic jams.

To get rid of this too complex global function, a solution is to build artificial systems for which the observed collective activity is not described in any part composing it but emerges from interactions between these parts. The common factor among all these systems lies in the emergent dimension of the observed behaviour. The two main properties of these systems are: the irreducibility of macro-theories to micro-theories [1] and the self-organising mechanisms which are the origin of adaptation and appearance of new emergent properties [17]. So, self-organisation seems to be a useful promising paradigm to design these systems.

As underlined by [27] using simulation tools to help MAS designers is still a challenge and the aim of this article is to study how simulation may be a help at the design stage of a self-organising system. Firstly the issue of designing self-organising systems through adaptive multi-agent systems and the use of the agent-oriented methodology ADELFE [3] is considered (Section 2). Once the proposed approach positioned relating to existing works (Section 3), a model of cooperative agent is described and implemented under SeSAm (Section 4). For the time being, in this paper, this model is focused on systems in which interactions are based on the environment, that is why this model is then applied to improve the behaviour of a simulated ecosystem in which agents are interacting in this way (Section 5).

2 Self-Organising Systems Design

The work presented in the paper consists in the enhancement of the existing ADELFE methodology, based on the AMAS theory and dedicated to the design of self-organising systems.

2.1 Self-Organising Systems and AMAS Theory

Self-organisation is a promising paradigm to take into account complexity and openness of systems and to reduce the complexity of the engineering process. A system is self-organising if it is able to change its internal organisation without explicit external control during its execution time [20], [12].

For developing these systems, designers have to define agents and the interactions between those agents, then the process achieved by the system results from the agents behaviour and interactions, making the global behaviour emerge. Consequently, the design is a bottom-up one and the difficulty is to find the right behaviour at the micro-level (i.e. agent level), in order to obtain a coherent, or adequate, function or behaviour at the macro-level (i.e. global system level).

The AMAS (Adaptive Multi-Agent Systems) theory provides a guide to design self-organising systems [7]. It is based on the observation that modifying interactions between the agents of the system modifies also the global function and makes the system adapt to changes in its environment. Agents have to be endowed of properties that make them locally change these interactions. According to the AMAS theory, interactions between agents depend on their local view and on their ability to "cooperate" with each other. Every internal part of the system (agent) pursues an individual objective and interacts with agents it knows by respecting cooperative techniques which lead to avoid Non Cooperative Situations (NCS) like conflict, concurrence etc. Faced with a NCS, a cooperative agent acts to come back to a cooperative state and permanently adapts itself to unpredictable situations while learning on others. An agent has a double role: an anticipative role and a repairing one. Roughly, applying the AMAS theory consists in enumerating, according to the current problem to solve, all the cooperation failures that can appear during the system functioning and then defining the actions the system must apply to come back to a cooperative state. However, building system enforcing this theory is not so easy and an agent-oriented methodology ADELFE was proposed to help designers.

2.2 ADELFE Methodology

ADELFE is an agent-oriented methodology dedicated to self-organising systems design and based on the AMAS theory [24], [3]. Its process is based on the RUP (Rational Unified Process) [21], modified to take into account characteristics of these systems, especially those concerning cooperation. The agent design and the interactions between agents and the environment form the essential steps of the design. The main points where non cooperative situations are detected and taken into account are located in the preliminary requirement phase and in the design phase.

Actually, during the preliminary requirements phase, designers must begin to think about the situations that can be "unexpected" or "harmful" for the system because these situations can lead to NCS at the agent level. ADELFE provides tools to express this in the use case diagrams.

During the design phase, it is possible to find if some deadlocks can take place within an interaction protocol, or if some protocols are useless or inconsistent. The protocol diagram notation has been extended to express these situations. Thus, the behaviour of several agents could be judged in accordance (or not) with the sequence diagrams described in the analysis phase. Then, the agent is defined and at this level of development, designers have to think about the different parts composing the behaviour of agents: what an agent knows about the domain enabling it to take actions (skills), how it reasons on this knowledge (aptitudes), how it communicates with others (interaction languages), how it represents its environment (representations) and most of all, how it stays cooperative towards others and itself (NCS) [7]. Once established, the behaviour of agents can be tested using the fast prototyping activity of the ADELFE design process.

However, the difficulty here for designers is twofold:
- On the one hand, all non cooperative situations have to be exhaustively found, and every agent provided with the right actions to come back to a cooperative situation,
- On the other hand, all possible actions have to be found for one situation, and every agent provided with a decision procedure to always choose the action which can be qualified as the best cooperative one.

It is the reason why, we propose to enhance ADELFE by using simulation during the design phase.

3 Simulation to Enhance Software Design

In this section, after reviewing other works using simulation for designing self-organising multi-agent systems, how improving the design phase of ADELFE through simulation is expounded.

3.1 Related Work

Simulation was already used in the development of agent-based systems or self-organising multi-agent systems.

Gardelli et al. assess that simulation could provide a substantial added-value when applied to support the development process of self-organising systems [15]. Simulation is used to detect abnormal agents' behaviours in a system in order to improve security. Authors took inspiration from the human immune system and exploit Pi-Calculus in the TucSoN infrastructure for simulating the security system. Unlike the aim of the work proposed here, the use of formal simulation is intended to let designers detect abnormal behaviours at the early stages of design before implementing any prototype.

In [25], Röhl and Uhrmacher propose a modelling and simulation framework called JAMES, based on the discrete event formalism DYNDEVS for supporting the development process of multi-agent systems. Agents' behaviour is validated on a model in a virtual environment and the real implementation starts when the model is mature.

Sierra et al. develop an Integrated Development Environment to design e-institutions [26]. Within this environment, a simulation tool SIMDEI is used to dynamically verify the specifications and the protocols to be implemented.

Fortino et al. propose a simulation–driven development process [13], [14] and enrich the PASSI (Process for Agent Societies Specification and Implementation) methodology with the simulation tool called MASSIMO (Multi-Agent System SIMulation framewOrk), a Java-based discrete event simulation framework [8]. Authors propose a semi-automatic translation of the agent implementation model provided by PASSI in the distilled statecharts specification of a multi-agent system needed by MASSIMO. At present, the simulation is used for validating the requirements of the system and evaluating some performances.

In the three last works, the simulation is more used to analyse the running system and to verify the agents and system behaviours. In our work, the simulation is also used to verify the global behaviour but the main objective is to help a designer to adjust and to build the agents behaviour.

De Wolf et al. combine agent-based simulations with scientific numerical algorithms for dynamical systems design [11]. In this approach, designers have to define the results that are expected from the analysis, the parameters of the simulation are then initialised and simulations are launched. Finally, results of the simulation are analysed and depending on the outcomes, the next initial values are determined. The analysis approach is integrated into the engineering process in order to achieve a systematic approach for building self-organising emergent systems.

3.2 Simulation to Improve the Design Phase of ADELFE

As already noticed, the difficulty for a designer in developing a self-organising system based on AMAS is to exhaustively find all non cooperative situations to obtain a coherent global behaviour. In order to help a designer to find all these situations, our idea is to enable him to observe a running system. Two main classes of multi-agent systems exist: the communicative ones, in which agents communicate by sending messages and the situated ones in which agents move in an environment and generally, communicate via the environment. As displayed by the variety of applications already implemented [7], the AMAS theory is able to deal with both types of MAS; however, in this paper and as a preliminary step, we chose to focus only on situated multi-agent systems. The main reason for such a choice was that the observation of the behaviour in an environment is implicitly associated with these systems.

By observing the system while it "lives", a designer can be aware of problems (NCS) that occur in the system and then, in order to solve them, means are provided to him for modifying the behaviour of agents. It is what we call the "living design" [16] which enables to design agents while the system is simulated. In this case, an agent can be partially designed and its capabilities of actions and reactions can be progressively improved by developers. The aim of this study is that developers design the behaviours of agents on a prototype or a simplified system with, for example, a simplified environment or a virtual one. Then when an agent behaviour becomes mature, it is validated. Eventually, once this task completed, the real system has to be implemented.

Observation of a Running System. Since simulation of the collective behaviour of agents acting in a n environment is the best means to observe it, a simulation platform has to be chosen. Our objective is not to build a new platform, but to find a platform enabling to easily modify the behaviour of an agent and to simulate a system. In order to find the most suitable one, the following existing and non-commercial platforms were considered: SeSAm[1] (Shell for Simulated Agent Systems) [22], Jade[2]

---

[1] http://www.simsesam.de
[2] http://jade.tilab.com/

(Java Agent DEvelopment Framework) [2], oRis[3] [19], Madkit[4] [18], NetLogo[5], AgentTool[6] [10], Zeus[7] [23].

Unlike Zeus, SeSAm is easy and simple to use. For instance, behaviours of the elements composing a simulation can be described by activity diagrams and using a graphical interface to model agents, resources or an environment is therefore possible. SeSAm enables to simulate situated agents (in a two-dimensional environment) or communicative agents whereas Jade and NetLogo focus only on one of the two categories. The behaviour adopted by the environment during a simulation is described by a world and various worlds can be associated with one environment. Situations enable also a modeller to define the initial state of a simulation and different situations can be given for one simulation. Therefore a given MAS can be tested under several different conditions. Moreover, SeSAm offers users tools to analyse a simulation and calibrate simulation parameters. Finally SeSAm is an open source piece of software which is still evolving, a beta version exists, and plug-ins can be created to add new functionalities (for example, to extend SeSAm agents with FIPA-compliant communication abilities [22]).

For these reasons, SeSAm was chosen to simulate a multi-agent system during the design process. Consequently a model of a cooperative agent, elaborated in the next section, was implemented under SeSAm to enable the observation of how an AMAS behaves while running.

Modification of an Agent Behaviour. It is clear that a minimal design and code of an agent must be provided before launching a simulation. Because simulation here is used as a tool for finding what the behaviour of an agent should be, perhaps at the end of this step, designers will have to code again the agents in a more efficient way.

Therefore, during the design and implementation phases, two challenges have to be taken up: firstly, designers must be allowed to easily modify the code of an agent, and secondly, these modifications have to be automatically done. Performances are not considered as the main objective yet and, for the time being, this work focuses on the former point. Thus, the agent behaviour must be necessarily expressed in a way that a designer could understand and modify easily. Accordingly, the second component of this work, expounded in section 5, proposes to use the subsumption architecture [5] to express an agent behaviour.

4 A Cooperative Agent Model to Simulate

According to the AMAS theory [7], it is possible to envisage that the less non cooperative situations occur, the better the global behaviour of an adaptive multi-agent is. If a designer has got a feedback about how many NCS have occurred during the execution of a simulated AMAS, he can therefore know where and when agents are not anymore cooperative and have the wrong behaviour. It is then more easy to try to find

---

[3] http://www.enib.fr/~harrouet/
[4] http://www.madkit.org
[5] http://ccl.northwestern.edu/netlogo/contact.shtml
[6] http://macr.cis.ksu.edu/projects/agentTool/agentool.htm
[7] http://labs.bt.com/projects/agents/zeus/

what must be added or modified in the behaviour of an agent to make it stay cooperative all along the simulation.

A model of a cooperative agent was thus implemented under SeSAm. Once instantiated, it is used to automatically detect NCS during the simulation of an AMAS. A designer can then improve the behaviour of the agents before launching again the simulation to judge how the global behaviour was enhanced. Moreover, this observation may also help him to find what is wrong and what should be improved in the system. In both cases, modifications of behaviours are still manually done at this stage of the study.

In SeSAm, an agent may have attributes and one or more activity diagrams are used to define its behaviour [22]. These activity diagrams are executed in a user-defined order and within an activity diagram, activities are executed, one after the other, following firing rules until the end node is selected or an activity with a "clock" is reached; in such a case the control is given to the next activity diagram. Agents are also synchronous and an agent is executed only when an activity with a clock is reached in the last activity diagram of the previous active agent.

Considering that NCS may occur during each phase of an agent's life cycle (perceive-decide-act), the model of a cooperative agent was implemented under SeSAm thanks to one activity diagram per life cycle phase.

4.1 Perception Phase

During the perception phase, non cooperative situations related to the interpretation an agent makes about what it perceives may be encountered: it does not understand its perceptions (incomprehension) or it has many ways to understand them (ambiguity). The perception phase is made up of four activities in the model we propose: perceiving the environment ("Perceive" in Fig. 1), interpreting those perceptions ("Interpret"), detecting incomprehension ("Incomprehension") and detecting ambiguity ("Ambiguity"). The last activity ("Go to Decision") is only here to make an agent change the phase.

An agent begins its cycle by memorizing what it perceives in a list of percepts related to the agent domain. Data manipulated by agents in SeSAm are strongly typed,

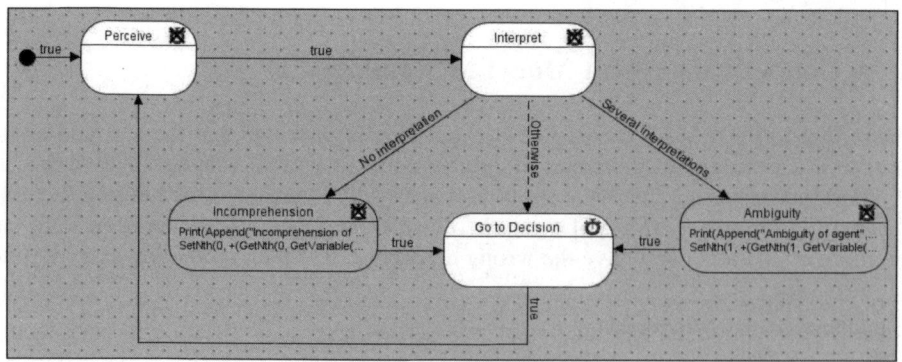

Fig. 1. Perception Phase

it is therefore impossible to create a perception model usable by any MAS; for that reason, designers have to implement the perception of their agents.

Interpreting perceptions consists in coding percepts in a way enabling reasoning of agents and is related to the MAS domain. Consequently, just like perceptions, designers have to implement interpretation.

In order to detect NCS related to interpretation, an agent has to verify the number of interpretations that are possibly associated with a percept. If no interpretation exists the Incomprehension activity is selected, if several interpretations are possible, the control is given to the Ambiguity activity.

4.2 Decision Phase

Once the perception phase completed, an agent begins its decision phase which consists in reasoning on its percepts by using its knowledge about some domains and the representation it has about its environment and itself. An activity enabling reasoning and an activity enabling to choose the actions it will do are required during this decision phase. Furthermore two more NCS may occur: interpretations done by an agent may not provide it something new (unproductiveness) or its reasoning does not lead to any conclusion (incompetence).

This phase of the model is then based on five activities in which an agent verifies whether its interpretations do not already belong to its knowledge ("Examine the Interpretation" in Fig. 2), processes the associated NCS if this is the case ("Unproductiveness") , reasons then on its knowledge and interpretations ("Reason"), deals with a NCS of incompetence if it cannot conclude ("Incompetence") and then decides what to do next depending on its situation ("Decide"). Once these decisions taken, control is given to the action phase.

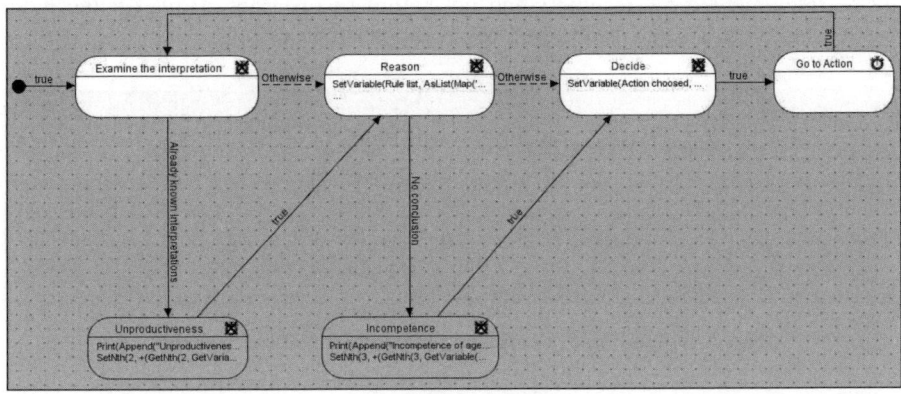

Fig. 2. Decision Phase

In the model, representations, knowledge, interpretations, conclusions and possible actions are stored in specific attributes. Various solutions exist to implement how an agent reasons and they are not especially associated with a specific domain. It is therefore possible to use the same reasoning engine for all MAS, for example a base of rules. In the same manner, finding what actions to do considering a given situation can be also implemented in different ways, for example by a subsumption architecture.

4.3 Action Phase

The last phase of a cooperative agent life cycle consists in carrying out the actions chosen during the previous phase. Three kinds of non cooperative situations may be encountered by an agent: its actions may prevent another agent from reaching its goal (conflict), it wants to do what another agent also intends to do (concurrence) or what it is going to do is not beneficial for others or itself (uselessness).

In order to detect these NCS, an agent has to code its actions in a (preconditions, additions, removals) form. Furthermore to be able to compare its goal with those of others, it must be able to imagine what are these other goals and code them in a similar way.

Firstly an agent combines the actions chosen in the previous phase ("Combine the chosen actions" in Fig. 3) in a recursive way starting from the first possible one and following the execution order of these actions. Formulae to obtain preconditions (Pre), additions (Add) and removals (Rem) for the ith action are:

$Pre \leftarrow Pre \cup \{ Pre_i \setminus Add \}$

$Add \leftarrow Add_i \cup \{ Add \setminus Rem_i \}$ and

$Rem \leftarrow Rem_i \cup \{ Rem \setminus Add \}$.

In order to imagine the goals of others ("Imagine goal of others" in Fig. 3), an agent takes into account the representations it has about them. These representations are closely related to the MAS domain and how an agent perceives others' goals can only be implemented by designers and not automatically done. However the model provides designers a way to store the imagined goals for an agent.

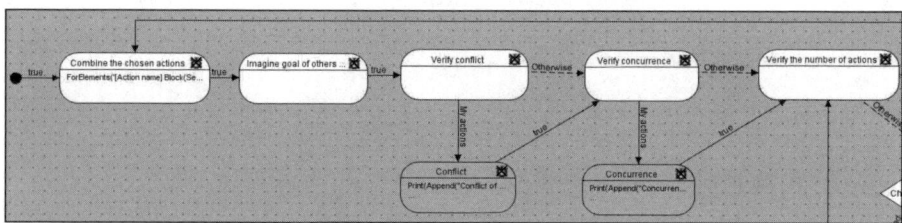

Fig. 3. Action Phase. Verification of conflicts and concurrences.

After that, verifications of conflicts or concurrences are done in the two following activities. A NCS of uselessness should occur when these actions do not increase the satisfaction of an agent or the satisfaction of others and actually should be detected afterwards actions are done. Furthermore this satisfaction should be evaluated for every goal an agent owns and finding the right computation would increase the workload of designers without ensuring that the results is the right one if NCS are wrongly detected. Consequently for the time being, no solution is given and this kind of NCS is not detected by an agent yet.

The decision phase provided an agent with the list of actions it can perform. In this phase, after having detected NCS, an agent verifies whether it has some actions to carry out ("Verify the number of actions" in Fig. 3). If no action exists, an agent starts its life cycle again or another agent completes its own. If at least an action exists, actions of the list are processed one after the other until none remains (the portion of the model enabling this part of the phase is not illustrated here).

5 Modification of an Agent Behaviour

During a simulation made up of cooperative agents based on this model, each time an activity related to NCS is triggered, the detected NCS is counted and a warning message is aimed at the designer. An agent memorises also the total number of NCS it meets during its life. Furthermore the total number of NCS that are encountered by all agents is stored thanks to a basic environment provided by the model (an example of code is given in Fig. 4). Consequently considering this information, a designer has the opportunity to modify the right portion of behaviour of his agents to try to progressively reduce the number of NCS and thus improve the global behaviour of the simulated MAS.

As it was said before, to reason an agent requires an inference engine and to decide a mechanism is needed, both have to be generic enough to be reused when simulating various MAS.

Answering the former issue has been done by implementing an ATMS (Assumption based Truth Maintenance System [9]) which is a system that maintains the consistency of a base of rules. A first plug-in to SeSAm provides then an activity Reason which enables an agent to reason on the interpretations it has using its knowledge and to store obtained results in a specific attribute. However, attributes under SeSAm are strongly typed and generic types cannot be used, it is therefore difficult to create an inference engine based on predicates and thus the provided plug-in only uses propositions. Using this plug-in, a designer has only to create the adequate knowledge base to make agents reason. Nevertheless he is not compelled to use this plug-in and can implement his own reasoning.

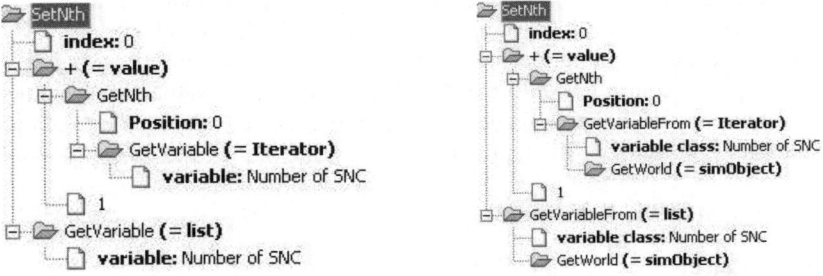

Fig. 4. Actions Counting NCS under SeSAm. On the left, an agent counts the number of NCS it encounters; on the right, the total number of NCS is incremented by every agent.

As introduced in section 3.2, a subsumption architecture implemented under SeSAm tries to answer the latter issue. A subsumption architecture (see Fig. 5 in section 6.2 for an example) is a way to express the behaviour of an agent by using (conditions, action) tuples. Such a tuple represents the action to do when the conditions are true. In a general way, conditions are linked to environmental perceptions of the agents. These pairs are ordered depending on the level of priority of an action relatively to another one. The action on the hierarchy top-level has then priority over the actions of the lower levels (in Fig. 5, the "Target the nearest fish" action has priority over the "Target this prawn" one if the "Near a fish" condition is true). Modifying, by re-ordering pairs, this architecture implies a modification of the behaviour of an agent and requires either to change the order between the tuples or to add new pairs. A plug-in to SeSAm was implemented to enable the user to modify the architecture at run-time. The main drawback is that, when modelling agents, a user does not exactly know how the architecture has to be modified to get the right behaviour. A further step would be to make agents modify it in an autonomous way depending on the actions they make and the results they get.

6 Applying the Model to an Ecosystem

Validation of the model was achieved by testing it on an application for which it was possible to automatically evaluate the emergent function. Actually this function was related to the performance of the built system and therefore could be measured.

6.1 Description of the Case Study

This application simulates an ecosystem made up of prawns, fishes and seaweeds. Prawns eat seaweeds and are able to increase their number by breeding. Fishes eat prawns and are also able to reproduce. The objective of prawns and fishes is to survive, they are autonomous and are implemented using the cooperative agent model described above. To comply with this model, agents do not communicate in a direct way but are endowed with perception abilities in the environment. Seaweeds just spread and have no real autonomy, they are thus considered as active objects which life cycle is also described under SeSAm with activity diagrams.

Prawns and fishes have similar behaviours and objectives, and here, only prawns are of interest knowing that what can be learnt about the behaviour of prawns can also be learnt for fishes. Taken as a whole, prawns form a MAS which can have various objectives; for example, either surviving the longest possible time or eating as many seaweeds as possible in a limited time. It is then possible to evaluate the rightness of the global function (or functional adequacy) of this MAS based on either the number of simulation cycles during which prawns survive or the number of eaten seaweeds.

6.2 Adjustment of the Behaviour

In order to evaluate the correlation between a decrease in the number of NCS encountered and the improvement of the functional adequacy of the MAS, a comparison was made between different versions of this application.

At the beginning prawns have no perceptions and NCS of incomprehension are detected. Once their perceptions improved (a prawn always knows what it perceives and does not learn anything), prawns do not encounter such NCS anymore. Thus, only NCS associated with conflict and concurrence can be detected by the model (and theoretically uselessness). If actions and a subsumption architecture (see Fig. 5) to decide which action to do next are given to prawns (random move, move towards a seaweed, eat a seaweed), the model indicates that conflicts are encountered.

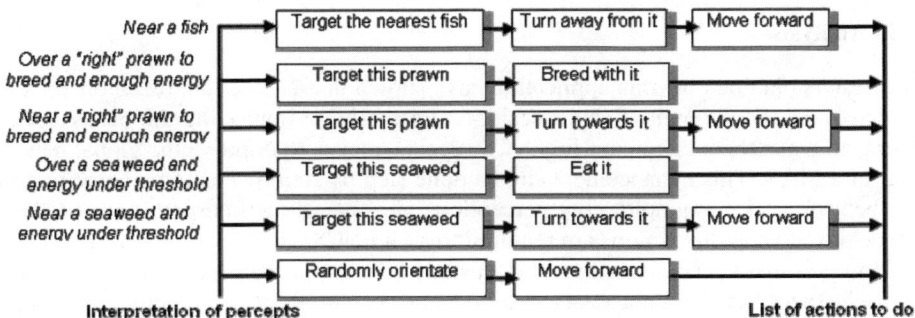

Fig. 5. Subsumption of a Prawn

By observing the simulation, it appears that these conflicts occur when two prawns want to eat the same seaweed or want to breed with the same prawn. The second conflict has the same nature than the first one on which we focus now. The left part of Fig. 6 shows the number of eaten seaweeds (top curve) and the number of conflicts (bottom curve), both curves have the same look which shows that (i) almost every time a prawn wants to eat a seaweed, there is a conflict with another prawn and (ii) prawns die at about the cycle 230 because all seaweeds are eaten.

Many possibilities exist to enhance the behaviour of a prawn and make it avoid conflicts; for example, eat a seaweed only if closer than the other prawn, do not eat if

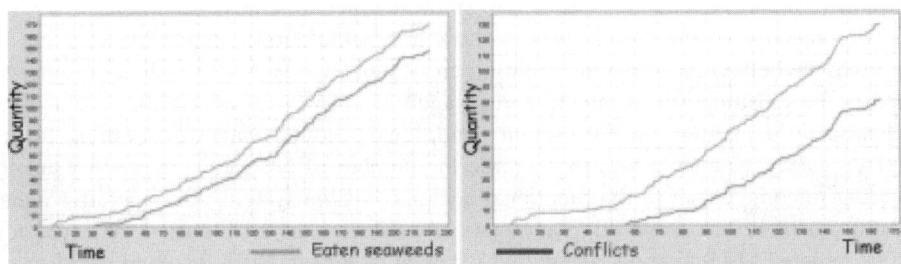

Fig. 6. Results Obtained in the Ecosystem Application. The top curve shows the number of eaten seaweeds, the bottom one shows the number of detected NCS. The left diagram shows results obtained when NCS are not processed, the right one shows results when NCS of conflicts are processed.

there is a conflict, evaluate the conflict probability for every seaweed etc. Simulation is a means to evaluate all this possibilities in order to choose the one that avoid the greatest number of conflicts. One solution was tested and results appear on the right part of Fig. 6: the number of conflicts has decreased over time, the number of eaten seaweeds has decreased and prawns survived a longer time (all remaining cycles are not shown on the diagram).

Best results are obtained when NCS about conflicts are considered by the designer and removed by agents.

6.3 Analysis

The results obtained for this application have shown that NCS could be automatically detected and then "repaired" by a designer. Furthermore improving the behaviour of agents to make them encounter less NCS seems a means to improve the global behaviour of a MAS. This enhancement can be done step by step by changing some part of the behaviour; for example, how perceptions are done, how interpretations of them are achieved and what actions are required to avoid NCS.

However many work is still needed to study how some NCS may be detected such as uselessness which is not taken into account yet. Some other situations related to cooperation have also to be studied; they are not directly related to NCS but could enable agents to be more cooperative or to avoid long-term NCS. For instance, a prawn could move towards a place where there is no prawns yet rather than in a random manner, that would avoid future conflicts.

7 Conclusion and Perspectives

The focus of this article was on self-organising systems in which agents have a cooperative social attitude making their relationships evolve to find the right system organisation and then the right collective function. The cooperation at the agent level is realised in a proscriptive way: an agent tries to avoid a certain number of situations that are judged, from its point of view, non cooperative. The main issue for designers of such systems is to find and enumerate all these non cooperative situations.

The objective of this article was to show how simulation could help designers to improve the behaviour of agents during their design. A way is to enable a designer to modify the subsumption architecture of an agent i.e. how an agent chooses the actions it has to do depending on the current conditions. Another way is, according to the AMAS theory, to enable him to decrease the number of NCS encountered by agents in order to come closer to the functional adequacy of the system. A cooperative agent model was therefore proposed to automatically detect some NCS during the execution of a simulated MAS to show where and when NCS appear. With this kind of information a designer is then able to enrich the behaviour of agents to let them detect NCS and act to remove or avoid them. Application of this model on a simple ecosystem has shown positive results and has enabled to improve its collective behaviour according to a global objective.

This first step in this work has then shown the feasibility of the approach for systems in which agents interact through the environment, nonetheless improvements are desirable.

First of all, the current detection of NCS could be improved and the model could be enhanced to detect other types of NCS like uselessness or NCS that could appear later on. It also could be envisaged to make an agent change its behaviour during a simulation. The last steady version of SeSAm was used to implement the model and did not enable this dynamical change, however this possibility exists under the beta-version 2 of SeSAm and could be used.

The autonomy of agents could be fully exploited by letting them change their behaviour in an autonomous way during simulation depending on what kind of NCS they encounter. An agent could, for instance, dynamically modify its subsumption architecture.

Finally, to fully comply with the AMAS theory, designers should be given the opportunity to design self-organising systems in which direct communication between agents is used and the proposed model should also be enhanced towards this perspective.

The main objective of this work is to include simulation tools into ADELFE to complete the life cycle of its development process. Indeed validation and testing are generally the neglected phases in agent-oriented methodologies either because tools are lacking or because agents concepts are less important at this level. Nevertheless we think that these phases are also required to help designers and that simulation could play an important role in their implementation. Actually modelling and formalisation of the systems considered here are hardly conceivable because of their emergent nature. Consequently, an empirical validation or testing is may be the only solution to overcome this complexity and that could be done by using simulation both at the individual and collective levels.

Acknowledgments. We would like to thank Nicolas Daures for the work done during his master of research training, especially for the model implementations under SeSAm.

References

[1] Ali, S., Zimmer, R., Elstob, C.: The Question Concerning Emergence: Implication for Artificiality. In: Dubois, D.M. (ed.) 1st CASYS'97 Conference (1997)

[2] Bellifemine, F., Poggi, A., Rimassi, G.: JADE: A FIPA-compliant Agent Framework. In: Proceedings of the Practical Applications of Intelligent Agents and Multi-Agents, pp. 97–108 (April 1999)

[3] Bernon, C., Camps, V., Gleizes, M.-P., Picard, G.: Engineering Adaptive Multi-agent Systems: the ADELFE Methodology. In: Henderson-Sellers, B., Giorgini, P. (eds.) Agent-Oriented Methodologies, pp. 172–202. Idea Group Pub. USA (2005)

[4] Bonabeau, E., Dorigo, M., Theraulaz, G.: Swarm intelligence - from natural to artificial systems. Oxford University Press, Oxford (1999)

[5] Brooks, R.A.: A robust control system for a mobile robot. IEEE Journal of Robotics and Automation, 2(1), 14–23 (1986)

[6] Camazine, C., Deneubourg, J.-L., Franks, N., Sneyd, J., Theraulaz, G., Bonabeau, E.: Self-organization in Biological Systems. Princeton University Press, Princeton, NJ (2002)

[7] Capera, D., Georgé, J.-P., Gleizes, M.-P., Glize, P.: The AMAS Theory for Complex Problem Solving Based on Self-organizing Cooperative Agents. In: Proc. 12th IEEE International Workshops on Enabling Technologies, Infrastructure for Collaborative Enterprises, Linz, Austria, 9-11 June 2003, pp. 383–388. IEEE Computer Society, Los Alamitos (2003)
[8] Cossentino, M.: From Requirements to Code with the PASSI Methodology. In: Henderson-Sellers, B., Giorgini, P. (eds.) Agent-Oriented Methodologies, pp. 79–106. Idea Group Pub. USA (2005)
[9] De Kleer, J.: An Assumption-Based TMS. Artificial Intelligence 28(2), 127–162 (1986)
[10] Deloach, S.A., Wood, M.F., Sparkman, C.H.: Multiagent Systems Engineering. Int. Journal of Software Engineering and Knowledge Engineering 11(3), 231–258 (2001)
[11] De Wolf, T., Samaey, G., Holvoet, T.: Engineering Self-Organising Emergent Systems With Simulation-Based Scientific Analysis. In: Brueckner, S.A., Serugendo, G.D.M., Hales, D., Zambonelli, F. (eds.) ESOA 2005. LNCS (LNAI), vol. 3910, pp. 138–152. Springer, Heidelberg (2006)
[12] Di Marzo Serugendo, G., Gleizes, M.-P., Karageorgos, A.: Self-Organization and Emergence in Muli-Agent Systems. In: Parsons, S. (ed.) The Knowledge Engineering Review, vol. 20(02), pp. 165–189. Cambridge University Press, Cambridge, UK (2005)
[13] Fortino, G., Garro, A., Russo, W., Caico, R., Cossentino, M., Termine, F.: Simulation-Driven Development of Multi-Agent Systems. In: Workshop on Multi-Agent Systems and Simulation (MAS&S'06), Palermo (2006)
[14] Fortino, G., Garro, A., Russo, W.: A Discrete-Event Simulation Framework for the Validation of Agent-based and Multi-Agent Systems. In: Proceedings of the Workshop on Objects and Agents (WOA), Camerino, pp. 14–16 (November 14-16, 2005)
[15] Gardelli, L., Viroli, M., Omicini, A.: On the Role of Simulations in the Engineering of Self-Organising MAS: the case of an intrusion detection system in TuCSoN. In: Brueckner, S.A., Serugendo, G.D.M., Hales, D., Zambonelli, F. (eds.) ESOA 2005. LNCS (LNAI), vol. 3910, pp. 153–166. Springer, Heidelberg (2006)
[16] Georgé, J.-P., Picard, G., Gleizes, M.-P., Glize, P.: Living Design for Open Computational Systems - 1st International Workshop on Theory And Practice of Open Computational Systems (TAPOCS'03). In: Proc. 12th IEEE International Workshops on Enabling Technologies (WETICE'03), Infrastructure for Collaborative Enterprises, Linz, Austria, 9-11 June 2003, pp. 389–394. IEEE Computer Society, Los Alamitos (2003)
[17] Goldstein, J.: Emergence as a Construct: History and Issues. Journal of Complexity Issues in Organizations and Management 1(1) (1999)
[18] Gutknecht, O., Michel, F., Ferber, J.: The MadKit Agent Platform Architecture, Research Report, LIRMM (April 2000)
[19] Harrouet, F.: oRis: s'immerger par le langage pour le prototypage d'univers virtuels à base d'entités autonomes, Thèse de Doctorat, Université de Bretagne Occidentale, Brest, France (December 8, 2000)
[20] Heyligen, F.: The science of self-organization and adaptivity – Knowledge Management, organizational Intelligence and Learning, and Complexity, The Encyclopedia of Life Support Systems (EOLSS), Publishers (2003)
[21] Jacobson, I., Booch, G., Rumbaugh, J.: The Unified Software Development Process. Addison-Wesley, London, UK (1999)
[22] Klügl, F., Herrler, R., Oechslein, C.: From Simulated to Real Environments: How to use SeSAm for Software Development. In: Schillo, M., Klusch, M., Müller, J., Tianfield, H. (eds.) Multiagent System Technologies. LNCS (LNAI), vol. 2831, pp. 13–24. Springer, Heidelberg (2003)

[23] Nwana, H., Ndumu, D., Lee, L.: ZEUS: An Advanced Tool-Kit for Engineering Distributed Multi-Agent Systems. In: Proceedings of the Practical Application of Intelligent Agents and Multi-Agent Systems, London, UK, pp. 377–392 (1998)

[24] Picard, G., Gleizes, M.-P.: The ADELFE Methodology – Designing Adaptive Cooperative Multi-Agent Systems. In: Bergenti, F., Gleizes, M.-P., Zambonelli, F. (eds.) Methodologies and Software Engineering for Agent Systems, pp. 157–176. Kluwer Publishing, Dordrecht (2004)

[25] Röhl, M., Uhrmacher, A.M.: Controlled Experimentation with Agents – Models and Implementations. In: Oza, N.C., Polikar, R., Kittler, J., Roli, F. (eds.) MCS 2005. LNCS, vol. 3541. Springer, Heidelberg (2005)

[26] Sierra, C., Rodriguez-Aguilar, J.A., Noriega, P., Esteva, M., Arcos, J.L.: Engineering Multi-Agent Systems as Electronic Institutions, Novatica 170, (July-August 2004)

[27] Uhrmacher, A.M.: Simulation for Agent-Oriented Software Engineering. In: Lunceford, W.H., Page, E. (eds.) First International Conference on Grand Challenges for Modelling and Simulation, San Antonio, Texas. SCS, San Diego (2002)

Adaptation of Autonomic Electronic Institutions Through Norms and Institutional Agents

Eva Bou[1], Maite López-Sánchez[2], and J.A. Rodríguez-Aguilar[1]

[1] IIIA - CSIC Artificial Intelligence Research Institute,
Campus UAB 08193 Bellaterra, Spain
{ebm,jar}@iiia.csic.es

[2] WAI, Volume Visualization and Artificial Intelligence, MAiA Dept.,
Universitat de Barcelona
maite@maia.ub.es

Abstract. Electronic institutions (EIs) have been proposed as a means of regulating open agent societies. EIs define the rules of the game in agent societies by fixing what agents are permitted and forbidden to do and under what circumstances. And yet, there is the need for EIs to adapt their regulations to comply with their goals despite coping with varying populations of self-interested external agents. In this paper we focus on the extension of EIs with autonomic capabilities to allow them to yield a dynamical answer to changing circumstances through norm adaptation and changes in institutional agents.

Keywords: Autonomic Electronic Institutions, Multiagent Systems, Adaptation.

1 Introduction

The growing complexity of advanced information systems in the recent years, characterized by being distributed, open and dynamical, has given rise to interest in the development of systems capable of self-management. Such systems are known as self-* systems [1] , where the * sign indicates a variety of properties: self-organization, self-configuration, self-diagnosis, self-repair, etc. A particular approximation to the construction of self-* systems is represented by the vision of autonomic computing [2], which constitutes an approximation to computing systems with a minimal human interference. Some of the many characteristics of autonomic systems are: it must configure and reconfigure itself automatically under changing (and unpredictable) conditions; it must aim at optimizing its inner workings, monitoring its components and adjusting its processing in order to achieve its goals; it must be able to diagnose the causes of its eventual malfunctions and repair itself; and it must act in accordance to and operate into a heterogeneous and open environment.

In what follows we argue that EIs [3] are a particular type of self-* system. When looking at computer-mediated interactions we regard Electronic Institutions (EI) as regulated virtual environments wherein the relevant interactions among participating agents take place. EIs have proved to be valuable to develop open agent systems [4]. However, the challenges of building open systems are still considerable, not only because of the inherent complexity involved in having adequate interoperation of heterogeneous agents, but also because the need for adapting regulations to comply with institutional goals despite varying agents' behaviors. Particularly, when dealing with self-interested agents.

The main goal of this work consists in studying how to endow an EI with autonomic capabilities that allow it to yield a dynamical answer to changing circumstances through the adaptation of its regulations. Among all the characteristics that define an autonomic system we will focus on the study of self-configuration as pointed out in [2] as a second characteristic: "An autonomic computing system must configure and reconfigure itself under varying (and in the future, even unpredictable) conditions. System configuration or "setup" must occur automatically, as well as dynamic adjustments to that configuration to best handle changing environments".

The paper is organized as follows. In section 2 we introduce the notion of autonomic electronic institution as an extension of the classic notion of electronic institution along with a general model for adaptation based on transition functions. Section 3 details how these functions are automatically learned. Section 4 details a case study to be employed as a scenario wherein to test the model presented in section 2. Section 5 provides some empirical results. Finally, section 6 summarizes some conclusions and related work and outlines paths to future research.

2 Autonomic Electronic Institutions

The idea behind EIs [3] is to mirror the role traditional institutions play in the establishment of "the rules of the game" –a set of conventions that articulate participants' interactions. The main goal of EIs is the enactment of a constrained environment that shapes open agent societies. EIs structure agent interactions, establishing what agents are permitted and forbidden to do as well as the consequences of their actions.

In general, an EI regulates multiple, distinct, concurrent, interrelated, dialogic activities, each one involving different groups of agents playing different roles. For each activity, interactions between agents are articulated through agent group meetings, the so-called *scenes*, that follow well-defined interaction protocols whose participating agents may change over time (agents may enter or leave). More complex activities can be specified by establishing networks of scenes (activities), the so-called *performative structures*. These define how agents can legally move among different scenes (from activity to activity) depending on their role.

Although EIs can be regarded as the computational counterpart of human institutions for open agent systems, there are several aspects in which they are nowadays lacking. According to North [5] human institutions are not static; they may evolve over time by altering, eliminating or incorporating norms. In this way, institutions can adapt to societal changes. Nonetheless, neither the current notion of EI nor the engineering framework in [6] support their adaptation so that an EI can self-configure. Thus, in what follows we study how to extend the current notion of EI to support self-configuration in order to be used in systems that need adaptation in their regulation (e.g. electricity market system).

First of all, notice that in order for EIs to adapt, we believe that a "rational" view must be adopted (likewise the rational view of organizations in [7]) and thus consider that *EIs seek specific goals*. Hence, EIs continuously adapt themselves to fulfill their goals. Furthermore, we assume that an EI is *situated* in some environment that may be either totally or partially observable by the EI and its participating agents.

With this in mind, we observe that according to [3] an EI is solely composed of: a dialogic framework establishing the common language and ontology to be employed by participating agents; a performative structure defining its activities along with their relationships; and a set of norms defining the consequences of agents' actions. From this follows that further elements are required in order to incorporate the fundamental notions of goal, norm configuration, and performative structure configuration as captured by the following definition of *autonomic electronic institution*.

Definition 1. *Given a finite set of agents A, we define an Autonomic Electronic Institution (AEI) as a tuple $\langle PS, N, DF, G, P_i, P_e, P_a, V, \delta, \gamma \rangle$ where:*

- *PS stands for a performative structure;*
- *N stands for a finite set of norms;*
- *DF stands for a dialogic framework;*
- *G stands for a finite set of institutional goals;*
- $P_i = \langle i_1, \ldots, i_s \rangle$ *stands for the values of a finite set of institutional properties, where $i_j \in \mathbb{R}$, $1 \leq j \leq s$ contains the value of the j-th property;*
- $P_e = \langle e_1, \ldots, e_r \rangle$ *stands for the values of the environment properties, where each e_j is a vector, $e_j \in \mathbb{R}^{n_j}$, $1 \leq j \leq r$ contains the value of the j-th property;*
- $P_a = \langle a_1, \ldots, a_n \rangle$ *stands for the values that characterize the institutional state of the agents in A, where $a_j = \langle a_{j_1}, \ldots, a_{j_m} \rangle$, $1 \leq j \leq n$ stands for the institutional state of agent A_j;*
- *V stands for a finite set of reference values;*
- $\delta : N \times G \times V \to N$ *stands for a normative transition function that maps a set of norms into a new set of norms given a set of goals and a set of values for the reference values; and*
- $\gamma : PS \times G \times V \to PS$ *stands for a performative structure transition function (henceforth referred to as PS transition function) that maps a performative structure into a new performative structure given a set of goals and a set of values for the reference values.*

Notice that with both the *normative transition function*, δ, and with the *PS transition function*, γ, our AEI definition has included the mechanisms to support their adaptation. Notice that a major challenge in the design of an AEI is to learn a *normative transition function*, δ, along with a *PS transition function*, γ, that ensure the achievement of its institutional goals under changing conditions. Next, we dissect the new elements composing an AEI.

2.1 Goals

Agents participating in an AEI have their social interactions mediated by the institution according to its conventions. As a consequence of his interactions, only the *institutional (social) state* of an agent can change since an AEI has no access whatsoever to the inner state of any participating agent. Therefore, given a finite set of participating agents $A = \{A_1, \ldots, A_n\}$ where $n \in \mathbb{N}$, each agent $A_i \in A$ can be fully characterized by his institutional state, represented as a tuple of observable values $\langle a_{i_1}, \ldots, a_{i_m} \rangle$ where $a_{i_j} \in \mathbb{R}$, $1 \leq j \leq m$. Thus, the actions of an agent within an AEI may change his institutional state according to the institutional conventions.

The main objective of an AEI is to accomplish its goals. For this purpose, an AEI will adapt. We assume that the institution can observe the environment, the institutional state of the agents participating in the institution, and its own state to assess whether its goals are accomplished or not. The temperature of a room can be an example of an environment property, the time an agent is playing in the institution can be an example of an agent' institutional property, and the number of scenes can be an institutional property. Thus, from the observation of environment properties(P_e), institutional properties (P_i), and agents' institutional properties (P_a), an AEI obtains the reference values required to determine the fulfillment of goals. Formally, the reference values are defined as a vector $V = \langle v_1, \ldots, v_q \rangle$ where each v_j results from applying a function h_j upon the agents' properties, the environmental properties and/or the institutional properties; $v_j = h_j(P_a, P_e, P_i)$, $1 \leq j \leq q$.

Finally, we can turn our attention to institutional goals. An example of institutional goal for the Traffic Regulation Authority could be to keep the number of accidents below a given threshold. In other words, to ensure that a reference value satisfies some constraint.

Formally we define the goals of an AEI as a finite set of constraints $G = \{c_1, \ldots, c_p\}$ where each c_i is defined as an expression $g_i(V) \lhd [m_i, M_i]$ where $m_i, M_i \in \mathbb{R}$, $\lhd$ stands for either $\in$ or $\notin$, and g_i is a function over the reference values. In this manner, each goal is a constraint upon the reference values where each pair m_i and M_i defines an interval associated to the constraint. Thus, the institution achieves its goals if all $g_i(V)$ values satisfy their corresponding constraints of being within (or not) their associated intervals.

2.2 Norm Transition

An AEI employs norms to constrain agents' behaviors and to assess the consequences of their actions within the scope of the institution. Although there is a plethora of formalizations of the notion of norm in the literature, in this paper we adhere to a simple definition of norms as effect propositions as defined in [8]:

Definition 2. *An effect proposition is an expression of the form*

$$A \; causes \; F \; if \; P_1, \ldots, P_n$$

where A is an action name, and each of $F, P_1, \ldots, P_n (n \geq 0)$ is a fluent expression. About this proposition we say that it describes the effect of A on F, and that $P_1, \ldots, P_n$ are its preconditions. If n = 0, we will drop if and write simply A causes F. Notice that since we use norms only to describe prohibitions, our norms are a particular case of regulative norms [9].

From this definition of norm, changing a norm amounts to changing either its pre-conditions, or its effect(s), or both. Norms can be parameterized, and therefore we propose that each norm $N_i \in N$, $i = 1, \ldots, n$, has a set of parameters $\langle p_{i,1}^N, \ldots, p_{i,m_i}^N \rangle \in \mathbb{R}^{m_i}$. Hence, changing the values of these parameters means changing the norm. In fact this parameters correspond to the variables in the *norm transition function* that will allow the institution to adapt under changing situations.

Notice that agents do not have the capability to change norms. In our approach we have external agents, internal agents and a mechanism of the institution to change norms. Thus, only the institution is entitled to change norms.

2.3 PS Transition

As mentioned above, an EI involves different groups of agents playing different roles within scenes in a performative structure. Each scene is composed of a coordination protocol along with the specification of the roles that can take part in the scene. Notice that we differentiate between institutional roles (played by staff agents acting as the employees of the institution) and external roles (played by external agents participating in the institution as users). Furthermore, it is possible to specify the number of agents than can play each role within a scene.

Given a performative structure, we must choose the values that we aim at changing in order to adapt it. This involves the choice for a set of parameters whose values will be changed by the PS transition function. In our case, we choose as parameters the number of agents playing each role within each scene. This choice is motivated by our intention to determine the most convenient number of institutional agents to regulate a given population of external agents.

Scenes can be parameterized, and therefore, we propose that each scene in the performative structure, $S_i \in PS$, $i = 1, \ldots, t$, has a set of parameters $\langle p_{i,1}^R, \ldots, p_{i,q_i}^R \rangle \in \mathbb{N}^{q_i}$ where $p_{i,j}^R$ stands for the number of agents playing role r_j in scene S_i.

3 Learning Model

Adapting EIs amounts to changing the values of their parameters. We propose to learn the *norm transition function* (δ) and the *PS transition function* (γ) in two different steps in an overall learning process. For the initial step, the AEI learns by simulation the best parameters for a list of different populations, exploring the space of parameter values in search for the ones that best accomplish goals for a given population of agents. Afterwards, in a second step in a real environment, the AEI will adapt itself to any population of agents. This second learning step involves to identify the current population of agents (or the most similar one) in order to use the learned parameters that best accomplish goals for this population (e.g., using Case-Based Reasoning (CBR) problem solving technique). This paper focuses on the first learning step, in how to learn the best parameters for a population.

We propose to learn the *norm transition function* (δ) and the *PS transition function* (γ) by exploring the space of parameter values in search for the ones that best accomplish goals for a given population of agents. In this manner, if we can automatically adapt an EI to the global behavior of an agent population, then, we can repeat it for a number of different agent populations and thus characterize both δ and γ.

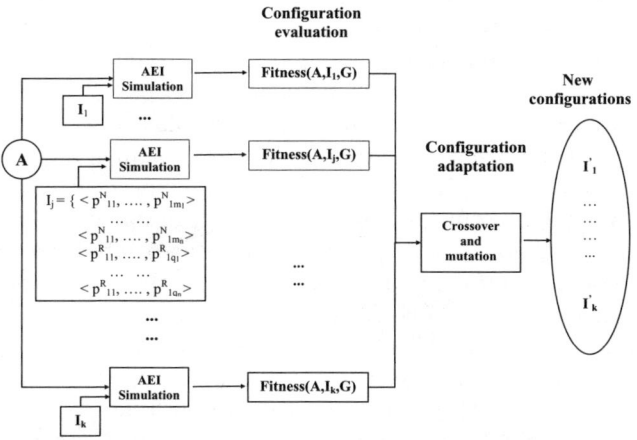

Fig. 1. Example of a step in EI adaptation using an evolutionary approach

Figure 1 describes how this learning process is performed for a given population of agents (A) using an evolutionary approach. We have an initial set of individuals $\langle I_1, \ldots, I_k \rangle$, where each individual represents the set of norm and role parameters defined above $\{\langle p^N_{1,1}, \ldots, p^N_{1,m_1} \rangle, \ldots, \langle p^N_{n,1}, \ldots, p^N_{n,m_n} \rangle, \langle p^R_{1,1}, \ldots, p^R_{1,q_1} \rangle, \ldots, \langle p^R_{t,1}, \ldots, p^R_{t,q_t} \rangle\}$. Each individual represents a specific AEI configuration, and therefore, the institution uses each configuration to perform a simulation with the population of agents A. The corresponding configuration can then be evaluated according to a fitness function that measures the satisfaction degree of institutional goals (*configuration evaluation*). Finally, the AEI compiles the evaluations of all individuals in order to breed a new generation from the best ones *configuration adaptation*. This process results with a new set of individuals (*New configurations*) to be used as next generation in the learning process. Since we are working with a complex system, we propose use an evolutionary approach for learning due to the fact that the institutional objective function can be naturally mapped to the fitness function and an evolutionary approach provides a solution good enough. Notice that the AEI does not learn any agent parameter, it learns the best parameters by simulation for a certain population of agents, that is whose values will be changed by the normative transition function and by the PS transition function.

4   Case Study: Traffic Control

Traffic control is a well-known problem that has been approached from different perspectives, which range from macro simulation for road net design [10] to traffic flow improvement by means of multi-agent systems [11]. We tackle this problem from the Electronic Institutions point of view, and therefore, this section is devoted to specify how traffic control can be mapped into Autonomic Electronic Institutions.

In this manner, we consider the Traffic Regulation Authority as an Autonomic Electronic Institution, and cars moving along the road network as external agents interacting inside a traffic scene through driving actions. Additionally, indirect communication is established by means of stop, rear and turn signal indicators. Considering this set-up, traffic norms regulated by Traffic Authorities can therefore be translated in a straight forward manner into norms belonging to the Electronic Institution. Norms within this normative environment are thus related to actions performed by cars (in fact, in our case, they are always restricted to that). Additionally, norms do have associated penalties that are imposed to those cars refusing or failing to follow them. On the other hand, institutional agents in the traffic scene represent Traffic Authority employees. In our case study, we assume institutional agents to be in charge of detecting norm violations so that we will refer to them as police agents. Notice that police agents are internal agents that play an institutional role. As opposed to other approaches like $\mathcal{M}OISE^{Inst}$ [12] where there is an institution agent middleware dedicated to the management of the organization and to the arbitration our police agents are not linked to any external agent. Each police agent is able to detect only a portion of the total number of norm violations that car agents actually do. Therefore, the number of police agents in the traffic scene directly affects the number of detected norm violations, and thus, the overall quantity of penalties imposed to car agents. Furthermore, our Electronic Institution is able to adapt both norms and the number of deployed police agents based on its goals – just as traffic authorities do modify them– and, therefore, it is considered to be autonomic.

Our AEI sets up a normative environment where cars do have a limited amount of credit (just as some real world driving license credit systems) so that norm offenses cause credit reductions. The number of points subtracted for each traffic norm violation is specified by the sanction associated to each norm, and this sanction can be changed by the regulation authority if its change leads the accomplishment of goals. Eventually, those cars without any remaining points are forbidden to circulate. On the other hand, we assume a non-closed world, so expelled cars are replaced by new ones having the total amount of points.

Getting into more detail, we focus on a two-road junction. It is a very restrictive problem setting, but it is complex enough to allow us to tackle the problem without losing control of all the factors that may influence the results. In particular, no traffic signals (neither yield or stop signals nor traffic lights) are considered, therefore, cars must only coordinate by following the traffic norms imposed by the AEI. Our institution is required to define these traffic norms based on general goals such as minimization of the number of accidents or deadlock avoidance.

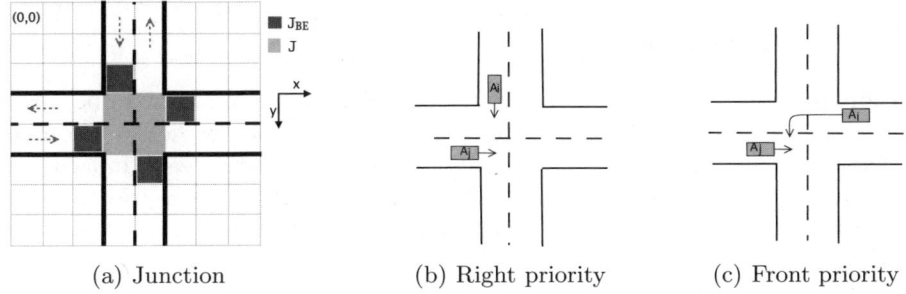

Fig. 2. (a)Grid environment representation of a 2-lane road junction. (b)Priority to give way to the right. (c)Priority to give way to the front.

We model the environment as a grid composed by road and field cells. Road cells define 2 orthogonal roads that intersect in the center (see figure 2(a)).

Discretization granularity is such that cars have the size of a cell. As section 4.2 details, our model has been developed with the Simma tool [13]. Although the number of road lanes can be changed parametrically, henceforth we assume the 2-lane case. Next subsections are devoted to define this "toy problem" and present our solution proposal in terms of it. But before that, we introduce some nomenclature definitions:

- A_i: an external agent i, agents correspond to cars.
- t: time step. Our model considers discrete time steps (ticks).
- (J_x, J_y): size in x, y of our road junction area.
- J: inner road junction area with (x_0^J, y_0^J) as top left cell inside it
  $J = \{(x,y) \mid x \in [x_0^J, x_0^J + J_x - 1], \ y \in [y_0^J, y_0^J + J_y - 1]\}$
  Considering the 4 J cells in the junction area of Figure 2(a):
  $J = \{(x_0^J, y_0^J), (x_0^J + 1, y_0^J), (x_0^J, y_0^J + 1), (x_0^J + 1, y_0^J + 1)\}$.
- J_{BE}: Junction Boundary Entrance, set of cells surrounding the junction that can be used by cars to access it. They correspond to cells near by the junction that belong to incoming lanes. Figure 2(a) depicts $J_{BE} = \{(x_0^J, y_0^J - 1), (x_0^J - 1, y_0^J + J_y - 1), (x_0^J + J_x - 1, y_0^J + J_y), (x_0^J + J_x, y_0^J))\}$.

Nevertheless, the concept of boundary is not restricted to adjacent cells: a car can be also considered to be coming into the junction if it is located one –or even a few– cells away from the junction.
- (x_i^t, y_i^t): position of car A_i at time t, where $(x, y) \in \mathbb{N} \times \mathbb{N}$ stands for a cell in the grid.
- (h_{ix}^t, h_{iy}^t): heading of car A_i, which is located in (x, y) at time t. Heading directions run along x, y axes and are considered to be positive when the car moves right or down respectively. In our orthogonal environment, heading values are: 1 if moving right or down; -1 if left or up; and 0 otherwise (i.e., the car is not driving in the axis direction). In this manner, fourth car's heading on the right road of figure 3 is (-1,0).

4.1 AEI Specification

Environment. As mentioned above, we consider the environment to be a grid. This grid is composed of cells, which can represent roads or fields. The main difference among these two types is that road cells can contain cars. Indeed, cars move among road cells along time. (Figure 2(a) depicts a 8×8 grid example) The top left corner of the grid represents the origin in the x, y axes. Thus, in the example, cell positions range from (0,0) in the origin up to (7,7) at the bottom-right corner.

We define this grid environment as:

$$P_e = \langle\ (x, y, \alpha, r, d_x, d_y)\ |\ 0 \leq x \leq max_x,\ 0 \leq y \leq max_y,\ \alpha \subseteq P(A) \cup \emptyset, \\ r \in [0,1],\ d_x \in [-1,0,1],\ d_y \in [-1,0,1]\ \rangle$$

being x and y the cell position, α defines the set of external agents inside the grid cell (x, y) (notice that $\alpha \subseteq A$), r indicates whether this cell represents a road or not, and, in case it is a road, d_x and d_y stand for the lane direction, whose values are the same as the ones for car headings. Notice that the institution can observe the environment properties along time, we use P_e^t to refer the values of the grid environment at a specific time t. This discretized environment can be observed both by the institution and cars. The institution observes and keeps track of its evolution along time, whilst cars do have locality restrictions on their observations.

Agents. We consider $A = \langle A_1, ..., A_n \rangle$ to be a finite set of n external agents in the institution. As mentioned before, external agents correspond to cars that move inside the grid environment, with the restriction that they can only move within road cells. Additionally, external agents are given an account of points which decreases with traffic offenses. The institution forbids external agents to drive without points in their accounts. The institution can observe the $P_a = \langle a_1, \ldots, a_n \rangle$ agents' institutional properties, where

$$a_i = \langle x_i,\ y_i,\ h_{ix},\ h_{iy},\ speed_i,\ indicator_i,\ offenses_i, \\ accidents_i,\ distance_i,\ points_i \rangle$$

These properties stand for: car A_i's position within the grid, its heading, its speed, whether the car is indicating a trajectory change for the next time step (that is, if it has the intention to turn, to stop or to move backwards), the norms being currently violated by A_i, wether the car is involved in an accident, the distance between the car and the car ahead of it; and, finally, external agent A_i's point account. Notice that the institution can observe the external agent properties along time, we use a_i^t to refer the external agent A_i's properties at a specific time t.

Reference Values. In addition to car properties, the institution is able to extract reference values from the observable properties of the environment, the participating agents as well as the institution. Thus, these reference values are computed as a compound of other observed values. Considering our road junction case study, we identity different reference values:

$$V = \langle col,\ crash,\ off,\ block,\ expel,\ police \rangle$$

where col indicates total number of collisions for the last t_w ticks ($0 \leq t_w \leq t_{now}$):

$$col = \sum_{t=t_{now}-t_w}^{t_{now}} \sum_{e \in P_e^t} f(e_{\alpha^t})$$

being P_e^t the values of the grid environment at time t, e_{α^t} the α^t component of element $e \in P_e^t$ and

$$f(e_{\alpha^t}) = \begin{cases} 1 & if\ |e_{\alpha^t}| > 1 \\ 0 & otherwise \end{cases}$$

Similarly, off indicates the total number of offenses accumulated by all agents during last t_w ticks ($0 \leq t_w \leq t_{now}$):

$$off = \sum_{t=t_{now}-t_w}^{t_{now}} \sum_{i=0}^{|A|} offenses_i^t \qquad (1)$$

$crash$ counts the number of cars involved in accidents for the last t_w ticks:

$$crash = \sum_{t=t_{now}-t_w}^{t_{now}} \sum_{i=0}^{|A|} accidents_i^t \qquad (2)$$

$block$ describes how many cars have been blocked by other cars for last t_w ticks:

$$block = \sum_{t=t_{now}-t_w}^{t_{now}} \sum_{i=0}^{|A|} blocked(a_i, t) \qquad (3)$$

where $blocked(a_i, t)$ is a function that indicates if the agent a_i is blocked by another agent a_j in time t.

$$blocked(a_i, t) = \begin{cases} 1\ if\ \exists\ e \in P_e^t\ |\ (e_{x^t} = x_i^t + h_{ix}^t\ \&\\ \quad e_{y^t} = y_i^t + h_{iy}^t\ \&\ |e_{\alpha^t}| \geq 1\ \&\\ \quad \exists\ a_j \in e_{\alpha^t}\ so\ that\ speed_j^t = 0) \\ 0\ otherwise \end{cases} \qquad (4)$$

being e_{x^t}, e_{y^t}, e_{α^t} the x^t, y^t, α^t components of element $e \in P_e^t$.

Furthermore, *expel* indicates the number of cars that have been expelled out of the environment due to running out of points, and finally, *police* indicates the percentage of police agents that the institution deploys in order to control the traffic environment.

Goals. Goals are in fact institutional goals. The aim of the traffic authority institution is to accomplish as many goals as possible. The institution tries to accomplish these goals by defining a set of norms and by specifying how many police agents should be deployed on traffic scene.

Institutional goals are defined as constraints upon a combination of reference values. Considering our scenario, we define restrictions as intervals of acceptable values for the previous defined reference values (V) so that we consider the institution accomplishes its goals if V values are within their corresponding intervals. In fact, the aim is to minimize the number of accidents, the number of traffic offenses, the number of blocked cars, the number of cars that are expelled from the traffic scene, as well as the percentage of deployed police agents. In order to do it, we establish the list of institutional goals G as:

$$G = \langle\ g(col) \in [0, maxCol],\ g(off) \in [0, maxOff],\ g(crash) \in [0, maxCrash],$$
$$g(block) \in [0, maxBlock],\ g(expel) \in [0, maxExpel],\ g(police) \in [0, maxPolice]\ \rangle$$

Having more than one institutional goal requires to combine them. We propose an *objective function* [14] that favors high goal satisfaction while penalizing big differences among them:

$$O(V) = \sum_{i=1}^{|G|} w_i \sqrt{f(g_i(V), [m_i, M_i], \mu_i)}$$

where $1 \leq i \leq |G|$, $w_i \geq 0$ are weighting factors such that $\sum w_i = 1$, g_i is a function over the reference values, $\mu_i \in [0,1]$ and f is a function that returns a value $f(x, [m, M], \mu) \in [0,1]$ representing the degree of satisfaction of a goal:

$$f(x, [m, M], \mu) = \begin{cases} \dfrac{\mu}{e^{k\frac{m-x}{M-m}}} & x < m \\[6pt] 1 - (1-\mu)\dfrac{x-m}{(M-m)} & x \in [m, M] \\[6pt] \dfrac{\mu}{e^{k\frac{x-M}{M-m}}} & x > M \end{cases}$$

Norms. Autonomic Electronic Institutions use norms to try to accomplish goals. Norms have associated penalties that are imposed to those cars refusing or failing to follow them. These penalties can be parameterized to increase its persuasiveness depending on the external agent population behavior.

Considering a road junction without traffic signals, priorities become basic to avoid collisions. We consider, as in most continental Europe, that the default priority is to give way to the right. This norm prevents a car A_i located on the

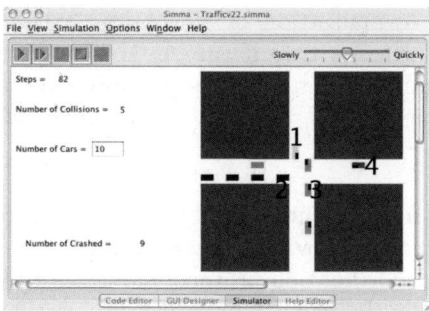

Fig. 3. Priority to give way to the right (Simma tool screenshot)

Table 1. Right and Front priority norms

	Right priority norm	Front priority norm
Action	$in(a_i, J_{BE}, t-1) \wedge$ $in(a_i, (x_i^{t-1} + h_{ix}^{t-1}, y_i^{t-1} + h_{iy}^{t-1}), t) \wedge$ $\neg indicator(a_i, right, t-1)$	$in(a_i, J_{BE}, t-1) \wedge$ $in(a_i, (x_i^{t-1} + h_{ix}^{t-1}, y_i^{t-1} + h_{iy}^{t-1}), t) \wedge$ $indicator(a_i, left, t-1)$
Pre-conditions	$right(a_i, a_j, t-1)$	$in(a_j, J_{BE}, t-1) \wedge$ $front(a_i, a_j, t-1)$
Consequence	$points_i^t = points_i^t - fine_{right}$	$points_i^t = points_i^t - fine_{front}$

Junction Boundary Entrance (J_{BE}) to move forward or to turn left whenever there is another car A_j on its right. For example, car 1 in figure 3 must wait for car 2 on its right, which must also wait for car 3 at the bottom J_{BE}. The formalization in table 1 can be read as follows: "if car A_i moves from a position in J_{BE} at time $t-1$ to its next heading position at time t without indicating a right turn, and if it performs this action when having a car A_j at the J_{BE} on its right, then the institution will fine A_i by decreasing its points by a certain amount" (see figure 2(b)).

Where the predicate $\mathbf{in}(a_i, Region, t)$ in table 1 is equivalent to $\exists (x, y, \alpha^t, r, d_x, d_y) \in P_e^t$ so that $(x, y) \in Region$ and $a_i \in \alpha^t$ and $\mathbf{right}(a_i, a_j, t)$ is a boolean function that returns true if car A_j is located at J_{BE} area on the right side of car A_i. For the 2-lane J_{BE} case, it corresponds to the formula: $(x_i^t - h_{iy}^t + h_{ix}^t J_x,\ y_i^t + h_{ix}^t + h_{iy}^t J_y) == (x_j^t, y_j^t)$.

Similarly, we define an additional norm that is somehow related to the previous 'right priority norm'. We name it 'front priority norm'. It applies when two cars A_i, A_j reach Junction Boundary Entrance areas (J_{BE}) located at opposite lines, and one of them (A_i in Figure 2(c)) wants to turn left. Car A_i turning left may interfere A_j's trajectory, and therefore, this norm assigns priority to A_j so that A_i must stop until its front J_{BE} area is clear. Otherwise A_i will be punished with the corresponding $fine_{front}$ fee.

Table 1 shows the formalization of this norm, where **front**(a_i, a_j, t) is a boolean function that returns true if car A_j is located in front of car A_i at time t. In an orthogonal environment, this function can be easily computed by comparing car headings $((h_{ix}^t, h_{iy}^t), (h_{jx}^t, h_{jy}^t))$ by means of the boolean formula $(h_{ix}^t h_{jx}^t + h_{iy}^t h_{jy}^t) == -1$.

Performative Structure. As introduced in 1, an AEI involves different groups of agents playing different roles within scenes in a performative structure. Each scene is composed of a coordination protocol along with the specification of the roles that can take part in the scene. Our case study particularizes the Performative Structure component so that we define it as being formed by a single traffic scene with two possible agent roles. On one hand, there is an institutional role played by police agents, whereas, on the other hand, the external role is played by car agents. Notice also that it is possible to specify the number of agents than can play each role within a scene.

4.2 Experimental Settings and Design

As a proof of concept of our proposal in section 3, we have designed an experimental setting that implements the traffic case study. In this preliminary experiment we consider four institutional goals related to *col*, *off*, *expel*, and *police* reference values; and both right and front priority norms in table 1. Institutional goals are combined with the objective function introduced in section 4.1, assuming corresponding weights are $4/10, 4/10, 1/10, 1/10$ so that the first two goals are considered to be most important. On the other hand, norms are parameterized through its fines (i.e., points to subtract to the car failing to follow the corresponding norm).

The 2-road junction traffic model has been developed with Simma [13], a graphical MAS simulation tool shown in Figure 3, in such way that both environment and agents can be easily changed. In our experimental settings, we have modeled the environment as a 16×16 grid where both crossing roads have 2 lanes with opposite directions. Additionally, the environment is populated with 10 cars, having 40 points each.

Our institution can observe the external agent properties for each tick and can keep a record of them in order to refer to past ticks. Institutional police agents determine traffic offenses by analyzing a portion of car actions along time. External agent actions are observed through consecutive car positions and indicators (notice that the usage of indicators is compulsory for cars in this problem set up). Furthermore, during our discrete event simulation, the institution replaces those cars running out of points by new cars, so that the cars' population is kept constant. Cars follow random trajectories at a constant 1-cell/tick speed and they collision if two or more cars run into the same cell. In that case, the involved cars do remain for two ticks in that cell before they can start following a new trajectory.

Cars correspond to external agents without learning skills. They just move based on their trajectories, institutional norms and the percentage of deployed agents on the traffic scene. Cars have local information about their environment (i.e., grid surrounding cells). Since the institution informs cars about changes in both norms and number of police agents, cars know whether their next (intended) moves violate some norms and the amount of the fine that applies to such violations. In fact, cars decide whether to comply with a norm based on four parameters: $\langle fulfill\_prob, high\_punishment, inc\_prob, police \rangle$; where $fulfill\_prob \in [0, 1]$ stands for the probability of complying with norms that is initially assigned to each agent; $high\_punishment \in \mathbb{N}$ stands for the fine threshold that causes an agent to consider a fine to be high enough to reconsider the norm compliance; $inc\_prob \in [0, 1]$ stands for the probability increment that is added to $fulfill\_prob$ when the fine threshold is surpassed by the norm being violated; and $police \in [0, 1]$ stands for the percentage (between 0 and 1) of police agents that the traffic authority has deployed on the traffic environment. In summary, agents decide whether they keep moving –regardless of violating norms– or they stop –in order to comply with norms– based on a probability that is computed as:

$$final\_prob = \begin{cases} police \cdot fulfill\_prob & fine \leq high\_punishment \\ police \cdot (fulfill\_prob + inc\_prob) & fine > high\_punishment \end{cases}$$

Our goal is to adapt the institution to agent behaviors by applying Genetic Algorithms (GA)[1] to accomplish institutional goals, that is, to maximize the objective function, which comprises the number of collisions, the number of offenses, the number of expelled cars and the percentage of police agents that should be deployed to control the traffic environment. We shall notice, though, that these offences do not refer to offences detected by police agents but to the real offences that have been actually done by car agents.

As section 3 describes, we propose to adapt the institution to different external agent population behaviors by running a genetic algorithm for each population. Therefore, institution adaptation is implemented as a learning process of the "best" institution parameters. In our experiments, Genetic Algorithms run 50 generations of 20 individuals. An individual corresponds to a list of a binary codifications of specific values for the following institution parameters: right norm penalty, front norm penalty, and percentage of police agents. Crossover among individuals is chosen to be singlepoint and a mutation rate of 10% is applied. The fitness function for individual evaluation corresponds to the objective function described above, which is computed as an average of 5 different 2000-tick-long simulations for each model setting (that is, for each set of parameters):

$$O(V) = \frac{4}{10} \cdot \sqrt{f(g(col), [0, maxCol], \frac{1}{2})} + \frac{4}{10} \cdot \sqrt{f(g(off), [0, maxOff], \frac{1}{2})} +$$

$$\frac{1}{10} \cdot \sqrt{f(g(expel), [0, maxExpel], \frac{3}{4})} + \frac{1}{10} \cdot \sqrt{f(g(police), [0, maxPolice], 0)}$$

---

[1] We use a genetic algorithm Toolbox [15].

where $g(col)$, $g(off)$, $g(expel)$ and $g(police)$ correspond to average values of each reference value averaged for 5 different simulations; and $f(x, [m, M], \mu) \in [0, 1]$ represents the goal satisfaction.

5 Results

From the experimental settings specified above, we have run experiments for five different agent populations. These populations are characterized by their norm compliance parameters, being $fulfill\_prob = 0.5$ and $inc\_prob = 0.2$ for the five of them whereas $high\_punishment$ varies from 5 for the first, to 8 for the second, to 10 for the third, to 12 for the fourth, up to 14 for the fifth (see table 2).

Since both right and front priority norms contribute to reduce accidents, our AEI must learn how to vary its fine parameters to increase its persuasiveness for agents, and eventually, to accomplish the normative goal of minimizing the total number of collisions. Nevertheless, it is also important for the AEI to reduce the total number of offenses, as well as, to a lesser extent, the number of expelled cars and the police deployment percentage. Each institutional agent has an associated cost, so that the AEI pursues the success of the traffic environment (i.e., a few collisions, agents respecting traffic norms and not having many expelled agents) at minimum cost. Thus, the AEI must learn what is the minimun percentage of police agents that should be deployed to control the traffic environment.

Table 2. Agent populations

Parameters	population1	population2	population3	population4	population5
$fulfill\_prob$	0.5	0.5	0.5	0.5	0.5
$high\_punishment$	5	8	10	12	14
inc_prob	0.2	0.2	0.2	0.2	0.2

Learning AEI parameters is a rather complex task because individual AEI's goals are interrelated and can generate conflicts (for example, increasing police helps with collisions but raises costs). Furthermore, goals are related to agents' behaviors. As explained before, agent's behavior is so that its probability of complying with norms is proportional to the percentage of police, and therefore, since norms contribute to reduce accidents, collisions increase when police decrease. Moreover, the more percentage of police is deployed, the more number of fines are applied, and thus, the higher number of cars are expelled. Nevertheless, agents generate less offences when the police percentage increases. Additionally, the number of expelled cars decreases proportionally, not only to the police percentage, but also to the amount of applied fines. On the other hand, it may be worth recalling that agent's behavior also increases its probability of complying with norms when the fine is larger than $high\_punishment$. Therefore, any fine value higher than the population's $high\_punishment$ value will have the same effect, and thus, will generate equivalent individual goal satisfaction

Table 3. Learning results for five different agent populations

Population	Learned $fine_{right}$	Learned $fine_{front}$	Learned $police$	Goal satisfaction
population1	15, 12, 7	8, 14, 13	0.93, 0.93, 0.93	0.699, 0.7, 0.691
population2	13, 13, 14	10, 11, 9	0.93, 0.93, 0.93	0.689, 0.694, 0.691
population3	15, 12, 15	14, 11, 15	0.93, 0.87, 0.93	0.685, 0.681, 0.685
population4	15, 13, 15	14, 13, 13	0.93, 0.93, 0.87	0.676, 0.686, 0.68
population5	15, 15, 15	15, 15, 8	0.93, 0.93, 0.93	0.668, 0.674, 0.677

degrees. As a result, the AEI must learn the best combination of parameters ($fine_{right}$, $fine_{front}$ and $police$) according to the 4-goal objective function and to the agents' behavior.

When learning, we have repeated tests for each setting three times –i.e., three separated learning runs for each agent population and setting–. Table 3 shows the learned parameters, where columns Learned $fine_{right}$, Learned $fine_{front}$, and Learned $police$ include the learned values for each corresponding parameter and agent population. Each cell in the table contains three values: one per repeated experiment. Thus, for example, considering population1, learned values for $fine_{right}$ are 15 for the first test, 12 for the second one and 7 for the third test. Notice that, due to the agent's behavior, any fine value higher that 5 ($high\_punishment$ value) will have the same effect. Table 3 also shows the goal satisfaction value obtained for each test (this value corresponds to the objective function value explained above $(O(V))$ using $maxCol = 150$, $maxOff = 200$, $maxExpel = 200$ and $maxPolice = 1$).

As it can be seen, learned fines are larger than the population's $high\_punishment$ value except for the third test in population5 (where the GA fails to find the maximum). Therefore, the institutional goals are successfully reached in fourteen of the fifteen tests. In this manner, we can rather state the AEI succeeds in learning the norms that better accomplish its goals. Relating to the police percentage, learned values are close to the 90%. This is due to its low associated weight in the objective function. Notice that the objective function is weighted in such a way that goals aiming to decrease the number of collision and offenses are considered to be significantly more important than those that pursue to decrease the number of expelled cars and the police percentage.

For the seek of clarity, figure 4 shows the overall goal functions for population1, population3 and population5 respectively. These 3D charts depict all the values of the goal function using only two parameters ($fine_{right}$ and $police$[2] but not $fine_{front}$), so that search space for the learning algorithm is kept 3 dimensional. The domain for each is $16 \times 16 \times 1$. The figure shows the dependency between both parameters: when the police percentage is 100% the effect of the norm fine ($fine_{right} > high\_punishment$) is greater than for smaller values of $police$, and becomes null when the police percentage goes down to 0%.

---

[2] Notice that the parameter $police$ is scaling to 15.

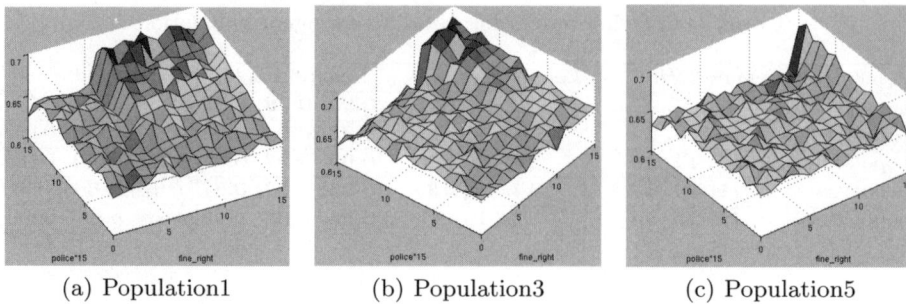

(a) Population1   (b) Population3   (c) Population5

Fig. 4. Objective functions with 4 goals: col,off,expel,police. (a)Population1 ($high\_punishment = 5$), (b)Population3 ($high\_punishment = 10$), (c)Population5 ($high\_punishment = 14$).

6 Discussion and Future Work

Within the area of Multi-Agent Systems, adaptation has been usually envisioned as an agent capability: agents learn how to reorganise themselves. Along this direction, works such as the one by Excelente-Toledo and Jennings [16] propose a decision making framework that enables agents to dynamically select the coordination mechanism that is most appropriate to their circumstances. Hübner et al. [17] propose a model for controlling adaptation by using the $\mathcal{M}$OISE+ organization model, and Gâteau et al. [12] propose $\mathcal{M}$OISE^{Inst} as an extension of $\mathcal{M}$OISE+ as an institution organization specification of the rights and duties of agents' roles. In both models agents adapt their MAS organization to both environmental changes and their own goals. In [18] Gasser and Ishida present a general distributed problem-solving model which can reorganize its architecture, in [19] Ishida and Yokoo introduce two new reorganization primitives that change the population of agents and the distribution of knowledge in an organization; and Horling et al. [20] propose an approach where the members adapt their own organizational structures at runtime. The fact that adaptation is carried out by the agents composing the MAS is the most significant difference with the approach presented in this paper. In our approach there is indeed a group of internal agents who can punish external agents but the reorganization is carried out by the institution, instead of by the agents.

On the other hand, it has been long stated [21] that agents working in a common society need norms to avoid and solve conflicts, make agreements, reduce complexity, or to achieve a social order. Boella et al. [9] approached the change of norms by using constitutive norms that make possible to create new norms. Our approach differs from their because we only modify norms, instead of creating new norms. Norm adaptation has been also considered from the individual perspective of agents within an agent society. Thus, in [22] agents can adapt to norm-based systems and they can even autonomously decide its commitment to obey norms in order to achieve associated institutional goals. Unlike this, we focus on adapting norms instead of adapting agents to norms.

Most research in this area consider norm configuration at design time instead of at run-time as proposed in this paper. In this manner, Fitoussi and Tennenholtz [23] select norms at design stages by proposing the notions of minimality and simplicity as selecting criteria. They study two basic settings, which include Automated-Guided-Vehicles (AGV) with traffic laws, by assuming an environment that consists of (two) agents and a set of strategies available to (each of) them. From this set, agents devise the appropriate ones in order to reach their assigned goals without violating social laws, which must be respected. Our approach differs from it becasue we do not select norms at design stages. Previously, Sierra et al. [24] used evolutionary programming techniques in the SADDE methodology to tune the parameters of the agent populations that best accomplished the global properties specified at design stages by the electronic institution. Their approach differs from our approach because they search the best population of agents by a desired institution and we adapt the institution to the population of agents.

The most similar work to ours is [25]. Their proposed approach to adapt organizations to environmental changes dynamically consists on translating the organizational model into a max flow network. Therefore, their purpose differs from ours because they only focus on adapting to environment fluctuation, and because their work is based on organizational models instead on norms.

Regarding the traffic domain, MAS has been previously applied to it [11] [26], [27]. For example, Camurri et al. [28] propose two field-based mechanisms to control cars and traffic-lights. Its proposed driving policy guides cars towards their (forward) destinations avoiding the most crowded areas. On the other hand, traffic light control is based on a linear combination between a distance field and the locally perceived traffic field. Additionally, authors combine this driving policy and traffic light control in order to manage to avoid deadlocks and congestion. Traffic has been also widely studied outside the scope of MAS, for example, the preliminary work by [29] used Strongly Typed Genetic Programming (STGP) to control the timings of traffic signals within a network of orthogonal intersections. Their evaluation function computed the overall delay.

This paper presents AEI as an extension of EIs with autonomic capabilities. In order to test our model, we have implemented a traffic AEI case study, where the AEI learns two traffic norms and the number of institutional agents in order to adapt the norms and the performative structure to different agent populations. Preliminary results in this paper provide soundness to our AEI approach. We are also currently performing the same experiments with other norms and with more goals. As future work, and since this basically represents a centralized scenario, we plan to develop a more complex traffic network, allowing us to propose a decentralized approach where different areas (i.e., junctions) are regulated by different institutions. Additionally, we are interested in studying how institutional norms and agent strategies may co-evolve. Nevertheless, this will require to extend the agents so that they become able to adapt to institutional changes. Nevertheless, we plan to extend both our traffic model and the institutional adaptation capabilities so that the AEI will not only learn the most appropriate norms for a given agent population, but it will be able to adapt to any change in the population.

Acknowledgments. This work was partially funded by the Spanish Science and Technology Ministry as part of the Web-i-2 project (TIC-2003-08763-C02-01) and by the Spanish Education and Science Ministry as part of the IEA (TIN2006-15662-C02-01) and the 2006-5-0I-099 projects. The first author enjoys an FPI grant (BES-2004-4335) from the Spanish Education and Science Ministry.

References

1. Luck, M., McBurney, P., Shehory, O., Willmott, S.: Agentlink Roadmap. Agenlink.org (2005)
2. Kephart, J.O., Chess, D.M.: The vision of autonomic computing. IEEE Computer 36, 41–50 (2003)
3. Esteva, M.: Electronic Institutions: from specification to development. IIIA PhD Monography, vol. 19 (2003)
4. Jennings, N.R., Sycara, K., Wooldridge, M.: A roadmap of agent research and development. Autonomous Agents and Multi-agent Systems 1, 275–306 (1998)
5. North, D.: Institutions, Institutional Change and Economics Perfomance. Cambridge U. P, Cambridge (1990)
6. Arcos, J.L., Esteva, M., Noriega, P., Rodríguez-Aguilar, J.A., Sierra, C.: Engineering open environments with electronic institutions. Engineering Applications of Artificial Intelligence, 191–204 (2005)
7. Etzioni, A.: Modern Organizations. Prentice-Hall, Englewood Cliffs, NJ (1964)
8. Gelfond, M., Lifschitz, V.: Representing action and change by logic programs. Journal of Logic Programming 17, 301–321 (1993)
9. Boella, G., van der Torre, L.: Regulative and constitutive norms in normative multiagent systems. In: Proc. of Int. Conf. on the Principles of Knowledge Representation and Reasoning (KR'04), Whistler (CA) (2004)
10. Yang, Q.: A Simulation Laboratory for Evaluation of Dynamic Traffic Management Systems. PhD thesis, MIT (1997)
11. Luke, S., Cioffi-Revilla, C., Panait, L., Sullivan, K.: Mason: A new multi-agent simulation toolkit. In: Proceedings of the 2004 SwarmFest Workshop, vol. 8 (2004)
12. Gâteau, B., Khadraoui, D., Dubois, E.: Moiseinst: An organizational model for specifying rights and duties of autonomous agents. In: 3rd European Workshop on Multiagent Systems (EUMAS'05), Brussels, Belgium (2005)
13. López-Sánchez, M., Noria, X., Rodríguez-Aguilar, J.A., Gilbert, N.: Multi-agent based simulation of news digital markets. International Journal of Computer Science and Applications 2, 7–14 (2005)
14. Cerquides, J., López-Sánchez, M., Reyes-Moro, A., Rodríguez-Aguilar, J.: Enabling assisted strategic negotiations in actual-world procurement scenarios. Electronic Commerce Research 2006 (to appear)
15. Chipperfield, A., Fleming, P.: The matlab genetic algorithm toolbox. IEE Colloquium Applied Control Techniques Using MATLAB, 10–14 (1995)
16. Excelente-Toledo, C.B., Jennings, N.R.: The dynamic selection of coordination mechanisms. Autonomous Agents and Multi-Agent Systems 9, 55–85 (2004)

17. Hübner, J.F., Sichman, J.S., Boissier, O.: Using the ℳoise+ for a cooperative framework of mas reorganisation. In: Bazzan, A.L.C., Labidi, S. (eds.) SBIA 2004. LNCS (LNAI), vol. 3171, pp. 506–515. Springer, Heidelberg (2004)
18. Gasser, L., Ishida, T.: A dynamic organizational architecture for adaptive problem solving. In: Proc. of AAAI-91, Anaheim, CA, pp. 185–190 (1991)
19. Ishida, T., Yokoo, M.: Organization self-design of distributed production systems. IEEE Trans. Knowl. Data Eng. 4, 123–134 (1992)
20. Horling, B., Benyo, B., Lesser, V.: Using Self-Diagnosis to Adapt Organizational Structures. In: Proceedings of the 5th International Conference on Autonomous Agents, pp. 529–536 (2001)
21. Conte, R., Falcone, R., Sartor, G.: Agents and norms: How to fill the gap? Artificial Intelligence and Law, 1–15 (1999)
22. López-López, F., Luck, M., d'Inverno, M.: Constraining autonomy through norms. In: AAMAS '02: Proceedings of the first international joint conference on Autonomous agents and multiagent systems, pp. 674–681. ACM Press, New York (2002)
23. Fitoussi, D., Tennenholtz, M.: Choosing social laws for multi-agent systems: Minimality and simplicity. Artificial Intelligence 119, 61–101 (2000)
24. Sierra, C., Sabater, J., Agustí-Cullell, J., García, P.: Integrating evolutionary computing and the sadde methodology (2003)
25. Hoogendoorn, M.: Adaptation of organizational models for multi-agent systems based on max flow networks. In: Proceedings of the Twentieth International Joint Conference on Artificial Intelligence. AAAI Press, Stanford, California, USA 2007 (to appear)
26. Dresner, K., Stone, P.: Multiagent traffic management: An improved intersection control mechanism. In: The Fourth International Joint Conference on Autonomous Agents and Multiagent Systems, pp. 471–477. ACM Press, New York (2005)
27. Doniec, A., Espié, S., Mandiau, R., Piechowiak, S.: Dealing with multi-agent coordination by anticipation: Application to the traffic simulation at junctions. In: EUMAS, pp. 478–479 (2005)
28. Camurri, M., Mamei, M., Zambonelli, F.: Urban traffic control with co-fields. In: Weyns, D., Parunak, H.V.D., Michel, F. (eds.) E4MAS 2006. LNCS (LNAI), vol. 4389, pp. 239–253. Springer, Heidelberg (2006)
29. Montana, D.J., Czerwinski, S.: Evolving control laws for a network of traffic signals. In: Genetic Programming 1996: Proceedings of the First Annual Conference, Stanford University, CA, USA, pp. 333–338. MIT Press, Cambridge (1996)

Managing Resources in Constrained Environments with Autonomous Agents

C. Muldoon[1], G.M.P. O'Hare[2], and M.J. O'Grady[2]

[1] Practice and Research in Intelligent Systems and Media (PRISM) Laboratory,
School of Computer Science and Informatics, University College Dublin (UCD),
Belfield, Dublin 4, Ireland
conor.muldoon@ucd.ie

[2] Adaptive Information Cluster (AIC), School of Computer Science and Informatics,
University College Dublin (UCD), Belfield, Dublin 4, Ireland
{gregory.ohare,michael.j.ogrady}@ucd.ie

Abstract. In the future electronic devices will permeate the environment where they will work invisibly and autonomously to deliver new and enhanced services that go far beyond the mandate of the desktop era. Intelligent agents will form the basis of many applications in this emergent ubiquitous domain. Agent Factory Micro Edition (AFME) is a framework that facilitates the construction of agent-based applications for computationally constrained devices, this paper outlines three enhancements introduced to AFME to enable resources to be managed more effectively, namely a new threading model, an extended rational decision making infrastructure, and a syntactic modification to the agent programming language that improves efficiency. The extended reasoning capabilities of AFME enable agents to choose the most appropriate course of action with respect to their finite resources in a social context.

1 Introduction

The true potential for information technology is in making it an integral, invisible part of the way people live their lives [1]. Ubiquitous computing prescribes a new model of computation, one that takes into consideration the natural human environment and removes computers from our direct focus into the objects that surround us. In the ubiquitous computing era intelligent agents will operate on 'smart' devices where they will manage and control dynamic ad-hoc networks, working both reactively and pro-actively to achieve individual and common goals. This paper concerns three features introduced to Agent Factory Micro Edition (AFME) [2], a framework for the construction of agents that operate on resource constrained mobile devices.

A new threading model, embedded within AFME, ensures that agents are sensitive to absolute time.[1] This enables agent response time values to be specified accurately. Rather than each agent creating their own thread agents share a thread

---
[1] When the application begins to execute a clock value is recorded in the scheduler. Subsequent timing values are relative to the initial clock value.

pool whereby they are scheduled to execute at regular intervals. To reduce the number of computational bottlenecks, and prevent agents from synchronising with one another, response time values are altered such that they are prime numbers.

To be able to solve problems collectively social structures must be present to enable agents to interact with each other. For agents to act as a team more is required than the synchronisation of individual isolated events. The group must act as a single agent that adopts beliefs, desires, and intentions of its own [3]. The manner by which collective decisions are made, however, is ultimately governed by the choices made by, the desires of, and the goals of the individuals that form the group. The extended rational decision making capabilities introduced to AFME enable agents to choose the most appropriate course of action with respect to their finite resources.

Belief labeling has been introduced to AFME to improve the efficiency of the reasoning process and to reduce development time. With belief labeling common sequences of predicates are only encoded and evaluated once.

The paper is organised as follows. Section 2 provides a broad overview of AFME. Section 3 describes the threading model. Section 4 describes the rational decision making infrastructure. Section 5 discusses belief labeling. An evaluation is provided in section 6. Some related work and a discussion that focuses on the social aspects of the system is provided in section 7.

2 AFME

AFME is loosely based on Agent Factory [4]. It uses a subset of the Agent Factory Agent Programming Language (AFAPL) [5] and augments it with a number of features specific to AFME. AFAPL is founded on a logical formalism of belief and commitment. Rules that define the conditions under which agents should adopt commitments are used to govern and encode agent behaviour. AFME is described elsewhere [6] [2]. This section provides a broad overview of AFME to put the work in context but the primary focus of this paper is on the three features introduced to AFME to enable resources to be managed more effectively.

An AFME platform comprises a scheduler, several platform services, and a group of agents (see figure 1). The scheduler is responsible for the scheduling of agents to execute at periodic intervals. Rather than each agent creating a new thread when they begin operating, agents share a thread pool.

AFME delivers support for the creation of BDI agents that follow a *sense-deliberate-act cycle*. The control algorithm performs four functions. (1) Preceptors are fired and beliefs are resolved within the belief resolution function. (2) The beliefs are used within the deliberation process to identify the agent's desired states. Agents are resource bounded and will be unable to achieve all of their desires even if their desires are consistent. (3) A subset is chosen, within the *intention selection process*, that maximises their self-interest with respect to their finite resources. (4) The final function of the control algorithm concerns commitment management. Depending on the nature of the commitments adopted various actuators are fired.

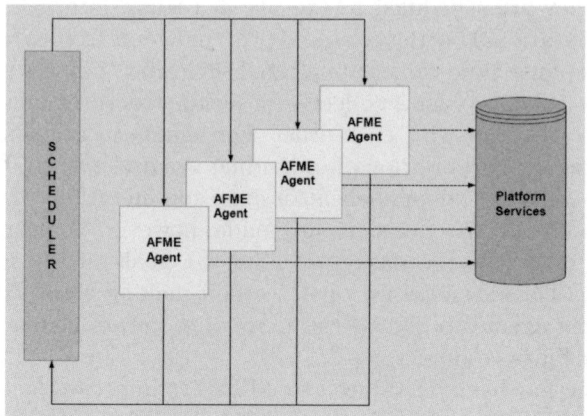

Fig. 1. AFME Architecture

AFME has been used in several applications [7] [8]. A discussion of these applications is beyond the scope of the paper. A key requirement for the development of an agent platform for constrained environments is that the platform shall manage resources efficiently and effectively. This is primary focus of this paper.

3 Threading Model

3.1 Thread Management

Agents in AFME are scheduled to execute at periodic intervals. In the original Agent Factory threading model when an agent was constructed its controller created a new thread and the agent began executing. A sleep time parameter[2] was passed to the controller and was used to determine the responsiveness of the agent and to facilitate cooperative multi-threading. The original system was somewhat limited in that all of the threads operated in an independent adhoc manner, there was no management or scheduling of the threads involved. Consider the use of the sleep time to indicate the responsiveness of the agent. The original system did not support the concept of fixed-rate execution, whereby subsequent executions take place at approximately regular intervals that are sensitive to absolute time. In the AFME threading model the agent's true response time is determined by a combination of the time taken to perform background activities (such as garbage collection), the agent's execution time, and the agents sleep time.

The objective of the new threading model is to introduce thread management and scheduling into AFME. The system is composed of a scheduler, a thread pool, and a task buffer. The threading model can be used to schedule any task

---

[2] The sleep time parameter specified the amount of time the agent's thread would sleep in between iterations of the control algorithm.

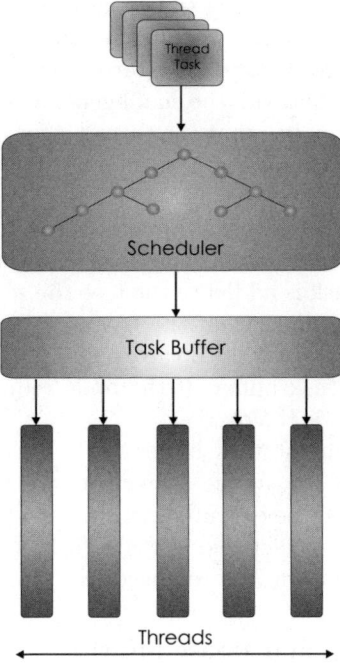

Fig. 2. Threading Model

but in AFME the task is typically an agent. The threading model is used in the collaborative agent tuning framework [9]. Figure 2 illustrates the architecture of the new threading model. Rather than starting a new thread when an agent is created, the agent's control algorithm is incorporated into a thread task object that is added to the scheduler. When the task is added the desired response time is specified. The task is then scheduled to be performed at approximately regular intervals with respect to absolute time. The amount of sleep time is varied so that the response time will be consistent. The accuracy of the timing is dependent on the clock underlying the *Object.wait()* method. The scheduler synchronises all response times with an initial clock value that is recorded when the application begins to operate. Task sleep times are constantly adjusted to be in synchronisation with the universal clock value[3] so that an agent's response time value will be regular and consistent with respect to (1) its previous execution times and (2) the relative execution times of the other agents.

The threading model is not limited to the execution of agents. Other internal platform or application tasks, such as IO operations, are scheduled to be performed at some point in the future either on a once off basis or periodically.

---

[3] It is possible to obtain a universal clock value as the threading model is only concerned with agents on the local platform. Agents on other platforms will be using different CPUs and therefore their computational load will not overlap with local agents.

These tasks are subject to the same timing criteria and scheduling algorithm. An example of where this is used is within the message transport service, which is required to periodically connect to an external server to receive incoming messages [2]. This is possible because the management logic of the thread pool is decoupled from the functionality that the thread is executing. A thread is simply viewed as a process. The management logic is not concerned about the specific task that the thread is performing.

The Scheduler contains an internal binary search tree to schedule tasks efficiently. The tree ensures that tasks are ordered in accordance to their scheduled execution time. When a task is added to the tree the scheduler thread is notified. The minimum node is obtained from the tree and its execution time is calculated. The scheduler thread waits until the execution time of the minimum node and then places it in the task buffer. If the node requires periodic execution it is rescheduled otherwise it is removed.

When placing tasks into the task buffer the scheduler first checks an active set of tasks. If the task is not in the active set it is added. The task buffer is composed of a set of tasks to be executed within one of the threads in the thread pool. The pooled threads extract and execute tasks from the task buffer. Once a task has completed execution it is removed from the buffer.

3.2 Primal Scheduling and Phase Shifting

To prevent the agents from synchronising with one another the agents' response times are altered to be prime numbers. For example consider two agents, one with a response time of 500 milliseconds, and the other with a response time of 1000 milliseconds. Both agents begin to execute at time 0 and finish executing at some later point. At time 500 the first agent begins to operate again and then finishes at some later point. At time 1000 because both agents are sensitive to absolute time they will both begin to execute at the same point because any variance incurred due to background tasks or executing times will have been removed. The cycle will repeat itself and both agents will begin executing at the same time point at 2000, 3000, 4000... milliseconds. The harmonics of the agents' response times cause the agents to synchronise with one another, the agents are literally in tune. This causes a computational bottleneck. It is undesirable, from a computational efficiency perspective, to have two or more agents beginning to execute at exactly the same time. The time between the last agent finishing to execute and the next agent or the next two agents beginning to execute is effectively wasted.

To prevent this problem agents' response times are altered to be prime numbers. Consider the case when the first agent's response time is changed to 499 and the second agent's response time is changed 1013. Rather than the first agent synchronizing with the second agent once every two iterations and the second agent synchronizing with the first agent on every iteration the first agent will synchronize with the second agent once every 1013 iterations whereas the second agent will synchronize with the first agent once every 499 iterations. The worst case scenario, whereby both agents begin executing at exactly the same time, only occurs once every 505487 (499*1013) milliseconds rather than once every 1000 milliseconds. The system ensures that the agents are out of tune with respect to each other.

The effect that this has is to average the agents' computational load over the available time range. The agents' computational overhead will still sometimes overlap but the time between agents completing execution and subsequently beginning to execute will not always be wasted. It should be noted that 1013 was chosen rather than a prime value, such as 997, closer to the original because it would cause the agents to come out of synchronization faster since there is a greater difference between the harmonics of 1013 and 499 than that of 997 and 499. The primes chosen are those that have the greatest harmonic difference within a particular threshold value. The threshold value is intended to keep the values close to the originals specified. If there are no primes within this range then the primes closest to the originals are chosen.

To further reduce computational bottlenecks the threading model phase shifts thread tasks with equal response times. Consider two agents with response times of 499. Rather than altering their responsiveness to be different primes such 491 and 509 the system phase-shifts one of the agents by 180 degrees. So the first agent will begin executing at time 0 whereas the second agent begins executing at time 249. Because the agents are sensitive to absolute time and have equal responsiveness values they will never begin executing at exactly the same point, there will always be approximately 249 milliseconds between their execution times. The number of degrees that the agents are phase shifted is equal to 360 divided by the number of agents with equal responsiveness multiplied by the agent's arbitrary ordinal number. Thus in the previous example the first agent is phase shifted by 249 * 0, i.e. it is not phased shifted, whereas the second agent is phase shifted by 249 * 1. The choice as to which agent is first, second, third... is capricious.

Another improvement to the efficiency of the threading model is that it randomizes start times. When applications begin to operate rather than having all tasks or agents begin to execute at time 0 the system staggers agents' and other scheduled tasks' start times.

4 Rational Decision Making

The requirement for agents to be rational necessitates that they act in a manner that maximises their self-interest or utility. The term self-interest was first popularised by Adam Smith [10] during the enlightenment period. In AFME Agents are rational. Their self-interest enables them to act in a consistent manner to achieve their objectives. For some agents their objectives necessitate selfish behaviour for others they do not.

AFME is consistent with the BDI model of agency [11]. The BDI model acknowledges that agents are resource bounded and thus will be unable to achieve all of their desires even if their desires are consistent. The agent must fix upon a subset of desires and commit resources to achieving them. This subset of desires is the agent's intentions. To be able to fix upon a subset of desires the agent must make a decision over a number of options. There must be some concept of utility or preference for an agent to make a decision[4]. Otherwise the agent would be unable to choose between the various alternatives. Even a random or arbitrary choice requires the concept of utility in that all utility values must be equal, thus no preference is specified. A random choice is rational in such circumstances because it returns the maximum possible benefit from the various alternatives.

This illustrates the difference between utility and preference. Utility specifies the benefit of an option whereas preference specifies a greater than relation between utilities. To say that the taste of a cake is preferred to the taste of some other cake does not imply that they both taste equally as nice. It implies that one tastes better than the other. If you prefer something, you believe it to be better than something else. This differs from weak stipulative definitions of preference [12] whereby two items can be equally preferred to each other. In the context of this work if the utility values of two items are equal, no preference is specified between those two items. Stipulative definitions cannot be considered correct or incorrect because they are not propositions. A definition can only be considered correct or incorrect when discussing usage, in such a case the definition is lexical.

The intention selection process has been realised within AFME by extending the Agent Factory Agent Programming Language (AFAPL) with additional constructs for rational decision making. In AFME the subset of commitments chosen maximises the agent's utility for a particular course of action. In AFAPL rules that define the conditions under which agents adopt commitments are used to govern an agent's behaviour. These rules are based on the agent's current model of the world, namely the agent's beliefs. In AFME's extended version of AFAPL the decision as to whether to adopt a commitment is contingent on the amount of resources an agent has available to it and the previous commitments that it has made. For example if an agent has a significant number of commitments it will decide not to adopt an additional commitment that it otherwise would have adopted if it did not already have such a heavy workload. If the agent does not have the requisite resources available the commitment is not adopted. If the agent can free additional resources by dropping previous commitments and the benefit of adopting the new commitment is greater than the loss of dropping the previous commitments then the agent drops the older commitments to make the requisite resources available. At a given point in time an agent will have, based on its model of the world, a number of commitments that it wishes to adopt. The agent chooses to espouse the subset of commitments that maximises its utility with respect to its finite resources. If an agent fails to achieve a commitment they must still incur the costs of the attempt. This is true for both human and computational agents.

---

[4] The utility is usually determined in the interpreter and is not specified within the logic components.

In BDI logics the concept of desire is a qualitative representation of utility. The intentions are chosen, within the interpreter, from among the desired states, using some metric that represents the actual utility value. In the AFAPL extension the metric for determining the utility is removed from the intention selection algorithm and is replaced within perceptors that generate beliefs about the costs and benefits of certain actions[5]. This is useful because it enables different metrics to be used for different commitments. The values generated by the perceptors are only potential utility. The chosen commitments must still be from among the desired states as determined by the commitment rules within the agent design. If a commitment is desired then the potential utility value will represent its actual utility value within the intention selection algorithm otherwise the commitment is not considered for selection. The desires are still a qualitative representation of utility.

Decoupling the utility metric from the intention selection algorithm has other advantages. It enables the developer to more easily alter system behaviour within the agent design. The benefit and cost of certain actions will be dependent on context. Variable beliefs provide a natural way of representing such data. The beliefs are not exclusive to the intention selection process and will sometimes be used by the agent to drive other system behaviour.

In the AFAPL extension the total amount of resources available to the agent is specified so that the agent is aware of its limitations or constraints. The following is an example of an agent design written with the addition constructs.

```
resources: ?res;

BELIEF(a) & BELIEF(costX(?cx)) & BELIEF(benefitX(?bx))
=> COMMIT(Self, Now, BELIEF(true), doX, ?cs, ?bx);

BELIEF(b) & BELIEF(costY(?cy)) & BELIEF(benefitY(?by))
 => COMMIT(Self, Now, BELIEF(true), doY, ?cy, ?by);

BELIEF(c) & BELIEF(costZ(?cz)) & BELIEF(benefitZ(?bz))
=> COMMIT(Self, Now, BELIEF(true), doZ, ?cz, ?bz);
```

In AFAPL terms preceded by the question mark symbol are variables. For illustrative purposes consider the case where the variables ?cx, ?bx, ?cy, ?by, ?cz, and ?cz assume the values 30, 11, 25, 5, 10, and 2 respectively. These variables represent the cost and benefits of adopting the commitments. Assume the resources variable, ?res, has a value of 20. If at a given point in time the agent adopts either *BELIEF(a)*, *BELIEF(b)*, or *BELIEF(c)*, along with the beliefs for costs and benefits, the commitments for *doX*, *doY*, *doZ* will be adopted respectively. The agent compares the commitment's cost to the available resources and if the cost is lower the commitment is adopted. If the agent adopts all three beliefs at

---

[5] The utility values do not necessarily have to be determined by a perceptor. They may be hard coded by the developer in the agent design but this will not usually be the case because it will often be difficult to determine utility values a priori.

the same discrete time point all three commitments are adopted because their combined cost is 18 which is less than the available resource allocation. Commitment resolution is non-deterministic, it does not make any difference what order the commitment rules are specified. The subset of commitments that maximise utility with respect to available resources at a particular time point will be chosen to be adopted.

Consider the case when all three beliefs are adopted but when the resource constraint is set at 15 rather than 20. The agent wishes to adopt all three commitments but does not have the resources to do this and must therefore make a decision as to what commitments to espouse. Considering the available resources the agent's options are (1) *doX* and *doZ*, or (2) *doY* and *doZ*. The agent chooses *doX* and *doZ* because it yields a greater utility.

The costs and benefits of temporal commitments will sometimes vary over time. If this is the case agents adopt beliefs about the cost and benefits of maintaining the commitments. Variables within the maintenance condition of a commitment are matched with the variables for cost and benefit.

The resource constraints specified are an abstract representation of resources rather than a direct reference to the CPU's computational overhead. There are a number of reasons why this approach has been adopted. The agent does not need to be aware of its low level processing performance to reason about its potential actions in much the same way that a person does not need to be aware of the exact number of joules they are going to use to perform a particular task. People have a more abstract notion of the amount of work involved and use this abstract concept, along with other factors such as their spatial and temporal constraints, previously adopted commitments etc., within their decision making process. The abstraction effectively hides such low level data. The tasks typically performed by intelligent agents are at a higher level of granularity than CPU scheduling. An agent's current load will only be required when an agent is making a decision as to whether or not to perform a future action. Processor performance is not useful in determining the overhead of future events because it can only be measured after the event has occurred. An agent's abstract representation of the cost of its commitments would be more useful in such situations. A developer could profile a particular application and use the data as a pre-emptive indication of a task's overhead. The data must still only be used as an approximation to overhead and be specified abstractly because it is dependent on the platform on which profiling was carried out. The abstract representation of resource usage facilitates the development of generic agents whose functionality is not coupled or dependent on particular processor capabilities. Java is a platform independent language and hides processor specific information from the developer. J2ME does not support the Java native interface so platform dependent code cannot be written to obtain such data in any case.

The metric for determining the amount of resources available could of course include the residual power of an embedded device along with its computational constraints. In such a case the power of the device would have an effect on the nature and degree of the commitments adopted.

The original AFAPL is a particular instance of the extended language specification whereby the benefit of adopting commitments is 1, the cost of adopting commitments is 0, and the resource allocation is non negative. In this case the agent will adopt all of its potential commitments because each commitment will increase its utility for free. In the extended version of the language if the values for a commitment's benefit and cost are not specified their default values are 1 and 0 respectively. If the resource allocation is not specified it has a default of 0. In this respect, it is consistent will the original specification. An agent written in the standard AFAPL will act in the same manner within an AFME environment as it does in the standard framework.

In reality when someone is considering adopting a non-trivial commitment their beliefs about the costs and benefits of the commitment along with an abstract concept of the amount of resources available to them form an integral part of their reasoning process. Over time their beliefs about the costs and the benefits of adopting certain commitments change because they are dependent on context.

5 Belief Labeling

Belief labeling has been introduced to improve the efficiency of the reasoning algorithm and reduce development time. It is syntactically different from AFAPL but the underlying semantics are equivalent. The following commitment rules

```
BELIEF(x) & BELIEF(y) is somelabel;
BELIEF(w) => COMMIT(...) requires somelabel;
BELIEF(z) => COMMIT(...) requires somelabel;
```

are an equivalent alternative to

```
BELIEF(x) & BELIEF(y) & BELIEF(w) => COMMIT(...);
BELIEF(x) & BELIEF(y) & BELIEF(z) => COMMIT(...);
```

similarly

```
BELIEF(z) => COMMIT(...) requires !somelabel;
```

is equivalent to

```
!BELIEF(x) &  BELIEF(y) & BELIEF(z) => COMMIT(...);
 BELIEF(x) & !BELIEF(y) & BELIEF(z) => COMMIT(...);
!BELIEF(x) & !BELIEF(y) & BELIEF(z) => COMMIT(...);
```

The second case is not logically equivalent to the single commitment rule

```
BELIEF(z) => COMMIT(...);
```

because in that case if both *BELIEF(x)* and *BELIEF(y)* were adopted the commitment would still be espoused, this is not what we want. In this example the agent will only ever adopt one commitment because each rule is mutually exclusive. Its clear, in this case, that considerably less reasoning is required than the original AFAPL approach. Additionally, the developer need only write one line of code rather than three (provided *somelabel* is already declared). This improves efficiency as the belief sequence is only evaluated once. The behaviour encoded is a negated *logical and*.

Agent designs that contain belief labels may be compiled into a format that is equivalent to the original AFAPL syntax so that the agent can run on the standard platform. When an agent migrates from an AFME environment it is converted to the alternative format.

The developer need not worry about writing the optimal agent design for performance. The compiler converts the code to the optimum automatically. Nevertheless, the developer can save time by doing so and improve the maintainability of their code by minimising redundancy and increasing reuse. Standard AFAPL design files can be compiled into the belief label format using the AFME compiler.

Belief labels can be embedded to further reduce redundancy. The *somelabel* belief set is embedded in the *otherlabel* belief set in the following example.

```
BELIEF(x) & BELIEF(y) is somelabel;
BELIEF(w) & BELIEF(z) is otherlabel requires somelabel;
```

It should be noted that cycles are prohibited. If the developer encodes cyclical dependencies the compiler will throw an error.

6 Evaluation

6.1 Evaluation of Threading Model

To evaluate the effectiveness of the threading model we conducted four experiments. In the first two experiments the unmanaged approach[6], the scheduled (timed) approach, the phase shifting approach, and the random start time approach were compared. Agents with equal average response time values[7] and random computational overheads were used. The second two experiments evaluated primal scheduling.

The agents summed the integers from 0 up to a pseudo random number chosen between 0 and 100000. Arithmetic series rules were not used. The same seed was used to generate the pseudo random numbers to ensure that the results were consistent in each case. In the new threading model agents share a thread pool rather creating their own thread. The number of threads in the pool is three. The results of the computational overhead experiment are illustrated in figure 3.

The scheduled (timed) approach is the least efficient. This is as we would expect in that by ensuring that agents' response time values are consistent we

---

[6] The unmanaged (original) approach refers to when scheduling is not used.
[7] The term response time cannot be used for agents using the unmanaged approach. In that case we are referring to the sleep time of the agent.

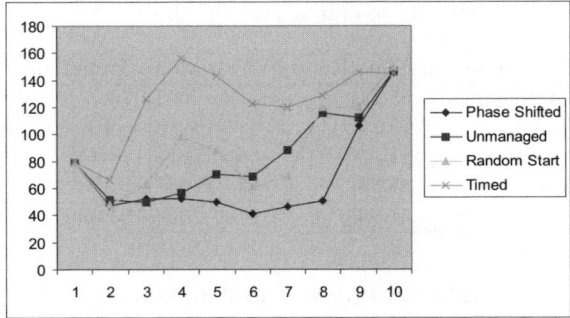

Fig. 3. Average time taken to process tasks with random computational overheads

ensure that agents are in tune with respect to each other. The random start time approach is sometimes better than the unmanaged approach other times it is not. The phase shifting approach is the most efficient. This is because it ensures that the agents never begin operating at the same time.

The experiment was repeated but this time the responsiveness of the agents was recorded rather than their computational overhead. The results of the responsiveness experiment are illustrated in figure 4.

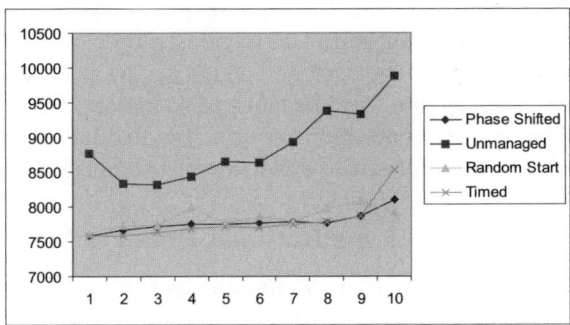

Fig. 4. Average processing duration for sixteen iterations for tasks with random overheads

The results indicate that Phase shifting is the most desirable. The problem is that phase shifting can only be used when the agents' response time values are equal. To improve the efficiency of agents that have different response times primal scheduling was developed.

To evaluate primal sceduling we conducted an experiment using four homogeneous agents with response times of 1000, 500, 250, and 125 milliseconds. Each of the agents has an equal computational overhead. The agents sum the integers from 0 to 100000. Again, arithmetic series rules were not used. The experiment was conducted with and without random start time values. The results for the

Table 1. Standard Scheduling Results

	Random Time	Random Duration	Time	Duration
125	55.11	2061	76.77	2689
250	65.11	3960	106	3866
500	81.22	7828	119.44	7619
1000	88.88	15585	115.33	15120
Average	71.33	7358	104.39	7324

Table 2. Primal Scheduling Results

	Random Time	Random Duration	Time	Duration
131	53.44	2103	60.44	2222
241	63.33	3804	60.55	3806
509	69.66	7993	69.77	7851
991	76	15393	71.77	15068
Average	65.61	7323	65.64	7237

computational processing time and responsiveness are given in table 1. These results illustrate that randomising the start times improves efficiency with the standard scheduling approach.

The experiment was repeated but with the response times altered, by the scheduler, to be 991, 509, 241, and 131. The results are given in table 2.

To enable the developer to specify the responsiveness of an agent accurately a scheduling procedure must be adopted. The accurate scheduling of agents causes problems because agents with harmonically similar response times become synchronized. These experiments indicate that by altering an agent's response time value to be a prime number the efficiency of the platform is improved.

6.2 Algorithm Analysis of the Rational Decision Making Extension

The task of determining the subset of commitments to espouse is a classic 0-1 knapsack problem. Given n items, with corresponding values and weights, the knapsack problem concerns the packing of some of these items in a knapsack of a specified capacity C, such that the profit sum of the included items is maximised. This is equivalent to the problem of an agent attempting to adopt the subset of commitments that maximise its utility with respect to its finite resources. It is called the 0-1 version of the knapsack problem because there are no fractions. The whole item must be either included in the knapsack or left out. Similarly, an agent cannot adopt half a commitment.

Since the pioneering work of Dantzig [13] the knapsack problem has been studied extensively in practice as well as in theory. No polynomial-time solution is known for the general case; it is non-deterministic polynomial-time hard[14]. The solution adopted in AFME uses a standard dynamic programming approach [15] and operates in pseudo-polynomial time. It has a run-time complexity of $O(nC)$. The complexity of dynamic programming solutions assure a much faster running time than other techniques, such as backtracking or brute-force. As noted in [14], "A pseudo-polynomial-time algorithm... will display 'exponential behavior' only when confronted with instances containing 'exponentially large' numbers, [which] might be rare for the application we are interested in." It is not anticipated that 'exponentially large' numbers will be encountered in AFME, nevertheless, future work will investigate the use of a greedy approximation algorithm.

6.3 Efficiency of Belief Labeling

Belief labeling reduces redundant processing. It enables developers to encode common sub sequences of predicates, which are only evaluated once. A depth first search is still used to match variables but the search space is considerably reduced. A depth first search has a runtime complexity of $\bigcirc(b^m)$, where b is the branching factor[8] and m is the maximum depth. To evaluate the efficiency of belief labeling we consider the worst case, in which the developer encodes a negated *logical and*. A negated *logical and* specifies the conditions under which a commitment will not be desired. Under all other circumstances it will be desired. To model this type of situation with the original approach the system would exhibit exponential behaviour. The developer would have to write, and the system would have to process, 2^n-1 rules for n conditions. With belief labeling the developer need only write, and the system need only process, 1 rule.

7 Discussion and Related Research

7.1 Agent Platforms

There have been several agent platforms developed for resource constrained environments reported in the literature. The LEAP [16] (Light Extensible Agent Platform) is a FIPA compliant agent platform capable of operating on both fixed and mobile devices. LEAP extends the JADE (Java Agent DEvelopment) architecture by using a set of profiles that allow it to be configured for various Java Virtual Machines (JVMs). The platforms supported are J2SE, Personal Java, and CLDC/MIDP. The architecture is modular and contains components for managing the life cycle of the agents and controlling the heterogeneity of communication protocols. The LEAP add on when combined with JADE replaces certain components of the standard JADE runtime environment to form a modified kernel that is referred to as JADE-LEAP or JADE powered by LEAP.

3APL-M [17] provides a platform that enables the fabrication of agents using the 3APL language for internal knowledge representation. It provides a scaled down version of the pre-existing language infrastructure, which was designed for a desktop environment. The 3APL-M architecture contains sensor and actuator modules, the 3APL machinery, and the communicator module. The sensor and actuator modules enable the agents to sense and to act upon their environment respectively. The 3APL machinery is a BDI reasoning engine. The communicator module provides the support for inter-agent communication. mProlog was

---

[8] The branching factor is the average number of children.

developed as a subcomponent of the 3APL-M project. It is a reduced footprint Java Prolog engine, optimized for J2ME applications.

Agilla [18] is an agent platform that has been designed specifically for wireless sensor networks where power consumption is an issue. Agilla agents are modelled as genetic algorithms and are not reflective.

MicroFIPA-OS is a minimised footprint version of the FIPA-OS agent toolkit [19]. The FIPA-OS was developed as an agent middleware environment to enable the creation of FIPA compliant agents. MicroFIPA-OS was constructed because the original FIPA-OS employed software engineering techniques, such as excessive object creation and XML parsing, that did not scale down well. The MicroFIPA-OS improves the efficiency of the system by avoiding or removing some of the additional overhead, such as mandatory XML parsing. It manages resources better and introduces thread and other resource pools that are shared among agents.

Though sharing the same broad objectives of these projects, AFME differs in a number of key ways. JADE-LEAP, and MicroFIPA-OS are frameworks for the development of agent technology but they do not contain reasoning capabilities. Any intelligence required must be written by the application developer. These systems therefore do not adhere to the same definition of agency as AFME. It is sometimes claimed that intelligent frameworks that have been developed for JADE will work with JADE-LEAP without making alterations to the code but this is in fact not the case. This would only work for the J2SE and perhaps Personal Java versions of JADE-LEAP. If an application were developed for JADE without considering the possibility of porting it to a CLDC/MIDP environment it would contain dependencies on standard Java classes and APIs not present or supported in the CLDC/MIDP specification.

3APL-M provides support for the construction of cognitive rational agents but it is an API not an agent development framework as such. This differs from AFME whereby the agent functionality is specified in a platform independent AFAPL design file and the Java code generated from the design through the use of the AFME compiler. The AFME development process supports the development of agents written in AFAPL through the use of visual debugging tools, a development methodology, and an integrated development environment.

The threading infrastructure differs from other systems in that agents are phase shifted and their response times are altered to be prime numbers so as to prevent computational bottlenecks.

The AFME migration process distinguishes itself from other approaches in that it hides platform idiosyncrasies. Generic agent designs are combined with platform specific functionality enabling agents to move between heterogeneous environments. This process is handled autonomically and is transparent from the agent's perspective.

7.2 Social Structure

Following on from section 4, for an action to be performed jointly more is required than just the union of simultaneous individual coordinated events. When a group decides to collaborate a team is formed that acts as a single agent with beliefs, desires, and intentions of its own above those of the individual team mates. [3] gives a formal model of the mental properties of teams and how joint intentions act, are affected by, and are reduced to the mental states of the team members.

In AFME agents are rational. Sometimes they collaborate other times they compete depending on the context. The decisions agents make are governed by their self interest[9]. If the developer wishes to guarantee that agents will always collaborate they encode utility values or use metrics such that it is always in an agent's interest to collaborate. Individually rational altruistic agents adopt common goals and help other agents at their own expense because they believe there is value in doing so even though it might never improve their personal welfare. The utility values adopted do not necessarily represent the personal welfare of the agent. In AFME some agents are selfish. Selfish agents benefit themselves but they also benefit society because to gain profit off one's own labours in an open market something must be provided that others value [10].

The manner by which collective decisions are made is ultimately governed by the choices made by, the desires of, and the goals of the individual agents that form the team. This is not to say that responsibility should end with the individual agent, but that it must start with the individual agent. Individuals will choose a course of action that maximises their self-interest. To illustrate the influence of variable utility values on the collective behaviour of agents consider a wireless sensor network with a number of interconnected nodes. Transmission in a wireless sensor network is quite a costly operation, for example the transmission of a single bit is equivalent to the execution of 1000 instructions with regards to power consumption on a typical node [20]. As the power level of the device decreases the utility of performing collective actions will decrease whereas the utility of performing an operation locally will increase. Different perceptors are required for generating the requisite utility values of collective and local actions. When the device's residual power is high the agent will communicate more often. When the power is low the agent will still sometimes communicate but not as much. The relationship between individual and collective action is not binary, it depends both on variable context values and the agent's mental state. In this case the power of the device has an effect on the nature and degree of the commitments adopted.

The behaviour of agents with similar goals is not the same as agents with common goals. For example consider two agents called Alice and Bob who both have a goal of getting to a particular football match, Alice and Bob have similar but not common goals. Alice's goal is for Alice to get to the football match. Bob's goal is for Bob to get to the football match. For Alice and Bob to have a common goal Alice would have to have a goal for Alice to get to the football match and Bob would also have to have a goal for Alice to get to the football match. In such a circumstance Alice and Bob would collaborate to ensure that Alice gets to the match. Sometimes developers will want agents to continue to collaborate even when resources are scarce. As noted earlier, in AFME this type

---

[9] It should be noted that term self-interest is not synonymous with selfishness.

of functionality is modeled through the use of fixed utility values that ensure that it is in an agent's interest to maintain common goals. If common goals are dropped agents will begin to compete.

To illustrate this point we shall consider the case when there are a number of agents operating on a node in a wireless sensor network. The residual power of the device is dropping. If all of the agents stay on the device there will not be enough resources to support them. In this scenario the agents must adopt a common goal regarding migration. The agents must negotiate in order to make a decision as to which agents should stay and which should leave as several agents want to stay on the device. It is essential that consensus is reached prior to the critical point beyond which the chances of a common goal being adopted are considerably reduced. The critical point is the time at which there is not enough power to transfer all of the agents. Beyond this point all of the agents will want to migrate because it becomes about survival.

There are a number of scenarios that can occur. If all of the agents are altruistic, they will have difficulty deciding which agents to save and which to leave behind. They identify this problem quite rapidly and begin a random select process. This is the virtual equivalent to the drawing of straws. The ultimate test of altruism is then for the agent that draws the short straw. It must be remembered that not all agents are altruistic. Malevolent agents will attempt to terminate other agents in order to guarantee their survival. Another scenario that can occur is that all agents are selfish. In this case all of the agents might attempt to migrate at the same time. This results in some, if not all, of them perishing. If there is a combination of altruistic and selfish agents, the selfish agents migrate and the altruistic agents perish. When programming agents the developer models high priority agents that operate in risky environments as selfish so as to increase their chances of survival[10]. It might well be the case that the agent is designed to alter its behaviour from selfishness when there is no resource shortage. That is, the agent is sensitive to context.

System designers often want agents to give up a certain amount of local autonomy to facilitate a global optimal usage of resources whereby all are better off. A balance must be struck, however, between societal authoritarianism and anarchy. In AFME the balance is in favour of local autonomy and decentralised decision making rather than enforced societal rules. To encode normative behaviour in AFME agents must adopt social roles. That is, the agents must be aware of the rules of the society in question. The decision as to whether they adhere to the rules, however, is made by the individual agents[11]. It depends on their (utility) values. It might well be the case that it is in an agent's interest to adhere to the rules. Nonetheless, when resources are scarce society tends to break down. Agents drop their social roles in order to survive. This is also true in the real world and can clearly be observed in disaster situations. If there is not enough resources for all, people tend to protect their own interests.

---

[10] This will only increase their chances provided that some of the others agents are not selfish.

[11] Future work will investigate the use of enforcement.

The problem of getting the balance right between local autonomy and societal authority is somewhat similar to what Bellman referred to as 'the macroscopic principle of uncertainty' in control theory [21]. In a similar vein to the microscopic principle of uncertainty in quantum mechanics, the macroscopic principle of uncertainty in control theory concerns getting the balance right between using a small amount of data and making an early choice, or taking a large amount of time to receive a significant amount of data before making a decision. This is a non-trivial problem. It led Bellman to conclude that "There is no way to control a large system perfectly". Either we accept errors due to the lack of information or we allow to system to carry on regardless as we are collecting data. There is an inherent cost in controlling a system. At one extreme anarchy prevails, there is no control, and at the other extreme the benefits of decentralised decision making are lost due to too much societal control. Various types of societies may be engineered in AFME; it depends on the requirements of the problem the developer is trying to solve. In AFME the societal structure is capable of dynamically altering itself at runtime so as to adapt to context and to handle emergent system requirements.

On working on approximations connected with neutron transport theory, as part of the Atomic Bomb project in Los Alamos, Bellman identified the need for powerful numeric techniques. This led him to work on the principles of optimality and invariance and ultimately to the theory of dynamic programming [15]. Dynamic programming forms the basis of the extended rational decision making capabilities introduced to AFME (see section 6.2). We would hasten to add that technology itself is not a good or a bad thing; it's what it is used for. Some might argue that the development of atomic weapons could never be justified but what if an asteroid[12] were coming? It is ironic that what once was considered to be the greatest threat to human civilisation might some day be what saves it. Our job as scientists is to find out what can be done. It's a question for society as to what should be done.

As in probabilistic decision theory models, such as those based on the Savage axioms [22], dynamic programming provides an efficient solution. The decision making capabilities of AFME differ from classical probabilistic decision theory. A number of flaws in Savage's arguments are identified in [23]. Beliefs cannot be represented by additive probability distributions [24].

8 Conclusion

This paper detailed three new features of AFME that were introduced to enable resources to be managed more effectively. The ability to manage resources in an intelligent and prudent manner is a key requirement in the deployment of intelligent agents on computationally constrained ubiquitous devices. The extended rational decision making capabilities enable agents to reason about the relationship between their abstract computational limitations and their commitments. Belief labeling improves the efficiency of the reasoning algorithm by ensuring

---

[12] It is likely that a controlled nuclear reaction would be used rather than a bomb.

that duplicated belief sequences are only evaluated once. It decreases development time by reducing the redundancy and improving the maintainability of the software.

The new threading model enables the response times of agents to be specified accurately. Agent response times are altered to be prime numbers to prevent harmonic synchronisation and to ensure that agents are out of tune with respect to each other. Agents with the same response times are phase shifted to reduce computational overlap.

AFME addresses the issue of attempting to get the balance right between local autonomy and societal authority. Agents alter their behaviour according to context and emergent system requirements. It is not claimed that this is a perfect solution to the problem. A perfect solution is impossible due to the macroscopic principle of uncertainty. Nonetheless, it is an adaptive solution.

Acknowledgements

We gratefully acknowledge the comments of the anonymous ESAW reviewers, which have significantly improved the quality of this paper. Gregory O'Hare and Michael O'Grady gratefully acknowledge the support of the Science Foundation Ireland under Grant No. 03/IN.3/1361.

References

1. Weiser, M.: The Computer for the Twenty-First Century, pp. 94–104. Scientific American (1991)
2. Muldoon, C., O Hare, G.M.P., Collier, R.W., O Grady, M.J.: Agent Factory Micro Edition: A Framework for Ambient Applications. In: Alexandrov, V.N., van Albada, G.D., Sloot, P.M.A., Dongarra, J.J. (eds.) ICCS 2006. LNCS, vol. 3993, pp. 727–734. Springer, Heidelberg (2006)
3. Levesque, H.J., Cohen, P.R., Nunes, J.: On Acting Together. In: Proceedings of AAAI-90, pp. 94–99 (1990)
4. Collier, R.W., O Hare, G.M.P., Lowen, T., Rooney, C.: Beyond prototyping in the factory of the agents. In: Mařík, V., Müller, J.P., Pěchouček, M. (eds.) CEEMAS 2003. LNCS (LNAI), vol. 2691. Springer, Heidelberg (2003)
5. Collier, R.W.: Agent Factory: A Framework for the Engineering of Agent-Oriented Applications. Ph.D. Thesis (2001)
6. Muldoon, C., O Hare, G.M.P., Bradley, J.F: Towards Reflective Mobile Agents for Resource Constrained Mobile Devices. In: AAMAS 07: Proceedings of the Third Internatioal Joint conference on Autonomous Agents and Multiagent Systems. ACM, New York (2007)
7. Chen, J., O Grady, M.J., O Hare, G.M.P.: Autonomy and intelligence - opportunistic service delivery in mobile computing. In: Gabrys, B., Howlett, R.J., Jain, L.C. (eds.) KES 2006. LNCS (LNAI), vol. 4253, pp. 1201–1207. Springer, Heidelberg (2006)
8. Keegan, S., O Hare, G.M.P.: Easishop - agent-based cross merchant product comparison shopping for the mobile user. In: Proceedings of 1st International Conference on Information and Communication Technologies: From Theory to Applications (ICTTA '04) (2004)

9. Muldoon, C., O Hare, G.M.P., O Grady, M.J.: Collaborative Agent Tuning: Performance Enhancement on Mobile Devices. In: Dikenelli, O., Gleizes, M.-P., Ricci, A. (eds.) ESAW 2005. LNCS (LNAI), vol. 3963, pp. 241–261. Springer, Heidelberg (2006)
10. Smith, A.: The Wealth of Nations. Book 1 (1776)
11. Rao, A.S., Georgeff, M.P.: BDI Agents: from theory to practice. In: Proceedings of the First International Conference on Multi-Agent Systems (ICMAS'95), pp. 312–319 (1995)
12. Kreps, D.: A Course in Microeconomic Theory. Princeton University Press, New Jersey (1990)
13. Dantzig, B.: Discrete Variable Extremum Problems. Operations Research, 266–277 (1957)
14. Garey, M.R., Johnson, D.S.: Computers and Intractability; A Guide to the Theory of NP-Completeness. W.H. Freeman & Co. New York (1990)
15. Bellman, R.E.: Dynamic Programming. Courier Dover Publications (1957)
16. Berger, M., Rusitschka, S., Toropov, D., Watzke, M., Schlichte, M.: The Development of the Lightweight Extensible Agent Platform. EXP in Search of Innovation 3(3), 32–41 (2003)
17. Koch, F., Meyer, J.J., Dignum, F., Rahwan, I.: Programming Deliberative Agents for Mobile Services: the 3APL-M Platform. In: Bordini, R.H., Dastani, M., Dix, J., Seghrouchni, A.E.F. (eds.) Programming Multi-Agent Systems. LNCS (LNAI), vol. 3862. Springer, Heidelberg (2006)
18. Fok, C.L., Roman, G.C., Lu, C.: Mobile agent middleware for sensor networks: An application case study. In: Proc. of the 4th Int. Conf. on Information Processing in Sensor Networks (IPSN'05), pp. 382–387. IEEE, New York (2005)
19. Tarkoma, S., Laukkanen, M.: Supporting software agents on small devices. In: AAMAS '02: Proceedings of the first international joint conference on Autonomous agents and multiagent systems, pp. 565–566. ACM Press, New York (2002)
20. Hill, J.L.: System architecture for wireless sensor networks. PhD thesis, University of California, Berkeley, Adviser-David E. Culler (2003)
21. Bellman, R.E.: Some Vistas of Modern Mathematics. University of Kentucky Press (1968)
22. Savage, L.: The Foundations of Statistics. J. Wiley, New York (1954) second revised edition (1972)
23. Shafer, G.: Savage revisited. In: Readings in uncertain reasoning, pp. 122–160. Morgan Kaufmann Publishers Inc. San Francisco, CA, USA (1990)
24. Ellsberg, D.: Risk, Ambiguity, and the Savage Axioms. Quarterly Journal of Economics 75, 643–669 (1961)

Towards a Computational Model of Creative Societies Using Curious Design Agents

Rob Saunders

Faculty of Architecture, University of Sydney, Australia
rob@arch.usyd.edu.au

Abstract. This paper present a novel approach to modelling creative societies using curious design agents. The importance of modelling the social aspects of creativity are first presented and a novel agent-based approach is developed. Curious design agents are introduced as an appropriate model of individuals in a creative society. Some of the advantages of using curious design agents to model creative societies are discussed. Results from some initial investigations into self-organisation within creative societies using the model are given. This paper concludes by discussing some related work and exploring possible directions for future work.

1 Introduction

Creativity is often described as the ability to produce work that is both novel and appropriate [1] and researchers generally acknowledge that creativity must be defined differently at the level of the individual and the level of society [2, 3] but the relationship between individual and social creativity is complex.

An individual may determine that their work is creative independently of the judgement of others, but for it to be generally recognised as a creative work, other members of the society must agree that it is significantly novel and appropriate for a particular domain. In addition, an individual's determination of what is creative is informed by their experiences that are in turn based in the social and cultural environment within which they are situated. Consequently, we can say that creativity, at whatever level it is determined, is ascribed through a dynamic process of interactions between an individual, their society and the domains within which they work. This dynamic process of interactions is nicely captured by Csikszentmihalyi's systems view of creativity [4], illustrated in Figure 1.

In Csikszentmihalyi's view, creativity can only be discussed in terms of the creative system that extends beyond any particular individual and includes the socio-cultural context within which the individual works. Csikszentmihalyi identified three important components of a creative system; firstly there is the person engaged in the creative work referred to as the *individual*, secondly there is a social component called the *field*, and thirdly there is a cultural component called the *domain*. Creativity can be characterised by the following cycle of interactions; an individual takes some knowledge from the domain and produces

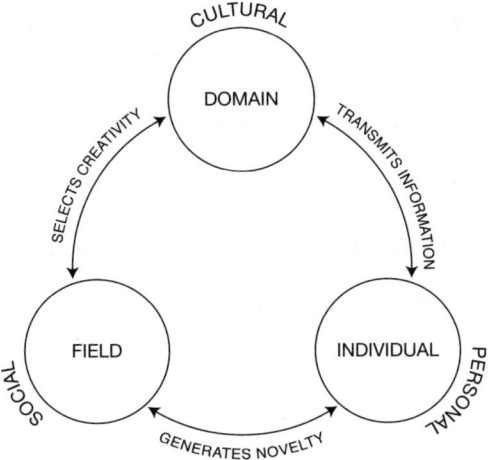

Fig. 1. Csikszentmihalyi's Systems View of Creativity

a work that is assessed by the field and if it is deemed to be creative the work, and any knowledge inherent in the work, is added to the domain.

The great majority of the research developing computational models of creativity has followed the lead of Newell et al. [5] and has focussed on developing computational models of creative processes such as divergent thinking, analogy making, and pattern recognition. Based the systems view of creativity, Csikszentmihalyi has questioned the validity of this approach [4], arguing that the these computational models cannot be said to model creativity without interaction with a field and its associated domain.

This paper presents a computational framework for studying the emergence of individual and social creativity within multi-agent systems based on the systems view of creativity. The goal of this research is to explore some of the interactive processes that occur within creative societies and how they might affect judgements of creativity.

2 A Framework for Modelling Creative Societies

The framework presented here provides an approach to developing models of social creativity based on Csikszentmihalyi's systems view. Previous work by Liu [6] recognised the need for a unified framework for modelling creativity. Liu's dual generate-and-test framework provided a synthesis of the personal and socio-cultural views of creativity in a single model. Liu proposed that existing computational models of personal creativity complemented computational models of social creativity by providing details about the inner workings of the creative individual missing from the models of the larger creative system. Liu proposed the dual generate-and-test model of creativity as a synthesis of Simon et al's generate-and-test model of creative thinking [5] and Csikszentmihalyi's systems view.

The dual generate-and-test model of creativity encapsulates two generate-and-test loops: one at the level of the individual and the other at the level of society. The generate-and-test loop at the individual level, illustrated in Figure 2(a), provides a model of creative thinking, incorporating problem finding, solution generation and creativity evaluation. The socio-cultural generate-and-test loop models the interactions among the elements of Csikszentmihalyi's systems view of creativity, as illustrated in Figure 2(b). In particular, it captures the role that the field plays as a social creativity test; ensuring that artefacts that enter into the domain are considered creative by more that just its creator. The combined dual generate-and-test model of creativity is illustrated in Figure 2(c).

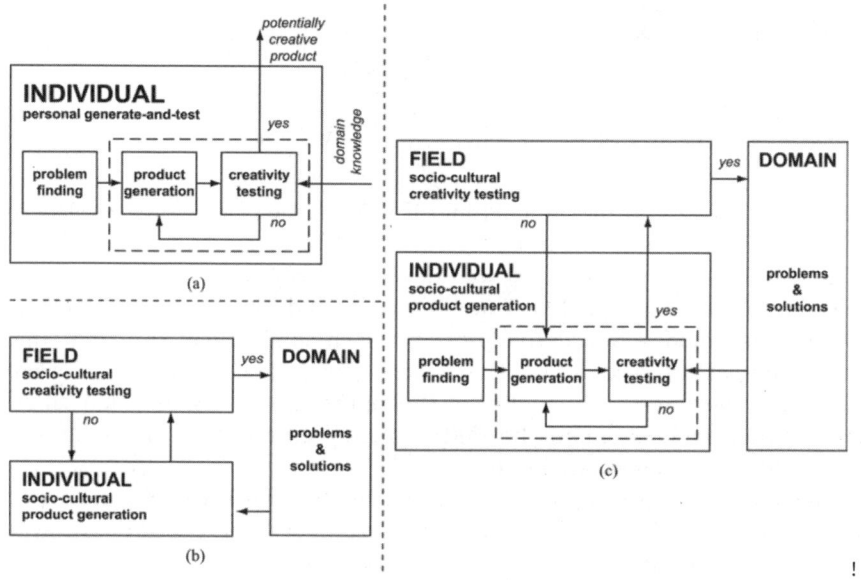

Fig. 2. Liu's Dual Generate-and-Test framework for building models of creative systems

A literal implementation of Liu's model requires separate processes to model the individual and social creativity test. This can be a pragmatic approach to adding a model of social factors to existing models of individual creativity and it is a viable solution for modelling some aspects of creativity, as demonstrated by the computational model developed by Gabora to study the memetic spread of innovations through a simulated culture Gabora [7].

The framework presented here takes a different approach, instead of implementing the social creativity test as a monolithic function, it distributes the social creativity test across all the individuals that constitute the field. The social creativity test is modelled through the communication of artefacts and evaluations of their creativity between individuals. An illustration of two individuals communicating artefacts and evaluations is given in Figure 3(a).

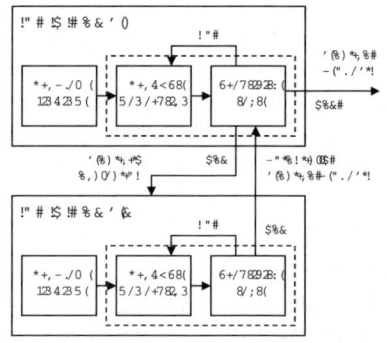 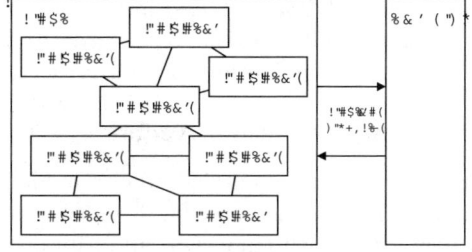

(a) The communication of artefacts and evaluations between agents.

(b) A model of a creative system using agents.

Fig. 3. The framework for modelling creative societies using agents

In the interaction illustrated in Figure 3(a), Agent A communicates an artefact that it considers to be creative, i.e. that passes its personal creativity test, to Agent B. Agent B evaluates the artefact according to its own personal creativity test and sends its evaluation back to Agent A. Each agent's evaluation of an artefact is affected by the traits of the individual, e.g. its preference for novelty, and its experiences, e.g. other artefacts it has evaluated.

Through the communication of evaluations, Agent B can affect the generation of future artefacts by Agent A by rewarding Agent A when it generates artefacts that Agent B considers to be creative. More subtly, Agent A can affect the personal creativity test of Agent B by exposing it to artefacts that Agent A considers to be creative, because the evaluation of creativity involves an evaluation of novelty, Agent A affects a change in Agent B's notion of creativity by reducing the novelty of the type of artefacts that it communicates. By exposing Agent B to artefacts that Agent A considers to be creative it can alter Agent B's evaluation of novelty and hence creativity.

To implement the social creativity test as a collective function of individual creativity tests a communication policy is needed. A simple communication policy would be for agents to communicate a product when their evaluation of that product is greater than some fixed threshold. To complete the implementation of the field as a collection of individuals, the individuals must be given the ability to interact with the domain according to some domain interaction policy. A simple domain interaction policy would follow the communication policy above and allow agents to add products of the generative process if the personal creativity evaluation is greater than a domain interaction threshold with the restriction that no individual is allowed to submit their own work to the domain. Thus, at least one other agent must find an individual's work creative before it is entered into the domain.

The individual, agent-centric, evaluations of creativity are key to the framework described here and permit the emergence of social definitions of creativity as the collective function of many individual evaluations. Without agent-centric evaluations of creativity, or at least interestingness, the collection of agents would simply represent parallel searches of the same design space. Curious design agents provide the necessary evaluations of creativity for this framework to be implemented.

2.1 Curious Design Agents

A curious design agent embodies a model of curiosity that uses a learning system called a novelty detector [8, 9]. A novelty detector can determine the novelty of a new input with respect to all of its previous inputs as a function of the errors generated when it attempts to classify the new input against one of its existing prototypes. Using a novelty detector, curious design agents are able to determine the novelty of new artefacts as they are produced. The novelty of each new work is measured as the distance between it and the nearest matching prototype, where the distance can be any measure of dissimilarity between a new work and an existing prototype, in the implementation that follows the distance is defined as the Euclidean distance between vectors representing a new work and the closest matching prototype.

The model of curiosity used by the curious design agents transforms the value of novelty determined by the novelty detector into a measure of interest by applying a "hedonic function". The hedonic functions used in the implementation are based on the Wundt Curve that Berlyne [10] used as a model for the typical reactions that animals and humans display in the presence of novel situations. A Wundt Curve for the determining the hedonic value, i.e. interest, from novelty is illustrated in Figure 4 as the combination of a reward function and a punishment function for discovering some novel work. Using a Wundt Curve to calculate interest, curious design agents favour works that are similar-yet-different to those

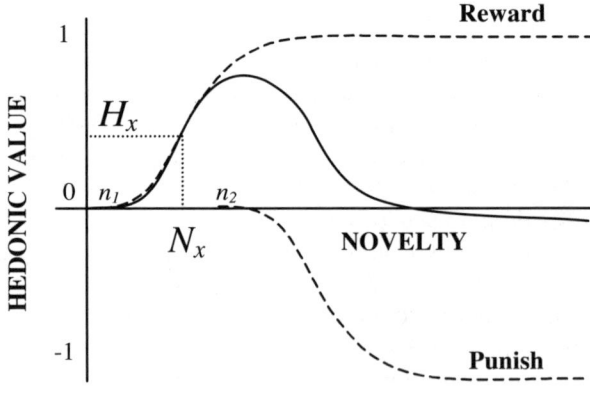

Fig. 4. The Wundt Curve

that have been experienced before. By changing the value of novelty at which the hedonic function is at its maximum, the agents can differ in how similar a new work must be for it to be considered interesting and therefore potentially creative.

The autonomy of curious design agents for determining what is interesting, and therefore potentially creative, is the key to adapting Liu's dual generate-and-test model to the study of emergent notions of creativity. This approach substitutes the monolithic social test of creativity found in Liu's model with a distributed agreement that emerges from the communication of individuals.

3 Experiments and Results

This section describes some results with an implementation of the framework described above. In this implementation the domain consists of "genetic artworks" [11]. Genetic artworks are images that are produced by evaluating an evolved program, typically a Lisp expression, at each (x, y) co-ordinate over the plane of the image. An example of a genetic artwork is shown in Figure 5?? together with the evolved Lisp expression that generated the image.

The curious design agents in this implementation use an interactive evolutionary art system based on the one developed by Witbrock and Reilly [12]. The images produced by the evolutionary system are converted into a vector that represents the contrast values of the pixels in the image. The vector is assessed using a novelty detector based on a self-organising map (SOM) [13] that provides a measure of each image's novelty whilst at the same time adapting the prototypes represented in the SOM to take into account the new images.

For the sake of simplicity, and to demonstrate the effects of different novelty evaluations on creative societies, all genetic artworks are assumed to be appropriate, i.e. any artworks that can be produced using the interactive evolutionary

(a) Genetic Artwork

```
(mod (iexp (mod (* (iexp (isin (* k x_iy_jx_ky))) (A1
(floor (iexp (conj golden))) (/ (/ x_iy (/ (exp (iexp
(/ x_iy (imax (iexp (rolL (iexp (* (conj golden) (normp
(exp (iexp (isin (/ j (* (floor x_iy_jx_ky) (+ i (conj
x_iy)))))))))))) (inv x_iy))))) (floor (exp (iexp (isin
(* k x_iy_jx_ky))))))) j))) (mod (iexp (conj golden))
(conj golden)))) (mod (* (/ (+ i (floor (/ j (/ (exp
(iexp (/ x_iy (imax (iexp (rolL (iexp (* (conj golden)
(normp (exp (iexp (isin (/ j (* (floor x_iy_jx_ky) (+
i (conj x_iy)))))))))))) (inv x_iy))))) (iexp (exp (iexp
(isin (* k x_iy_jx_ky)))))))) j) (inv x_iy)) (/ golden
(/ (/ x_iy (/ (exp (iexp (/ x_iy (imax (iexp (rolL (iexp
(* (conj golden) (normp (exp (iexp (isin (/ j (* (floor
x_iy_jx_ky) (+ i (conj x_iy)))))))))))) (inv x_iy)))))
(iexp (exp (iexp (isin (* k x_iy_jx_ky))))))) j))))
```

(b) Lisp Expression

Fig. 5. A genetic artwork and the Lisp expression that was evaluated at every (x, y) co-ordinate in the image to produce it where the x co-ordinates and y co-ordinates are in the range -1 to 1

3.1 The Law of Novelty

In "The Clockwork Muse" [14] Martindale presented an extensive investigation into the role that individual novelty-seeking behaviour played in literature, music, visual arts and architecture. He concluded that the search for novelty exerts a significant force on the development of styles. Martindale illustrated the influence of the search for novelty by individuals in a thought experiment where he introduced "The Law of Novelty". The Law of Novelty forbids the repetition of word or deed and punishes offenders by ostracising them. Martindale argued that The Law of Novelty was merely a magnification of the reality in creative fields. Some of the consequences of the search for novelty are that individuals that do not innovate appropriately will be ignored in the long run and that the complexity of any one style will increase over time to support the increasing need for novelty.

The following experiments were designed to study the effects of the search for novelty in creative societies modelled as curious agents that have hedonic functions with different preferred novelty values. The preferred novelty of each agent is expressed as a value N that indicates the amount of novelty associated with peak interest in the agent's hedonic function. In this implementation, N ranges from 0 to 32; this is equal to the range of the potential classification error generated by the novelty detectors used.

In the first experiment a group of 12 agents were created. Ten of the agents, agents 0–9, shared the same hedonic function, i.e. the same preference for novelty

Agent ID	Preferred Novelty (N)	Attributed Creativity
0	11	5.43
1	11	4.49
2	11	4.50
3	11	3.60
4	11	4.48
5	11	1.82
6	11	6.32
7	11	8.93
8	11	10.72
9	11	5.39
10	3	0.00
11	19	0.00

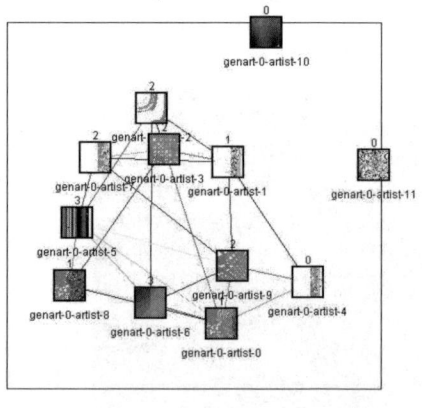

(a) The attributed creativity between agents.

(b) Screenshot of the running simulation.

Fig. 6. The Law of Novelty simulated within a single field of agents with different preferences for novelty

(N=11). Two of the agents were given quite different novelty preferences. One, agent 10, was given a preference for low amounts of novelty (N=3) and the other, agent 11, was given a preference for high amounts of novelty (N=19).

Figure 6(b) is a screenshot of the running simulation; the squares represent agents, the images in each square shows the currently selected genetic artwork for that agent, the number above each genetic artwork shows its attributed creativity, and the lines between agents indicate the communication of rewarded works between pairs of agents. Figure 6(b) shows how the network of communication links developed between agents that communicate artworks and evaluations on a regular basis excludes the two agents with different hedonic functions.

The results of the simulation are presented in Figure 6(a). The results indicate that the agents with the same preference for novelty to be somewhat creative according to their peers, with an average attributed creativity of 5.57. However, neither agent 10, with a preference for low amounts of novelty, nor agent 11, with a preference for high degrees of novelty, received any credit for their artworks. Consequently none of the artworks produced by these agents were saved in the domain for future generations. When these agents expired nothing remained in the system of their efforts.

The results of this experiment appear to show the emergence of the Law of Novelty in models of creativity societies that have agents with different preferences for novelty. One explanation for this may be that agents with a lower novelty preference tend to innovate at a slower rate than those with a higher hedonic preference and while an agent must produce novelty to be considered creative, it must do so at a pace that matches its audience. There is no advantage in producing many highly novel artefacts if the audience cannot appreciate them.

3.2 The Formation of Cliques

In this second experiment, the behaviour of groups of agents with different hedonic functions is investigated. To do this a group of 10 agents was created, half of them had a hedonic function that favoured novelty close to N=6 and the other five agents favoured novelty values close to N=15. Figure 7(a) shows the payments of creativity credit between the agents in recognition of interesting artworks sent by the agents.

Two areas of frequent communication can be seen in the matrix of payment messages shown in Figure 7(a). The agents with the same hedonic function frequently send credit for interesting artworks amongst themselves but rarely send them to agents with a different hedonic function. There are a large number of credit messages between agents 0–4 and agents 5–9, but only one payment between the two groups – agent 4 credits agent 5 for a single interesting artwork.

The result of putting collections of agents with different hedonic functions in the same group appears to be the formation of cliques: groups of agents that communicate credit frequently amongst themselves but rarely acknowledge the creativity of agents outside the clique. As a consequence of the lack of communication between the groups the style of artworks produced by the two cliques also remains distinct.

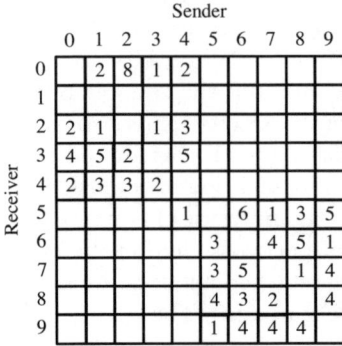
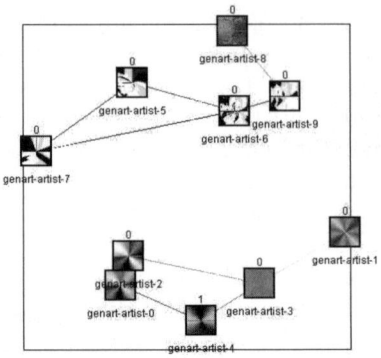

(a) A matrix of the number of positive creative evaluations sent between agents.

(b) A screenshot of a simulation showing two non-communicating cliques.

Fig. 7. The formation of cliques between agents with different hedonic functions

Figure 7(b) is a screenshot of the running simulation that has formed two cliques. To help visualise the emergent cliques, the distances between agents are shortened for agents that communicate frequently. The different styles of the two groups can also be seen, with agents 0–4 producing smooth radial images and agents 5–9 producing fractured images with clearly defined edges.

The results of this experiment show that when a population of agents contains subgroups with different hedonic functions, the agents in those subgroups form cliques. The agents within a clique communicate credit frequently amongst themselves but rarely to outsiders. The stability of these cliques depends upon how similar the individuals in different subgroups are and how often the agents in one subgroup are exposed to the artworks of another subgroup. Further research is needed to determine whether other factors that can affect judgements of interestingness can similarly affect the social structure.

Communication between cliques is rare but it is an important aspect of creative social behaviour. Communication between cliques occurs when two individuals in the different cliques explore design subspaces that are perceptually similar. Each of the individuals is then able to appreciate the other's work because they have constructed appropriate perceptual categories. The transfer of artworks from a source to a destination clique will introduce new variables into the creative processes of the destination clique, the two cliques can then explore in different directions, just as two individuals do when they share artworks. Cliques can therefore act as "super-artists", exploring a design space as a collective and communicating interesting artworks between cliques.

3.3 Domains of Complexity

To investigate the relationship between the search for novelty and the complexity of the resulting artworks an experiment was conducted to compare agents

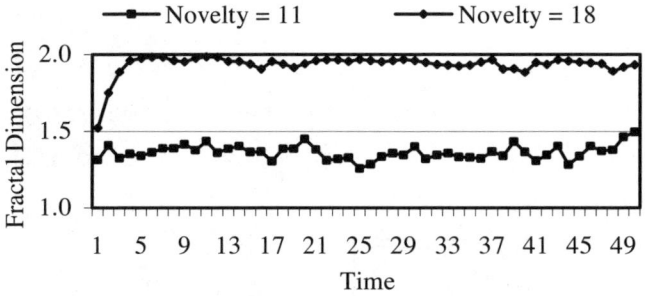

Fig. 8. The complexity of genetic artworks produced by two groups of agents with different preferences for novelty

with different preferences for novelty. To measure the complexity of the images the fractal dimension of selected images was calculated using the box counting method [15]. For any two-dimensional image, a measure of its fractal dimension will produce a value between 0.0 and 2.0, depending on how much of the space is filled in the image at different levels of detail.

To investigate the relationship between the preferred degree of novelty and the fractal dimension of the resulting images, two types of agents were used. One type preferred novelty values of $N=18$ and the other type favoured novelty values of $N=11$. Three agents of each type were allowed to explore the space of genetic artworks for 50 time steps. Figure 8 shows how the complexity of the images produced by the two groups of agents quickly diverge and then remain at a constant level. For the group with the higher preference for novelty, the results appear to confirm Martindales hypothesis that the search for novelty promotes increased complexity over time [14], at least up to some limited level of complexity.

To investigate the relationship between a field's preference for novelty and the complexity of the artefacts produced by its members, 19 test groups were created consisting of 3 agents in each group. In each group the agents favoured the same novelty value, across the 19 tests the groups favoured novelty values in the range $1 \leq N \leq 19$. Figure 9 shows that the relationship between the preferred value of novelty and the average fractal dimension of the resulting images is almost linear for the large proportion of values for preferred novelty. In other words, agents with a preference for greater novelty produce images with higher fractal dimensions.

How can we explain this relationship between the preferred novelty of an agent and the fractal dimension of the resulting images? One explanation is that the curious exploration of the space of genetic artworks drives the agents towards subspaces that have an appropriate amount of local variability to continually satisfy the need for novelty. Consequently, agents that prefer novel forms will tend towards areas of the design space that produce more complex images, as there is a great deal more variability between complex images than between simple ones.

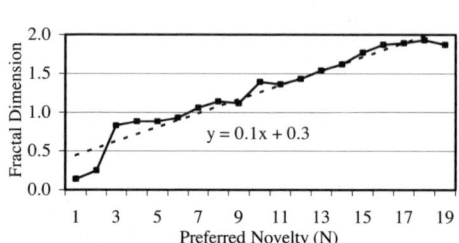
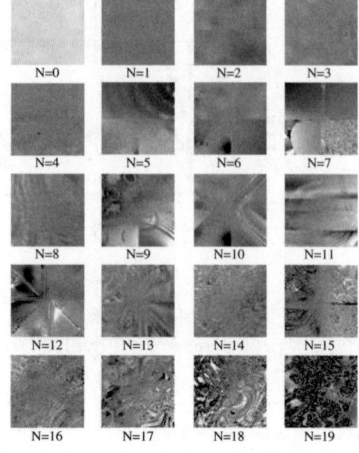

(a) A comparison of the average fractal dimension against a range of peak hedonic values.

(b) A small gallery of genetic artworks evolved by the curious design agents.

Fig. 9. The relationship between preferred novelty and the complexity of the genetic artworks evolved by the curious agents with different preferences for novelty $0 \leq N \leq 19$

4 Discussion

Curious agents have been developed by a number of other researchers. Schmidhüber created curious agents that competed against each other to determine what was interesting [16]. Marsland et al [17] produced curious robots that explored environments for novelty as a way of generating maps of the space. Interest in intrinsically motivated agents, like curious agents, is increasing as researchers discover the benefits of self-motivated learning in both modelling and applications [18, 19].

Other computational models based on Csikszentmihalyi's system view of creativity have also been developed [20] that demonstrate the important role that authority figures, or gatekeepers, play in creative fields. The contribution of the framework presented here is the bringing together of curious agents and the creative systems to support an approach to computationally modelling creative societies at multiple levels.

The work presented here is still in its early stages of development and there are many ways that it can be extended to improve the models or investigate other features of creative societies. Future work using this framework will aim to extend the experimental possibilities at both the individual and social levels of creativity. Three possible directions for future work are:

Integrating Evaluations of Appropriateness. One of the obvious limitations of the work presented here is the lack of an explicit test for the appropriateness of artefacts. To apply the computational model of more significant domains, future work will integrate domain-specific knowledge so that the test for creativity can include a test for appropriateness within a domain.

Integrating Alternative Models of Creative Processes. The curious design agents presented in this paper use an evolutionary design tool to explore a design space. Integrating alternative models of creative processes including analogy-making [21] could provide a useful framework for evaluating the effectiveness of such creative processes within a social and cultural context.

Modelling Individuals with Intrinsic Motivations other than Curiosity. Curiosity is not the only intrinsic motivation for creative individuals, although it is one of the most persistent [14]. Other motivations for exploring a design space can be computationally modelled in design agents, e.g. competency [19].

Modelling Large Creative Societies. The ability to simulate larger creative societies will permit the study of the spread of innovations and styles. It may also facilitate the emergence of new fields as cliques attain a critical size. Spatial and topological relationships will become more important issues in large population models.

Modelling Non-Homogenous Societies. There are several other important players in creativity societies besides the producers of innovations including, e.g. consumers, distributors, critics, etc. Each has their own role to play in creative societies; consumers evaluate products, distributors distribute products widely, and critics distribute their evaluations widely. Convincing other people that you've had a creative idea is often harder than having the idea in the first place. In non-homogenous societies of agents, the selection of which agents to communicate with becomes an important strategy for agents seeking recognition as a creative individual.

Modelling More Complex Social Interactions. Simulations of technological innovation in industry show that the consideration of the costs of innovation in decision-making can lead to complex behaviour [22]. Simulating similar costs in the design process may provide a better understanding of the economics of creative design in creative societies and the strategies needed to manage creativity with limited resources.

Modelling Domain-Specific Symbol Systems. Domains in the real world contain much more than examples of previously produced artefacts. Creative domains often include symbol systems, e.g. languages, that are specific to the knowledge held in the domain. These symbol systems can present opportunities for domains to differentiate as they present barriers to the flow of information between domains.

Modelling the Evolution of Domains. Domains and the symbol systems they contain evolve over time through use by the field. Computational models of the evolution of language [23] may provide a useful technique for developing computational models of domain-specific languages that evolve over time.

The aim of this paper has been to present a framework for computationally modelling creative societies using curious design agents and to show some of the research opportunities that exist using models developed using this framework.

By using curious design agents as models of individuals within creative fields, the framework provides a flexible basis for developing multi-agent systems that can be used to study the interaction between personal and social judgements of creativity.

References

[1] Sternberg, R.J., Lubart, T.I.: The concept of creativity: prospects and paradigms. In: Sternberg, R.J. (ed.) Handbook of Creativity, pp. 3–15. Cambridge University Press, Cambridge, UK (1999)
[2] Boden, M.A.: The Creative Mind: Myths and Mechanisms. Cardinal, London, UK (1990)
[3] Gardner, H.: Creating Minds: An Anatomy of Creativity Seen Through the Lives of Freud, Einstein, Picasso, Stravinsky, Eliot, Graham and Gandhi. Basic Books, New York (1993)
[4] Csikszentmihalyi, M.: Society, culture and person: a systems view of creativity. In: Sternberg, R.J. (ed.) The Nature of Creativity: Contemporary Psychological Perspectives, pp. 325–339. Cambridge University Press, Cambridge (1988)
[5] Newell, A., Shaw, J.C., Simon, H.A.: The process of creative thinking. In: Gruber, H., Terrell, G., Wertheimer, M. (eds.) Contemporary Approaches to Creative Thinking, pp. 63–119. Atherton Press, New York (1962)
[6] Liu, Y.T.: Creativity or novelty? Design Studies 21, 261–276 (2000)
[7] Gabora, L.: Meme and variations: A computer model of cultural evolution. In: Nadel, L., Stein, D. (eds.): 1993 Lectures in Complex Systems. Addison-Wesley, Reading (1995)
[8] Markou, M., Singh, S.: Novelty detection: A review, part i: Statistical approaches. Signal Processing 83, 2481–2497 (2003)
[9] Markou, M., Singh, S.: Novelty detection: A review, part ii: Neural network based approaches. Signal Processing 83, 2499–2521 (2003)
[10] Berlyne, D.E.: Conflict, Arousal and Curiosity. McGraw-Hill, New York (1960)
[11] Sims, K.: Artificial evolution for computer graphics. Computer Graphics 25, 319–328 (1991)
[12] Witbrock, M., Reilly, S.N.: Evolving genetic art. In: Bentley, P.J. (ed.) Evolutionary Design by Computers, pp. 251–259. Morgan Kaufman, San Francisco, CA (1999)
[13] Kohonen, T.: Self-Organizing Maps. Springer, Heidelberg (1995)
[14] Martindale, C.: The Clockwork Muse. Basic Books, New York (1990)
[15] Mandelbrot, B.B.: The Fractal Geometry of Nature. W. H. Freeman & Co. New York (1977)
[16] Schmidhuber, J.: Curious model-building control systems. In: Proceedings of the International Joint Conference on Neural Networks, Singapore, vol. 2, pp. 1458–1463. IEEE, New York (1991)
[17] Marsland, S., Nehmzow, U., Shapiro, J.: Novelty detection for robot neotaxis. In: International Symposium On Neural Computation (NC'2000), pp. 554–559 (2000)
[18] Singh, S., Barto, A.G., Chentanez, N.: Intrinsically motivated reinforcement learning. Neural Information Processing 18 (2004)
[19] Merrick, K., Maher, M.L.: Motivated reinforcement learning for non-player characters in persistent computer game worlds. In: ACM SIGCHI International Conference on Advances in Computer Entertainment Technology (ACE 2006), Los Angeles, USA (2006)

[20] Sosa, R., Gero, J.S.: A computational study of creativity in design: the role of society. In: AIEDAM (2005)
[21] Falkenhainer, B., Forbus, K.D., Gentner, D.: The structure-mapping engine: Algorithm and examples. Artificial Intelligence 41, 1–63 (1989)
[22] Haag, G., Liedl, P.: Modelling and simulating innovation behaviour within microbased correlated decision processes. Journal of Artificial Societies and Social Simulation 4 (2001)
[23] Steels, L.: A self-organizing spatial vocabulary. Artificial Life 2, 319–332 (1995)

Privacy Management in User-Centred Multi-agent Systems

Guillaume Piolle[1], Yves Demazeau[1], and Jean Caelen[2]

[1] Laboratoire Leibniz-IMAG, Université Joseph Fourier and CNRS
46, avenue Felix Viallet, F-38031 Grenoble Cedex, France
{Guillaume.Piolle,Yves.Demazeau}@imag.fr
[2] Laboratoire CLIPS-IMAG, CNRS
385 rue de la Bibliothèque, F-38041 Grenoble Cedex 9, France
Jean.Caelen@imag.fr

Abstract. In all user-centred agent-based applications, for instance in the context of ambient computing, the user agent is often faced to a difficult trade-off between the protection of its own privacy, and the fluidity offered by the services. In existing applications, the choice is almost never on the user's side, even though the law grants him a number of rights in order to guarantee his privacy. We examine here different technical works that seem to be as many interesting ways of dealing with privacy policies. The problems already solved will be identified, as well as remaining technical challenges. Then we will propose directions of research based on the most interesting aspects of the underlined approaches.

Keywords: Privacy, User-Centring, Personal Agents, Trusted Computing.

1 Introduction

Much work has been done in multi-agent systems about inter-agent communication and agent-based knowledge acquisition and sharing ([5], [7]). More and more researchers believe now that the stress should be put on the security concerns intrinsic to information disclosure, and specifically the privacy-related issues. Privacy becomes important above all in applications involving personal assistants, and in general agents having access to personal data. It is the case in ambient computing of course, where personal agents can be embedded in mobile or nomad user devices, or in more conventional agent society architectures involving contracts, payment or delivery, like agent-based web services usage. Let us illustrate it with a short example.

When it comes to ambient computing technologies, it has become very popular to illustrate one's work with idyllic usage scenarios. In such stories, one can follow the day of a salesman or a researcher evolving in a fluent "information society" in which the purchase of a return flight from his computer, with the mediation of his personal user agent, guarantees that his favourite movies will be available on board, a room will be pre-booked in a good hotel in Paris and he will receive the menus of the restaurants close to the conference venue via text messages on his cell phone. All the service agents knowing exactly the preferences of the character, and willing to facilitate all his actions, all these wonders are offered by pervasive computing technologies, service composition techniques, user agents evolving in ad-hoc personal networks and automatically contracting on behalf of the user, and highly collaborative service agents. Such scenarios always arouse enthusiasm, but "ordinary people" as well as ambient computing researchers sometimes also find that in order to get that fluidity in the services, the user has to loose control over all the personal and professional information involved in the transactions, i.e. he has to accept that his name, address, usage profile, and maybe his payment preferences and history might be communicated by one service agent to another. In other words, he renounces a part of his privacy.

This sacrifice may be acceptable, but only in a given context and in a limited measure, which should be defined and consented by the user himself. Laws often protect the privacy rights of the citizens, but it is actually difficult to ensure that they are respected: how could we know what a service agent does with the information that the user agent has given to it in the past? To what extent is it possible to give a proof to the consumer that a good use will be made of the personal data he communicates? Few work has been done so far in the domain of privacy applied to multi-agent systems (for instance [3], [10], [14]), mainly because it is not an intrinsically multi-agent-based problem. However, we think that the multi-agent paradigm may facilitate the understanding of the privacy concepts at a low level, by providing a cognitive and social layer. Multi-agent systems may thus provide us with interesting approaches, as we will see. In this study, we would like to explore the tools we have, be they from the multi-agent field or not. After having identified the contribution and drawbacks of several general privacy management approaches, we will propose directions of research based on that analysis and on multi-agent systems, as well as some evaluation methodology principles.

In the next section, we will define the different components of the concept of privacy, and precise the context in which we will use them. In the third part we will have an overview of the interesting works undertaken in or around the domain of privacy, and how each one of them concentrates on a different aspect of the problem. We will then propose directions for future work in the domain, identify our priorities and present our conclusions.

2 Context

2.1 What is Privacy?

The term of privacy encompasses a number of individual notions that are also put together under the denomination "protection of personal (or nominative) information". In [17], Alan Westin defines privacy as

> The right of the individuals to determine by themselves when, how and what private information is disclosed.

This is quite a general definition, which has adaptations in different fields. However, the main idea is still valid in all of them. In our applications, we will find appropriate precisions in the texts related to computing and electronic communications. The founding principle is that the actions that users undertake in the community must not force them to publish personal information: only the necessary information should be transmitted to the concerned people, and for a reasonable duration. One could note that "privacy" is usually used for "respect of privacy". By studying the different approaches quoted here, we have been able to define six components for the notion of privacy. These six sub-notions, depending on which restrictions and obligations are built on them, define the privacy context.

The components of privacy are the following:

- The **information** given to the user about data collection and processing;
- The user's **consent** regarding data collection and processing;
- The **goal** of the data processing and the **justification** of the data collection;
- The ability of the user to **modify** and **retract** the collected information;
- The right that the service processing the data has (or not) to **communicate** the collected and/or processed data to third parties;
- The **data retention duration**.

The risk about privacy only refers to nominative information: totally anonymous information, for statistics for instance, are not a threat for privacy. Usually a distinction is made between *directly* nominative information and *indirectly* nominative information[1].

Our problematics about privacy is finding a means of instantiating the components of privacy in the personal agents that interface and represent the user, for instance by contracting on his behalf.

2.2 Privacy as User-Centring: Illustration in Several Domains

Privacy management is a key feature for ambient computing applications involving multiple service agents, for they would gain great benefit from sharing (or selling to each other) their information about the customers' usage profiles. It is the case in the introduction scenario, where the traveller can be virtually tracked by service agents from one company to another, leaving a profile and preferences that could identify him, through his personal agent, almost as surely as a login. Here, in theory, each service agent should have informed the user agent that they wanted to collect the profile, for what goal and for how long they would keep it, and to whom they would forward it. The user should have given his consent individually to every one of them, he should have a means of updating his profile with each provider. And of course every service agent should have respected those engagements, which is virtually impossible to check for the user. It is obvious here that there is a trade-off between properly coping with privacy, and ensuring a fluent service to the user.

Another sensitive field is the integration of electronic medical files, or more specifically medical surveys, as exposed in [8], where the patients' anonymity is vital, but where specialists may want to ask further questions to an identifiable subset of the survey pool (those having answered in a specific way, for instance). How is it possible to

---

[1] The first (direct) case is when appears in the dataset an information like the name, the address, the social security number of the user. In the second (indirect) case, there is a less obvious way to identify the user, like an IP address, a pseudo or an email address used on the internet, the professional title or function of the user... The identification of the user can be made possible by intercorrelation of various indirectly nominative datasets.

allow that without threatening the patients' privacy? Here, complex asymmetric cryptographic protocols must be designed, and strong privacy policies must be enforced. The balance is here between the protection of the patients, and the general health interest. The same kind of issue arises in electronic voting systems [6]. Once again, part or all of the constraints can be moved from the user to his personal agent.

2.3 The European Legal Context

In our study, we will take as a reference the European Directive on privacy [15]. This document imposes a number of restrictions that must be implemented in the national laws of the member states. Thus it constitutes a good common denominator for European requirements in the matter of privacy. The general principles correspond to the different components of privacy, defined earlier:

- The user must be provided with clear information about the data collected, the purpose of the processing, the duration of the storage, and to whom this information will be delivered;
- The user has the right to refuse a personal data collection and/or processing. This might mean that the service cannot be performed;
- The access to the service may be subject to such acceptance only if the collected information are used "for a legitimate purpose";
- The users must be able (free of charge) to correct the collected information, and in some cases (directories) to remove it;
- Information forwarding to third parties cannot be done without the consent of the user;
- The data cannot be kept once they are not needed any more.

All these considerations about privacy also apply when a personal agent, instead of a human user, is acting. Since the agent, even autonomous, acts in accordance with the user's intention, the same principles should be taken into consideration.

3 Comparative Reading of Existing Propositions

We will now compare some works by analysing how they deal with the different components of privacy: user information, user consent, user ability to update data, control of the service-side data processing (storage, usage in relation to the original goals, data forwarding). We will expand the frame idea to service agents dealing with personal artifacts provided by user agents.

3.1 W3C Platform for Privacy Preferences

Platform for Privacy Preference (P3P, [18]) is a W3C work group whose purpose is to deal with a specific part of the privacy concept: Informing the user. The specifications describe a system for websites, allowing them to publish a privacy policy in a normalized form (an XML document). On the user side, the client application compares the policy with a set of requirements previously defined by the user. If they match, then the website is allowed to collect information, for instance in the form of a client-side cookie.

The information enclosed in the XML format are the following: identity of the service responsible for data collection, collected information details, purpose of the processing, what data will be shared, with whom, whether users can make changes in how their data is used, the legal jurisdiction of the processing, policy for data conservation, and a pointer to a human-readable policy. Website managers can specify part or all of it, possibly through the configuration of the service agents.

The W3C made it clear that no minimal level of privacy is assumed or required, the P3P protocol only provides information about the policy. Besides, it does not guarantee that the service will actually comply with the policy. P3P only addresses a limited and specific part of the problem, and does it properly. Several tools (policy editors, checkers, browser plugins...) are available at the moment, and in a general way P3P can be implemented as an integrated component in a privacy-compliant architecture. The XML format could also be used as is, as a common agent formalism for describing privacy policies.

3.2 IETF IDsec

IDsec [9] is a project of the Internet Engineering Task Force. Its goal is to manage and protect virtual identities and profiles online. In this approach there are three actors, the Profile Owner (the user agent), the Profile Requester (the service agent) and the Profile Manager (an independent entity). The user is previously registered (in a secure way) with the Profile Manager, on which is stored the profile, with Access Control Lists (ACL) attached to each element of it.

Figure 1 (quoted from the IDsec sourceforge directory) describes what happens when a Profile Requester asks for user information. After authentication, the Profile Owner gets a Session Certificate (acting as an access token) from the Profile Manager, and forwards it to the Profile Requester. The Requester sends this Certificate and its own Profile Requester Certificate (proof of identity, provided by an external certification authority or, more likely, by the Profile Manager itself) to the Profile Manager. The Profile Manager then sends back the parts of the profile that the Requester is entitled to get, given its Certificate and the ACLs of the profile.

This protocol has several weaknesses. First of all, the user agent has to totally trust the Profile Manager [4], because it is the one in charge of the storage, the integrity and the confidentiality of the profile. Besides, Profile Managers, storing personal data (maybe including addresses and banking information) of many users, would become an interesting target for attacks. The Profile Manager would then concentrate the security issues in a single point of failure. A second problem is that the user agent has to trust the requester itself, about what is done with the collected data. Indeed, no guarantee can be provided by IDsec about information storage, processing and forwarding after the profile has been delivered. This trust will be transcribed in the ACLs, but a Profile Requester can easily be punctually malicious and provoke a leak. Even audits from certification authorities will not be able to prevent that. This problem is a recurrent one in the design of a privacy protocol.

Why could not one act as one's own Profile Manager? The interest here of having a separate entity is that a nomad user does not have to keep the same terminal to access a personalized service. In ambient computing, this is an obvious advantage. And

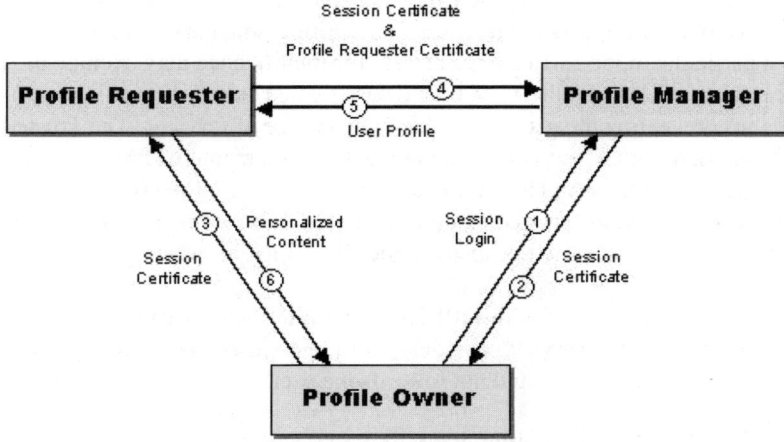

Fig. 1. IDsec general mechanism

of course, should the user be his own Profile Manager, most of the IDsec protocol becomes useless.

3.3 A Privacy Architecture Using Trusted Computing Platform Architecture

The system described in [10] is based on the Trusted Computing Platform (TCP) architecture, designed by the Trusted Computing Group [16]. A TCP is a platform in which has been integrated a hardware component, the Trusted Computing Module (TPM), typically in a chip on the motherboard. The TPM is a cryptographic component able to securely store private keys, data, small parts of software... The data in the TPM can be unlocked only in a certain execution state, i.e. if no problem has occurred during the boot phase, the right OS has been booted and the right program is running. A platform as a whole can be certified by a Certification Authority (called Privacy Certification Authority, or PCA, by the Trusted Computing Group), so that a distant platform can be certain that a *certified* program is running on a *certified* platform.

This capability is used in [10] to certify a client platform, on which user profiles are generated and attached to (certified) anonymous identities. In this approach the profile is generated and stored by client-side agents (located on the TPM, so that the process can be certified), and the cryptographic certification is for the server's use, so that it can be sure of the authenticity of the profile and the virtual identity (without having to know the actual identity of the user). Here the architecture is oriented towards the certification of the user platform, but by reversing the problem the same tools could eventually provide a certification to the user that the distant service is behaving in a specified way with the provided profile.

This paper has been published by Hewlett Packard, which is one of the founding societies of the Trusted Computing Group (with Microsoft, Intel, IBM, AMD...) and they seem to be very involved in the TCPA project (see [11]). However, some reservations apply for the privacy architectures involving Trusted Computing technologies.

Detractors of TCPA have exposed the fact that the architecture would allow PC builders to forbid (by imposing certification authorities) some operating systems or programs (in particular, open software are at risk, for their licence may exclude them from paying certification), so that the TPM would be activated only if the right OS/software is running, thus preventing the user from accessing a range of services. The Trusted Computing Group denies that, but it is possible that an implemented TPM would impose a given certification authority. The European "article 29 data protection working party" expose their worries about Trusted Computing in [1]. It is also interesting to notice that the TPM has been designed to encapsulate the Digital Rights Management capabilities of the platform.

The point of view on TCPA in [10] looks quite utopian (especially regarding the goals and interests of the service providers) but the architecture is clearly oriented towards the commercial service, and not towards the user.

3.4 User-Centred Profiling and Client-Side Privacy Policy

In [12] and [13], Brebner and Riché present a system whose main interests are the distribution of the profile over a user-centric ad-hoc network of devices, and the distribution of the privacy and consistency policy as well.

Here the profile is distributed on the client side, so that the user is in control of his own profile. The profile is organized as a tree, each leaf being a profile item. To each item is attached a replication policy (depending on the consistency need, which is for example strong for a password and weak for a display preference) and a diffusion policy (based on the credentials presented by the requester agent).

In [13], the authors describe their user-centric replication protocol, based on a dynamic migration of the master authority for each profile item. They make use of a trust management system on the network: all the devices are not equally trusted for replication, they have trust values based on their security capabilities and disconnection rate. Their replication model has the advantage of avoiding user intervention.

In order to enforce privacy at running time, they use a model which stands between a client-side execution model (mobile code running on the client device in a secure environment) and the standard server-side stored profile. It is closer to client-side personalization, where the matching of the distant service requirements with the user preferences is made on the user side, the server receiving only the necessary and relevant parts of the profile. Furthermore, since we are in a distributed environment, instead of using a centralized profile data store, the device-located data (replicated, more or less up-to-date depending on the on-demand access consistency level) is used as the user profile base.

3.5 Synthesis

We have seen a short selection of approaches here, each one addressing part of the privacy problem, in a way that involves personal agents for managing privacy on the client side, or that allows the integration of such autonomous agents. Let us have a second look at the six components of privacy identified in 2.1. By **information**, we now mean that the user must be warned by the service agent of which data will be collected, by whom, for what purpose, for how long. It is done here by displaying information on a terminal, or only

Table 1. Basic privacy components in the different approaches

Approach	Components of Privacy Checking of:					
	Information	User Update	Consent	Data retention	Data Usage	Transmission
P3P [18]	●		●			
IDsec [9]			●			
TCPA-based [10]	●		●	○	○	○
Client-Side Profile Storage [12], [13]	○		●			

matching the information against patterns previously given by the user. The **goal** of the data processing is enclosed in the information, or given to the user by the context of the transaction (e.g. when visiting a website, the user may be given sufficient information about the service). **Consent** is collected by the means of direct interaction with the user agent, but can also be a prior consent: by defining access rules with his personal agent, the user says he gives his consent to any data collection complying with this policy. This last approach has the advantage of a higher fluidity. The **modification and suppression** ability could be implemented in a very simple way in the form of an order from the user to the service agent, or by a direct control of the user over the distant data (for instance by the means of a web interface for a database). Control over **data communication** to third parties means here that a user agent should have a way to be sure that the service agent will not sell his profile, or correlate it with another one. The idea could involve a system similar to the "key" or "lock" icon in the web browser, which guarantees the user that the connection is secure. The same kind of guarantee could address the problem of **data retention duration**. Those last issues are technically difficult, and we have only tried to describe what kind of solution could be acceptable.

Table 1 makes a synthetic summary of which components of privacy are managed by each approach. An empty circle means that the problem has been partially addressed, or could be addressed with minor adaptations of the architecture. Here "User update" means the ability for the user to make corrections to the already collected information (and/or to delete it).

Information is now a solved problem here, since the P3P project provides a standard protocol, and tools that can be integrated in any architecture. Consent is also easy to address. Regarding user update, it could be quite surprising that no approach has presented a mechanism to notify changes in an already collected profile. The reason is that most approaches have focused on client-side profiles, and not on the data retrieved by remote agents. Of course the user can update profile information on the client side, but it is not the point here. Regarding the checking of the distant properties, this is particularly difficult to do from the client side: how can the user be sure of what the service agent does with the profile, without explicitly trusting it? Only the TCPA-based approach provides tools that could do that in theory, but this idea is not even mentioned. We believe that this part is the current technological challenge of privacy management. This problem of guaranteeing distant properties (data retention, usage and transmission) can be reduced to a trust issue, just like the authentication problem [4]. Without authentication protocol, you have to trust your correspondent about his identity, and it is a problem since the only person on which the trust lies, is the one that could eventually cheat. When you introduce an authentication protocol, your trust is now in a cryptographic algorithm and/or a certification authority, and not any more in the possible cheater. By this mechanism you lower the global risk or the transaction. A logic for the formalisation of trust in such authentication issues is presented in [2]. In our problem, the user agent has to trust the service agent about its privacy policy, whereas it could be tempted not to respect it. The challenge here is to find a way (similar to the introduction of an authentication protocol) to move the user's trust from the service to a certification authority or a well-known protocol.

Table 2 compares the approaches with regards to a few additional features that look helpful in managing the privacy of user profiles. They put the user in control of his data and give him more information and security. One must notice that server authentication can always be done by cryptographic encapsulation. The first feature is the distribution of the profile over a number of devices or agents. The second one is the quite common choice of storing the profile on the client side, rather than on the service side. Anonymous identities allow the user agent to manage several profiles without revealing his identity. Server authentication is a quite obvious security feature, and the client-side privacy policy is the way by which the user specifies with his agent the requirements regarding the privacy of his profile. By "legislation independence", we mean that the designed systems are not bound to one legal system, and are flexible enough to allow adaptation to any (foreseen) kind of restriction.

None of the four approaches would pose a problem for adaptation. Apart from that remark, we can notice that all approaches have a system allowing the user to define a privacy policy on the client side (either an "acceptable policy" pattern, or an access

Table 2. Additional features in profile and privacy management

Approach	Additional Features					
	Distributed Profile	User-side Profile	Anonymous Identities	Server Authentication	Client-Side Policy	Legislation-independent
P3P [18]		•		○	•	•
IDsec [9]			•	•	•	•
TCPA-based [10]	•	•	•	•	•	•
Client-Side Profile Storage [12], [13]	•	•		○	•	•

control policy). As seen earlier, it is a simple way to implement a prior consent system. Only the last approach has investigated privacy on distributed profiles, and the TCPA approach alone introduces the concept of anonymous identities, which would be so convenient for protection of indirectly nominative data.

4 Further Work

The study of these approaches puts a stress on a few problems that should be addressed, if we want to build a system fully guaranteeing privacy. Here are the orientations we are going to give to our work in the domain.

4.1 Software Cryptographic Certification

We think that Pearson's approach has a few shortcomings that could be avoided. First, hardware protection of the private keys [10] may not be absolutely necessary in our privacy perspective (because hacking of the profiling agent on the client machine may not be the greater risk here). It would be very interesting though to derive the ideas exposed by Pearson, but using a software certification of the profile management process. Moreover, the necessary presence of the TPM on the motherboard could possibly prevent the use of Trusted Computing on the service side: it looks very difficult to deploy TCP on a load-balancing multi-server system. Even distributed file systems could cause problems. Besides, the system would be very sensitive to hardware failure and modifications. That is why we think that TCPA has been designed to certify the user, and in no way the service. Despite these drawbacks, Pearson's approach brings very interesting ideas that we would like to adapt. Above all, there is the certification of an agent and a process: We will study the feasibility of replacing the hardware protection by a software certification, in order to eliminate some of the restrictions we have identified. It is currently the only way we foresee that could lead us to a guarantee, provided (by computational means) to the user, that a distant service proceeds in a given way.

In theory, should this approach be developed as desired, we would be able to certify that the personal information provided by the user to the service will not be transmitted to another one, for instance. But let us imagine a situation where the privacy policy on which the user and service agents have agreed allows the latter to forward some information to one identified third party. A limit must be defined for the control over the distant process, since it might be possible, in some cases, to check that the data is sent to the right third party, but the control cannot follow the data. In any way, this approach will be limited by the computational burden put on the user and on the service machines.

4.2 Monitoring the Privacy Compliance of a Service

Subirana and Bain [14] think that the application of a privacy policy (a legal privacy or a service-defined privacy) must be associated to a monitoring of the data processing. In the cases where no strong control can be established on the process, it should be possible to "test" the way in which the company complies with the privacy policy. We think that a profitable effort can be made here, by studying and integrating the existing ways of testing the privacy compliance of the services and the different kinds of "privacy probes" that can be used.

In the context of reasoning on the compliance of a service, we think that it would be a good thing to integrate the P3P [18] standards in the different components of our future platform, since they provide us with efficient tools for communicating about privacy policies. Extensions to the standard formats could be developed in order to encapsulate the other dimensions of our future approach.

4.3 Anonymous and Distributed User Profile

We have seen that the work in [10] contain very interesting ideas about the use of virtual identities in privacy-compliant profile management. For some kinds of problems, dissociation may transform directly nominative information into indirectly nominative information, and complexify the intercorrelation between sets of data. Distribution of the profile over a network, for its part, helps the user with keeping the control over his private data while gaining in dynamism and openness. Consequently, we believe that our future platform should be able to deal with (and facilitate) the profile management principles proposed in [9] and [12], for they will help improving the overall privacy level of the system.

4.4 Profiling in Other Domains

We have focused on the context of ambient computing commercial applications, but other fields use multi-agent based technologies in their distributed processes, and are concerned with privacy issues. In particular, ideas could be retrieved from the medical field, where the implementation of distributed medical records brings interesting problems to light [8]. All the areas where distributed profiles are present, like information research, may have such problems. We will identify the specificities of those sector-based privacy demands, so that our propositions can be as generic as possible.

4.5 Implementing and Testing Privacy Management Solutions

In order to test the different approaches of agent-based privacy management, we plan to build a platform on which simple scenarios could be run, and the models be evaluated. The scenario chosen for the first sets of evaluations is rather simple, but adequate: a service agent asks a personal (user) agent for a personal, nominative information (a simple numerical value, in our application), in order to give it as a parameter (possibly several times) to a processing function, internal to the service agent. This function represents the service itself, for which the user agent explicitly sent the personal information. After a given time, the service agent has to destroy the data. Before that, the user agent can ask the service agent to modify or destroy its personal information. The platform environment will be able to evaluate the following properties (related to the six components of privacy) in the transaction:

- **Information and Consent:** Has the service agent properly informed the user agent? Has the user agent given its explicit consent to the transaction?
- **Goal:** Has the service agent given the value to the right internal function and not to another one?

- **Update and Suppression:** Did the service agent modify or destroy the data after being told to do so?
- **Data forwarding:** Has the service agent transmitted the data to a third-party agent?
- **Data retention:** Has the service agent destroyed the data after the declared data retention limit?

Those properties constitute the indicators of the benchmark platform. They are directly related to the components of privacy we have identified, so that we can make a systematic evaluation of our different privacy management models.

In order to properly evaluate those properties, the agents implemented on the platform will be the following:

- A personal agent, without any privacy management (for reference),
- One personal agent per privacy management model implemented,
- A benevolent service agent,
- A third-party agent, which only purpose is to receive forwarded data,
- A certification authority agent, able to deliver cryptographic certificates, and which authority is accepted by all agents on the platform.

With these first we will be able to check that the tested models are working properly in a "normal" context. The first personal agent is a very obedient one, replying honestly to all requests, and not worrying about privacy issues. It is an image of a "naive" user, a potential target for dishonest service agents. With this agent, all privacy breaches are believed to be successfully exploited by the different service agents. Each of the other personal agents will implement a privacy management model, and its results will be compared to the ones of the obedient agent. The next step is to implement treacherous agents, one for each property we want to check:

- **Information and Consent:** A service agent that does not properly inform the user agent, and/or that does not explicitly ask for its consent,
- **Goal:** A service agent that gives the data as a parameter to another internal function,
- **Update and Suppression:** A service agent that does not modify or destroy the data when told to do so,
- **Data forwarding:** A service agent that forwards the data to a third-party agent,
- **Data retention:** A service agent that does not destroy the data after expiration of the data retention duration.

This second set of service agents is designed so that each of them tries to infringe one of the privacy principles. It is possible that derived service agents must be implemented, in order to cope with protocols imposed by specific privacy management models (for instance, implementation of a cryptographic certification system). This makes it possible for the privacy management models to be tested in a more real-world, "aggressive" environment. This set of experiments will provide us with actual measurement indicators for the quality of the different models.

This platform architecture will be the base for all implementation of our future work in privacy management. It will allow us to evaluate with clear and rational indicators the improvements brought by one theoretical model or another.

5 Conclusion

Our study gives an overview of the state of the art in the domain of privacy enforcement. We have shown what has been done, or could be done with little effort, given the existing works. We have identified the technological challenge of guaranteeing or certifying a distant process, and examined the steps already taken in that direction. Working on those points will allow us to build tools that would help us with putting the user at the centre of ambient computing and heterogeneous agent/human societies, by taking control over his personal data and their privacy. These theoretical and technical tools should be the basis of a possible evolution towards a privacy managed by autonomous cognitive agents. Even though no "computational guarantee" exists at the moment that could quantify the privacy compliance of a process, the knowledge management components of distributed artificial intelligence could integrate the notions we have identified, in order to deal with privacy at a cognitive level.

The very process of knowledge acquisition could actually be adapted to a privacy-compliant context, where the obtaining of directly or indirectly nominative information would be subject to a number of checks: the user agent must be given information, it must give its consent... The knowledge representation model, for its part, will consider the limitation of data retention to the duration agreed with the user agent. The knowledge representation engine should also be able to control access to the sensitive data, so that it could not be used for unexpected processing, or transmitted to non explicitly authorized third parties. Actually the whole knowledge processing model of the agent should be designed while keeping in mind all the possible requirements of a privacy policy, so that the policy itself could be part of the agent's cognitive contextual organization (be it hard-coded or dynamic, or even hybrid with a hard-coded "legal" part, and a dynamic part for optional privacy policies). Such a cognitive model would allow the agents to be easily adaptable to different national legislations, and then able to operate in the international e-commerce market, for instance.

This integration of the privacy mechanisms into cognitive multi-agent systems would mean significant progress in the direction of a better protection of the user in ambient computing contexts. Indeed such an approach would ally the fluidity of autonomous personal agents, with all the provable guarantees that could be gained from advances in pure privacy management techniques.

References

1. Article 29 Data Protection Working Party: Working Document on Trusted Computing Platforms and in particular on the work done by the Trusted Computing Group (TCG group), The European Commission (2004)
2. Burrows, M., Abadi, M., Needham, R.: A Logic of Authentication. In: Proceedings of the Royal Society of London, Series A, London, UK, vol. 426, pp. 233–271 (1989)
3. Bygrave, L.A.: Electronic Agents and Privacy: A Cyberspace Odyssey. International Journal of Law and Information Technology 9(3), 275–294 (2001)
4. Castelfranchi, C., Falcone, R.: Principles of Trust for Multi-Agent Systems: Cognitive Anatomy, Social Importance, and Quantification. In: International Conference of Multi-Agent Systems (ICMAS'98), Paris, pp. 72–79 (1998)

5. Castelfranchi, C.: Trust Mediation in Knowledge Management and Sharing. In: Jensen, C., Poslad, S., Dimitrakos, T. (eds.) iTrust 2004. LNCS, vol. 2995, pp. 304–318. Springer, Heidelberg (2004)
6. Cranor, L.F.: Electronic Voting: Computerized Polls May Save Money, Protect Privacy. In: Crossroads, vol. 2(4), pp. 12–16. ACM Press, New York (1996)
7. Finin, T., Fritzson, R., McKay, D., McEntire, R.: KQML as an Agent Communication Language. In: Proceedings of the Third International Conference on Information and Knowledge Management (CIKM '94), Gaithersburg, MD, USA, pp. 456–463 (1994)
8. Huberman, B.A., Hogg, T.: Protecting Privacy While Revealing Data. Nature Biotech. 20, 332 (2002)
9. Internet Engineering Task Force: IDSec: Virtual Identity on the Internet, http://idsec.sourceforge.net/
10. Pearson, S.: Trusted Agents that Enhance User Privacy by Self-Profiling. In: Falcone, R., Barber, S., Korba, L., Singh, M.P. (eds.) AAMAS 2002. LNCS (LNAI), vol. 2631, pp. 113–121. Springer, Heidelberg (2003)
11. Pearson, S.: Trusted Computing Platforms: TCPA Technology in Context. Prenctice Hall PTR, Upper Saddle River, New Jersey, USA (2002)
12. Riché, S., Brebner, G.: Client-Side Profile Storage. In: Gregori, E., Cherkasova, L., Cugola, G., Panzieri, F., Picco, G.P. (eds.) NETWORKING 2002. LNCS, vol. 2376, pp. 127–133. Springer, Heidelberg (2002)
13. Riché, S., Brebner, G.: Storing and Accessing User Context. In: Chen, M.-S., Chrysanthis, P.K., Sloman, M., Zaslavsky, A. (eds.) MDM 2003. LNCS, vol. 2574, pp. 1–12. Springer, Heidelberg (2003)
14. Subirana, B., Bain, M.: Legal Programming: Designing Legally Compliant RFID and Software Agent Architectures for Retail Processes and Beyond. Springer, USA (2005)
15. The European Parliament and the Council: Directive 2002/58/EC of the European Parliament and of the Council of 12 july 2002 Concerning the Processing of Personal Data and the Protection of Privacy in the Electronic Communications Sector. Official Journal of the European Communities (2002)
16. Trusted Computing Group (2006), https://www.trustedcomputinggroup.org/home
17. Westin, A.: Privacy and Freedom. Atheneum, New York, USA (1967)
18. World Wide Web Consortium: Platform for Privacy Preferences Specification 1.0, http://www.w3.org/P3P/

Effective Use of Organisational Abstractions for Confidence Models*

Ramón Hermoso, Holger Billhardt, Roberto Centeno, and Sascha Ossowski

Artificial Intelligence Group - DATCCCIA
University Rey Juan Carlos, Madrid (Spain)
{ramon.hermoso,holger.billhardt,roberto.centeno,sascha.ossowski}@urjc.es

Abstract. Trust and reputation mechanisms are commonly used to infer expectations of future behaviour from past interactions. They are of particular relevance when agents have to choose appropriate counterparts for their interactions as it may also happen within virtual organisations. However, when agents join an organisation, information about past interactions is usually not available. The use of organisational structures can tackle this problem and can improve the efficiency of trust and reputation mechanisms by endowing agents with some extra information to choose the best agents to interact with. In this context, we present how certain structural properties of virtual organisations can be used to build an efficient trust model in a local way. Furthermore, we introduce a testbed (TOAST) that allows to analyse different trust and reputation models in situations where agents act within virtual organisations. We experimentally evaluate our approach and show its validity.

1 Introduction

The concept of organisation is of significant importance to MultiAgent Systems (MAS). Particularly, organisational structures are often used in research about Agent-oriented Software Engineering [26]. In fact, organisational concepts are commonly used as abstract pieces that help designers to build more complex models in MAS design processes [14].

Organisational abstractions can be used to impose some structure on a society of agents and can endow MAS with certain behaviours. Agents joining an organisation play specific *roles* in different *interactions* and they are supposed to act conforming to the prescriptions of these concepts. Furthermore, these prescriptions may be complemented by a more general set of *norms* [23] and some kind of mechanisms that make it difficult for agents to transgress norms (e.g. by providing specific "governor" agents [4], by integrating "filtering" mechanisms [13], or by using protocols of sequential actions). We will call MAS with such organisational structures as *Virtual Organisations* (VOs) [18].

VOs can be considered as limiting the freedom of choice of agents because they regulate the interactions within a MAS. However, especially within low

---

* The present work has been funded by the Spanish Ministry of Education and Science under projects TIC2003-08763-C02-02 and TIN2006-14630-C03-02.

regulated organisations, agents will still have to tackle the problem of choosing appropriate counterparts for their interactions according to their own beliefs and goals. Within this scenario, trust and reputation mechanisms can be integrated into VOs providing support to agents' decision-making processes.

Trust and reputation systems have received a lot of attention in the last years. Such systems use past experiences to provide agents with hints about the future behaviour of their acquaintances [9,2,17]. Nevertheless, most of the research is based on the importance of distributed trust and on the exchange of information among agents (e.g. reputation values about third parties) in poorly structured systems.

Trust and reputation systems are not only useful for VOs. VOs can also be useful for trust and reputation mechanisms. The structure provided by a VO can be used to construct more effective trust mechanisms. In particular, the structural elements defined in a VO (e.g., roles and interactions) provide a certain notion of similarity which allows agents to infer the expected behaviour of acquaintances within totally new situations by analysing their past behaviour within *similar* situations. This property is especially useful in situations where agents can not count on their own past experiences, e.g., when they just have joined an organisation, or within very volatile environments.

In this paper we continue our previous work [7] on trust and reputation mechanisms for VOs. We present some experiments that show how the use of organisational abstractions can effectively improve trust and reputation mechanisms. The experiments have been carried out using a testbed called *TOAST* (Trust Organisational Agent System Testbed), which we have developed for testing trust and reputation models. In Section 2 we briefly summarise our model and show how it can guide an agent's decision-making within a VO. Section 3 introduces the testbed. In Section 4 we present experimental results comparing different models. We summarise related work in Section 5, and we present some conclusions and future lines of work in Section 6.

2 Confidence and Trust for Organisational Structures

In this section we summarise our previous work [7] on a trust model for VOs. We first show how basic trust mechanisms can be integrated in VOs. Afterwards, we explain how an agent can use knowledge about the organisational structure to infer confidence in an issue if no previous experience is available.

2.1 Basic Local-Based Trust Model for Virtual Organisations

As outlined in [7], it is natural to consider that agents participating in a VO play some roles in different interactions. In addition, we assume that the agents know the organisational structure, e.g., they know the existing roles and interaction types, the roles that participate in each interaction type, as well as the roles other agents are playing within the organisation.

Similarly to other approaches [10,24,16,15], we build our trust model on the idea of *confidence* and *reputation*. Both are ratings agents use in order to evaluate the trustworthiness of other agents in a particular issue (e.g., playing a particular role in a particular interaction). *Confidence* is a local measure that is only based on an agent's own past experiences, while *reputation* is an aggregated value an agent gathers by asking its acquaintances about their opinion regarding the trustworthiness of another agent. Thus, reputation can be considered as an external or *social* measure. We define *trust* as a rating resulting from combining *confidence* and *reputation* values.

A typical scenario for the use of a trust model is the following. An agent A wants to evaluate the trustworthiness of some other agent B – playing the role R – in the interaction I. This trustworthiness is denoted as $t_{A \to \langle B,R,I \rangle}$, with $t_{A \to \langle B,R,I \rangle} \in [0..1]$, and it measures the trust of A in B (playing role R) being a "good" counterpart in the interaction I. When evaluating the trustworthiness of a potential counterpart[1], an agent can combine its local information (confidence) with the information obtained from other agents regarding the same counterpart (reputation).

Confidence, $c_{A \to \langle B,R,I \rangle}$, is collected from A's past interactions with agent B playing role R and performing interactions of type I. We call LIT, *Local Interaction Table*, the agents' data structure dedicated to store confidence values for past interactions with other agents. Each entry corresponds to an *issue*: an *agent* playing a specific *role* in a particular *interaction*. LIT_A denotes agent A's LIT. An example is shown in Table 1.

Table 1. An agent's local interaction table (LIT_A)

$\langle X, Y, Z \rangle$	$c_{A \to \langle X,Y,Z \rangle}$	$r_{A \to \langle X,Y,Z \rangle}$
$\langle a_9, r_2, i_3 \rangle$	0.2	0.75
$\langle a_2, r_7, i_1 \rangle$	0.7	0.3
$\vdots$	$\vdots$	$\vdots$
$\langle a_9, r_2, i_5 \rangle$	0.3	0.5

Each entry in a LIT consists of: i) the Agent/Role/Interaction identifier $\langle X, Y, Z \rangle$, ii) the confidence value for the issue ($c_{A \to \langle X,Y,Z \rangle}$), and iii) a reliability value ($r_{A \to \langle X,Y,Z \rangle}$). The confidence value is obtained from some function that evaluates past experiences on the same issue. We suppose $c_{A \to \langle X,Y,Z \rangle} \in [0..1]$ and higher values to represent higher confidence.

Each direct experience of an agent regarding an issue $\langle X, Y, Z \rangle$ changes its confidence value $c_{A \to \langle X,Y,Z \rangle}$. In this sense, we suppose that the agents have some kind of mechanism to evaluate the behaviour of other agents they interact with. Let $g_{\langle X,Y,Z \rangle} \in [0..1]$ denote the evaluation value an agent A calculates for a particular experience with the agent X playing role Y in the interaction of type Z. In our work, we use the following equation to update confidence:

$$c_{A \to \langle X,Y,Z \rangle} = \epsilon \cdot c'_{A \to \langle X,Y,Z \rangle} + (1 - \epsilon) \cdot g_{\langle X,Y,Z \rangle}, \qquad (1)$$

---

[1] By potential counterpart we mean an agent which is a candidate to interact with.

where $c'_{A\to\langle X,Y,Z\rangle}$ is the confidence value in A's LIT before the interaction is performed and $\epsilon \in [0..1]$ is a parameter specifying the importance given to A's past confidence value. In general, the aggregated confidence value from past experiences will be more relevant than the evaluations of the most recent interactions.

Reliability ($r_{A\to\langle X,Y,Z\rangle}$) measures how certain an agent is about its own confidence in an issue. We suppose $r_{A\to\langle X,Y,Z\rangle} \in [0..1]$. Furthermore, we assume that $r_{A\to\langle X,Y,Z\rangle} = 0$ for any tuple $\langle X,Y,Z\rangle$ not belonging to LIT_A. We calculate reliability by using the approach proposed by Huynh, Jennings and Shadbolt [9,10]. This approach takes into account the number of interactions a confidence value is based on and the variability of the individual values across past experiences.

An agent may build trust directly from its confidence value or it may combine confidence with reputation. Reputation will be particularly useful if an agent has no experience about an issue or if the reliability value for the confidence is not high enough. Although we will not deal with it in this paper, social reputation may be obtained by asking other agents about their opinion on an issue. Agents that have been requested for their opinion will return the corresponding confidence and reliability ratings from their LIT. The requester might then be able to build trust by calculating a weighted mean over its own confidence value and the confidence values received from others, as it is represented in equation (2):

$$t_{A\to\langle B,R,I\rangle} = \begin{cases} c_{A\to\langle B,R,I\rangle}, & if\ r_{A\to\langle B,R,I\rangle} > \theta \\ \dfrac{\sum_{X\in AA\cup\{A\}} c_{X\to\langle B,R,I\rangle} \cdot w_{X\to\langle B,R,I\rangle}}{\sum_{X\in AA\cup\{A\}} w_{X\to\langle B,R,I\rangle}} & otherwise \end{cases} \quad (2)$$

$\theta \in [0..1]$ is a threshold on the reliability of confidence. If the reliability is above θ then an agents own confidence in an issue is used as the trust value. Otherwise trust is build by combining confidence and reputation. AA is a set of acquaintance agents that an agent asks for their opinion about the issue $\langle B,R,I\rangle$. Within a VO, the structural abstractions may provide hints for the proper selection of such a set of acquaintance agents. For instance, in some scenarios it may be useful to ask other agents that play the same role as A, since they may have similar interests and goals.

The weights $w_{X\to\langle B,R,I\rangle}$ given to the gathered confidence values is composed of the corresponding reliability value and a constant factor α that specifies the importance given to A's own confidence in the issue, as it is shown in the following equation:

$$w_{X\to\langle B,R,I\rangle} = \begin{cases} r_{X\to\langle B,R,I\rangle} \cdot \alpha, & if\ X = A \\ r_{X\to\langle B,R,I\rangle} \cdot (1-\alpha), & otherwise \end{cases} \quad (3)$$

2.2 Confidence Inference Using Organisational Structure Similarities

In this section we propose a local way for building trust on an issue when no past interactions have been performed and without relying upon social reputation. In [7] we proposed to use the agent/role confidence $c_{A\to\langle B,R,_\rangle}$ (or the agent confidence $c_{A\to\langle B,_,_\rangle}$) as an estimation for $c_{A\to\langle B,R,I\rangle}$ if agent A has no reliable experience about issue $\langle B, R, I \rangle$. This approach relies on the hypothesis that, in general, agents behave in a similar way in all interactions related to the same role. We argue that, exploiting this idea, the more similar I' and I are, the more similar will be the values $c_{A \to \langle B,R,I' \rangle}$ and $c_{A \to \langle B,R,I \rangle}$. The same applies to roles. Using this assumption, confidence ratings accumulated for similar agent/role/interaction tuples may provide evidence for the trustworthiness of the issue $\langle B, R, I \rangle$. Based on this idea, we propose to build trust by taking into account all the past experiences an agent has, focusing on their degree of similarity with the issue $\langle B, R, I \rangle$. In particular, we calculate trust as a weighted mean over all the confidence values an agent has accumulated in its LIT. This is shown in the following equation:

$$t_{A \to \langle B,R,I \rangle} = \frac{\sum_{\langle X,Y,Z \rangle \in LIT_A} c_{A \to \langle X,Y,Z \rangle} \cdot w_{A \to \langle X,Y,Z \rangle}}{\sum_{\langle X,Y,Z \rangle \in LIT_A} w_{A \to \langle X,Y,Z \rangle}} \quad (4)$$

$w_{A \to \langle X,Y,Z \rangle}$ is the weight given to agent A's confidence on issue $\langle X, Y, Z \rangle$. The weights combine the confidence reliability with the similarity of the issue $\langle X, Y, Z \rangle$ to the target issue $\langle B, R, I \rangle$ in the following way:

$$w_{X \to \langle X,Y,Z \rangle} = r_{A \to \langle X,Y,Z \rangle} \cdot sim(\langle X, Y, Z \rangle, \langle B, R, I \rangle) \quad (5)$$

The similarity function $sim(\langle X, Y, Z \rangle, \langle B, R, I \rangle)$ is computed as the weighted sum of the similarities of the individual elements (agent, role and interaction) as it is shown in the following equation:

$$sim(\langle X, Y, Z \rangle, \langle B, R, I \rangle) = \begin{cases} \beta \cdot sim_R(R, Y) + \gamma \cdot sim_I(I, Z), & if \ B = X \\ 0, & otherwise \end{cases} \quad (6)$$

where $sim_R(R, Y), sim_I(I, Z) \in [0..1]$ measure the similarity between roles and interactions, respectively, and β and γ, with $\beta + \gamma = 1$, are parameters specifying the sensibility regarding the individual similarities.

We suppose that organisational models include taxonomies of roles and/or interactions from which role and interaction similarity measures can be derived. In this case, $sim_R(R, R')$ and $sim_I(I, I')$ can be implemented by *closeness functions* that estimate the similarity between two concepts on the basis of their closeness in a concept hierarchy.

Equation (4) can be used as an alternative way to build trust. Especially if an agent has no reliable experience about a particular agent/role/interaction issue, this model can be used to estimate trust without the necessity to rely on the opinions of other agents. Thus, the proposed model makes agents less dependent from others, which is an important issue, in particular in VOs that do not provide mechanisms to keep their members from cheating.

3 Trust Organisational Agent System Testbed (TOAST)

In this section we present TOAST, a tool we have developed to evaluate trust models by showing the influence of these models on the evolution of the overall utility of an agent or a society of agents. This testbed simulates a virtual organisation where agents have to interact with others in order to to achieve their goals. TOAST is based on the following simplifications:

- TOAST does not consider the problem of finding appropriate interactions that help to achieve an agent's goals, nor it considers the task of locating possible candidates that can act as counterparts in the interactions an agent wants to perform. We suppose that both of these problems are resolved, e.g., the corresponding information is fully available to each agent within the organisation. The only problem the testbed actually addresses is the selection of appropriate counterparts out of a set of possible candidates.
- All interactions are binary, e.g., exactly two agents are required in order to perform an interaction.
- Agents are always willing to participate in an interaction if they have been chosen by others. That means, we do not consider the problem of the selected agent to decide whether or not to participate in an interaction.
- Interactions in TOAST are not actually carried out; they are simulated. In practice, this means that agents participating in an interaction do not actually evaluate the behaviour of the others in that interaction. Instead, the agents receive these evaluation values directly from the testbed. In order to generate these values, the system uses the notion of capability. Capability here indicates the "goodness" of agent's A behaviour in playing role R in an interaction of type I. Capability is represented through a normal probability distribution with constant mean and variance and is assigned to each tuple $\langle A, R, I \rangle$. The evaluation value an agent receives after performing an interaction is drawn from the corresponding capability distribution of the counterpart agent. The use of this schema implies the following simplifying assumptions:
  • An agent's capability playing a specific role within a particular interaction is stable with some variations. That is, each time an agent plays the same role in the same type of interaction it will behave in a similar way. The assumption of stability is actually the basis for any trust and reputation model. Nevertheless, in real cases it is likely that an agent's behaviour changes over time.
  • Agents don't cheat. That is, agents participating in an interaction do the best they can; they do always behave in the prescribed way (corresponding to their capabilities).

3.1 Setting Up a VO in TOAST

The testbed allows the user to create virtual organisations of agents. The elements that compose a virtual organisation and the relationships among them are represented in a class diagram in Figure 1.

- *Agents* are the entities that participate in a VO. They can play different *roles* depending on their aptitudes. Within the VO, agents aim to achieve their goals. TOAST does not fix the type of agents. That is, users can implement

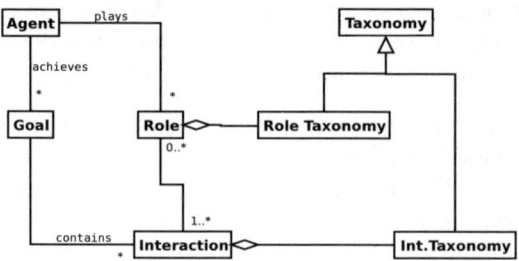

Fig. 1. Organisational elements in TOAST

different types of agents (agents with different behaviours) in order to make the VO more heterogeneous.
- *Roles* describe the functionality of the organisation. They will be played in specific interactions. The roles that are defined within an organisation have to be specified in a *role types taxonomy*. This taxonomy also specifies a conceptual hierarchy for the defined roles.
- *Interactions* are the actions that the VO allows to perform between agents. They are defined in an *interaction types taxonomy*, which also defines the roles that are involved in each interaction as well as the hierarchical relationships between interactions.
- *Goals* have to be achieved by agents. An agent may have several, different goals. A goal will be achieved by means of performing a specific interaction. The relation between goals and the interactions that help to achieve them has to be specified in a XML file which is loaded into the testbed. The following is an example of a fragment of such a file:

```
<Goal Name="Select Lecturer">
    <Interaction Name="Teach">
        <Role Name="Academic Staff">
        </Role>
        <Role Name="Student">
        </Role>
    </Interaction>
</Goal>
<Goal Name="Select Assignment Partner">
    <Interaction Name="Do Assignments">
        <Role Name="Student">
        </Role>
        <Role Name="Student">
        </Role>
    </Interaction>
</Goal>
. . .
```

After defining a virtual organisation, agents have to be added to the organisation. In particular, the user has to select the number of agents of each type that

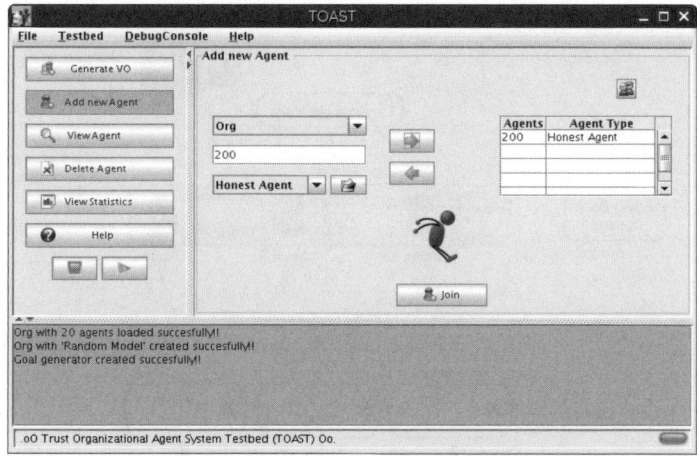

Fig. 2. Add Agents panel in TOAST's GUI

will be added to the organisation as it is shown in Figure 2. In our experiments we have only used agents of one type (with the same behaviour).

After agents have been added to the VO and according to the organisation specification described above, roles have to be assigned to agents and capability values for Agent/Role/Interaction tuples have to be generated. This is done automatically as follows:

- **Agent/Role assignment.** Roles are assigned randomly to agents. An agent may play different roles. For each role, the user can choose the percentage of agents to which the role will be assigned. This selection should be done with care. If some roles are played by only very few agents, then the scenario may not be appropriate for evaluating trust mechanisms because of the lack of sufficient candidates out of which an agent can choose its counterparts in its interactions.
- **Agent/Role/Interaction capability values.** As we have mentioned before, in order to evaluate the behaviour of an agent in a particular situation, the system uses a normal probability distribution that model the capability of that agent in this situation. Such distributions are assigned to each tuple $\langle A, R, I \rangle$ at startup by selecting the mean and standard deviation randomly such that $\mu \in [0..1]$ and $\sigma \in [0..0.5]$. The capabilities are correlated according to the type of interaction, the type of role and the agent which is performing the action; that is, similar normal probability distributions will be assigned to the same agent playing similar roles within similar interactions[2].

3.2 Executing Experiments with TOAST

As we have mentioned before, agents will deal with the problem of selecting appropriate counterparts to interact with in order to achieve their goals. The

---

[2] Similar distributions means similar mean values.

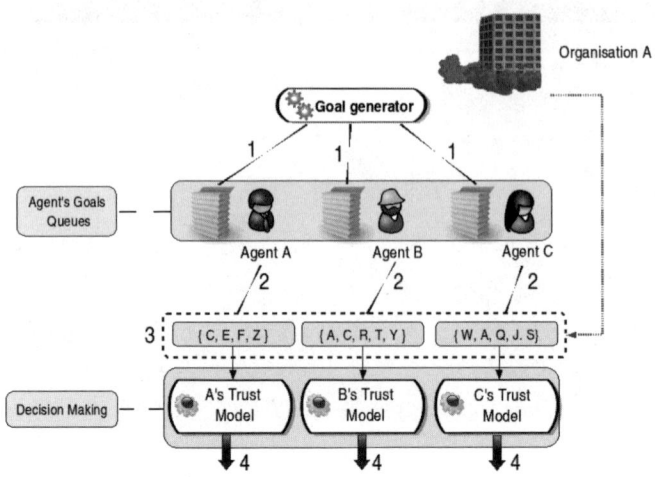

Fig. 3. Interaction simulation in TOAST

basic process that is repeated when the testbed is executed is summarised in Figure 3.

The process is the following:

1. A goal is generated for an agent. Each goal has an associated interaction that the agent must perform in order to achieve that goal. Each agent has a goal queue where goals are stored until they can be processed.
2. Once the agent has identified the interaction that eventually helps to achieve the goal, a list with potential agents that are possible counterparts in the required interaction is obtained from the organisation. As we mentioned before, in order to simplify the experiments all interaction types are binary, e.g., require exactly two agents playing particular roles.
3. The agent uses its trust model to select the agent that is expected to behave best in the required interaction and playing the specified role.
4. The interaction is simulated. As we have mentioned before, this step consists of sending to each agent the evaluation values ($g_{\langle X,Y,Z \rangle}$) for the other agent that participated in the interaction. These values are generated from the corresponding capability distributions. Finally, each agent uses the received evaluation values to update the confidence and reliability values in its LIT as described in section 2.1.

4 Experimental Results

We have used TOAST to experimentally evaluate our confidence inference approach. In this section, first we describe the scenario we have chosen to evaluate our assumptions. Then, we describe the different trust models we have tested and, finally, we show the obtained results.

4.1 The University Scenario

As a test scenario we use a "School of Computer Science" organisation, whose members play roles out of the taxonomy shown in Figure 4. Furthermore, the social functionalities provided by the organisation are summarised in the interaction taxonomy illustrated in Figure 5.

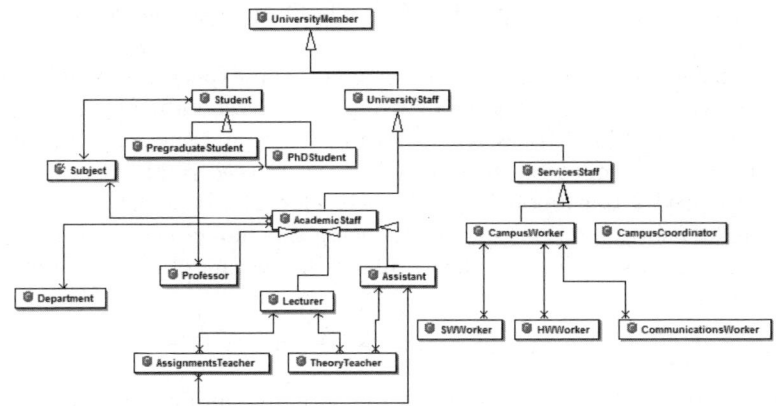

Fig. 4. Fragment of role taxonomy provided by University organisation

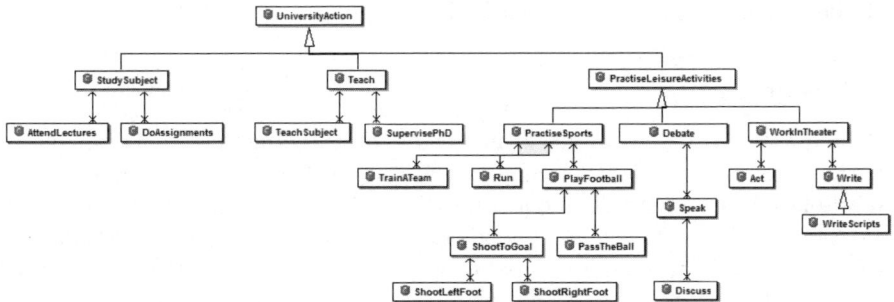

Fig. 5. Fragment of interaction taxonomy provided by University organisation

In this scenario, for example, a typical situation could be that an agent playing the role of a student needs to find a partner for some kind of assignment in a specific subject. The student will use its LIT to select the best partner for the assignment (e.g., another student) according to his/her own experiences about past interactions with other students.

4.2 Different Trust Models

We tested and compared three different *local* trust models in the sense that they do not use social reputation to compute trust and are based only on different confidence evaluation approaches. Thus, each model defines a different way how an agent uses its own past experience in order to select the best counterparts for its interactions out of the set of possible candidates. The models are the following:

- *Random Model*: in this model agents choose the counterparts for their interactions randomly among the potential agents provided by the organisation. Thus, the selection does not take into account any experience on past interactions.
- *Basic Model*: in this model, agents evaluate the expected behaviour of the potential candidates for an issue by using the corresponding confidence value stored in their LITs. If no entry exists about an issue, e.g., no previous experiences are available, the counterpart is selected randomly.
- *Inference Model*: this model implements our local trust model, as described in Section 2.2 using equation (4). Agents using this model will use the following simple formula to calculate the similarity between roles and interactions, respectively:

$$sim_R(x,y) = sim_I(x,y) = 1 - \frac{h}{h_{MAX}} \qquad (7)$$

where x, y are either roles or interactions, h is the number of hops between x and y in the corresponding taxonomy, and h_{MAX} is the longest possible path between any pair of elements in the hierarchy tree. Equation (7) is a very simple formula to measure the similarity between concepts in a taxonomy. Other functions have been described in [11,6].

4.3 Results

We tested and compared the performance of the three models in the university scenario by using exactly the same conditions in each case. In particular we used an organisation with 20 agents and the goal generator generated randomly 40000 goals for those 20 agents. The generated goals, the agent/role assignment and the agent/role/interaction capability generation was exactly the same for each model. In the *inference model* we used the similarity weights $\beta = 0.8$ and $\gamma = 0.2$. Furthermore, we repeated each experiment five times using a different random seed. The given results represent the average over the five runs.

Figure 6 shows the evolution of the overall system utility over the number of interactions. The overall system utility is calculated as the average of the utilities of all individual agents. As utility values we use the evaluation values (about its counterparts) an agent receives after performing interactions. As it can be observed, both trust models perform clearly better than a random agent selection. Moreover, the utility improves as the agents gain more experience, that is, as more interactions take place. The *inference model* obtains better outcomes as compared to the *basic model*, since agents' utility curves grow faster. Hence, with the *inference model* agents are able to find faster "good" counterparts to interact with. This confirms the hypothesis that the *inference model* improves the agents decision making processes when agents have none or only very few

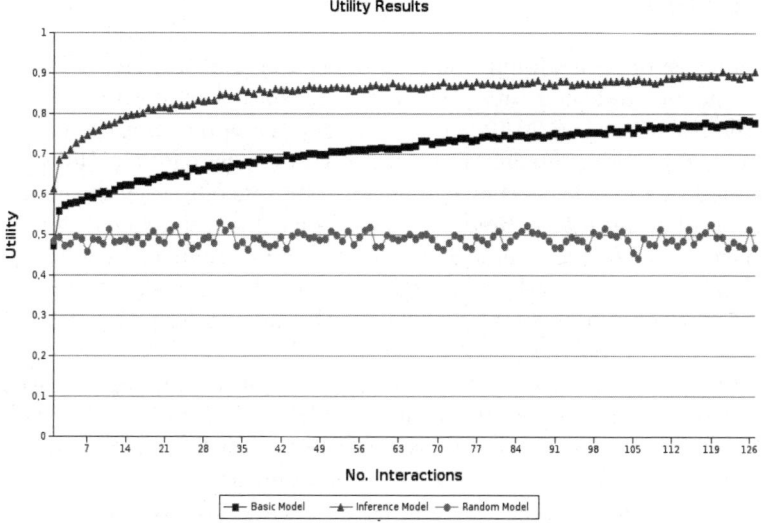

Fig. 6. Overall system utility

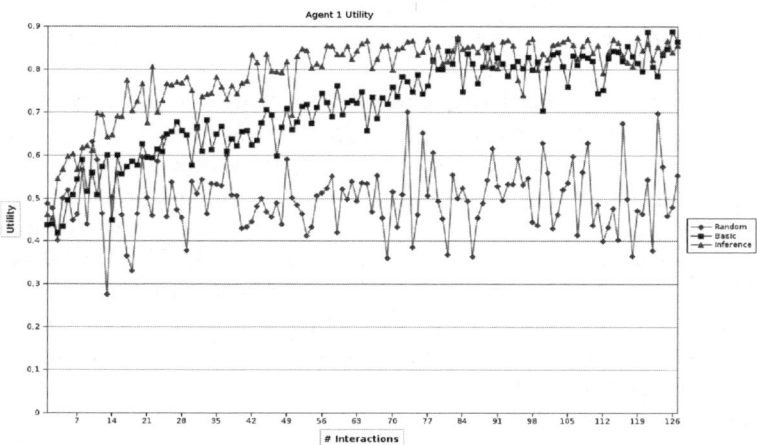

Fig. 7. Example of the evolution of an agent's individual utility

experiences from past interactions. In this case, the *inference model* makes use of past experiences about similar roles and interactions in order to estimate the expected utility of an unknown $\langle X, Y, Z \rangle$ tuple.

This evolution is similar if we consider individual agents. Figure 7 gives an example of how the utility curve of an individual agent evolves over time. Also here, the *inference model* curve grows faster that the others, because the agent is able to find good counterparts faster.

5 Discussion

A wide range of organisational (meta-)models aimed at describing basic organisational concepts and their interrelation in the context of MAS have been developed in the last years [5,8,20,12]. It is commonly accepted that the notion of *role* is central for putting agents and organisational models together. Roles are sometimes defined by the *actions* they can perform, but usually they are characterised by the types of *social interactions* to which they contribute. The latter term does not primarily refer to the interaction protocols that agents engage in, but rather to the social functionality that such interactions shall achieve. In this sense, we assume that VOs define roles and specify the interactions (functionalities) in which each role can participate.

Several meta-models allow for *specialisation* relations among essential organisational concepts. In the organisational model underlying the FIPA-ACL, for instance, *information exchange* interactions are a special kind of *request* interaction, where the requested action is a communicative action of type *inform* (e.g., [21]). In much the same way, the *informer* role involved in this interaction can be conceived as a specialisation of the *requester* role. In summary, organisational models often contain *taxonomies* of concept types, e.g., for roles or for interactions. Such taxonomies can be provided to the agents participating in an organisation – for instance, as an organisational service – and can be used to define the similarities among roles and interactions as it is described in this paper.

Trust and reputation mechanisms have been widely studied, recently above all in *peer-to-peer* systems in general (e.g. [25,2]), and in MAS in particular (e.g. [9,24,17]). In contrast to other approaches to trust systems (most of them based on reputation distribution – reputation values exchange about third parties), we have presented a way of evaluating trust at a local level that focuses on the experience of agents obtained in past interactions. The FIRE model proposed by Huynh, Jennings and Shadbold [9] is also related to *interaction trust* and *role-based trust*. As in our approach, the former is built from direct experiences of an agent, while the latter is the rating that results from role-based relationships between agents. However, the FIRE model does not consider inference on VO structures.

Sensoy and Yolum [19] deal with the problem of distributed service selection in an e-commerce setting where consumers are allowed to capture their experiences with the service providers. This approach is similar to the witness reputation approach presented in [9] in the sense that agents can locate others by making use of other agents' past experiences. Nevertheless, their approach does not consider inference on organisational structures. Instead, it uses ontologies to match between required and provided services and not in order to better approximate expected agents' behaviours in a local way.

The model proposed by Sabater and Sierra [17] also exploits ontologies to make up trust values (*ontological dimension of reputation*). Nevertheless, it does not consider organisations as a whole issue, and thus it does not take into account organisational structures. Teacy et al. [22] deals with a similar approach, where trust is obtained from using probability theory (*beta distributions* taking into account an agent's own past experiences if they exist, and information gathered from third parties (with reputation techniques) otherwise. Although the approach considers agents living within VOs, it does not deal with VO's internal structures – as our approach does. While our approach tackles the problem of improving agent skills to decide appropriate counterparts based only upon local experiences, they mainly focus on assessing reputation source accuracy.

Abdul-Rahman and Hailes [1] propose a trust model for virtual communities but use qualitative ratings for estimating trust. They focus on evaluating trust from past experiences and reputation coming from *recommender agents* without considering explicitly VO structures.

6 Conclusion

In this paper we have presented results of our work, aimed at integrating trust mechanisms into virtual organisations. We have tackled the problem of locally calculating trust, that is, finding "good" counterparts, even if only very few previous experiences are available and without the need of using reputation information obtained from external sources. The proposed model takes into account key concepts of organisational models, such as *roles* and *interactions*. It has confidence inference capabilities exploiting *taxonomies* of concept types provided by VOs. We have tested our model, confirming that the use of organisational structures makes agents' decision-making easier and more efficient, in particular when agents join an organisation and, thus, can not count on their own previous experiences. Furthermore, we have presented TOAST, a testbed we have developed to evaluate our assumptions.

In future work, we plan to extend our model with social reputation capabilities, and to study the effects of dishonest and non-cooperative agents. Furthermore, we will focus on developing an extension for TOAST that will allow us to evaluate trust models in non-stable and high-scalable environments, where agents join and leave organisations frequently. We are also focusing on finding more accurate similarity functions based on organisation taxonomies.

References

1. Abdul-Rahman, A., Hailes, S.: Supporting trust in virtual communities. In: HICSS'00, Proceedings of the 33rd Hawaii International Conference on System Sciences, vol. 6, pp. 1769–1777 (2000)
2. Aberer, K., Despotovic, Z.: Managing trust in a peer-2-peer information system. In: CIKM '01: Proceedings of the tenth international Conference on Information and Knowledge Management, pp. 310–317. ACM Press, New York (2001)
3. Esteva, M., Rodriguez, J.A., Sierra, C., Garcia, P., Arcos, J.L.: On the formal specifications of electronic institutions. In: Dignum, F., Sierra, C. (eds.) CP-WS 1996 and CDB 1997. LNCS (LNAI), vol. 1191, pp. 126–147. Springer, Heidelberg (1996)

4. Esteva, M., Rosell, B., Rodríguez-Aguilar, J.A., Arcos, J.Ll.: AMELI: An agent-based middleware for electronic institutions. In: Proceedings of the Third International Joint Conference on Autonomous Agents and Multiagent Systems, vol. 1, pp. 236–243 (2004)
5. Ferber, J., Gutknecht, O.: A meta-model for the analysis of organizations in multi-agent systems. In: Demazeau, Y. (ed.) Proceedings of the Third International Conference on Multi-Agent Systems (ICMAS'98), Paris, France, pp. 128–135. IEEE Press, New York (1998)
6. Ganesan, P., Garcia-Molina, H., Widom, J.: Exploiting hierarchical domain structure to compute similarity. ACM Trans. Inf. Syst. 21(1), 64–93 (2003)
7. Hermoso, R., Billhardt, H., Ossowski, S.: Integrating trust in virtual organisations. In: Noriega, P., et al. (eds.) COIN 2006 Workshops. LNCS (LNAI), vol. 4386, pp. 19–31. Springer, Heidelberg (2007)
8. Fred Hübner, J., Simão Sichman, J., Boissier, O.: Using the moise+ for a cooperative framework of mas reorganisation. In: Bazzan, A.L.C., Labidi, S. (eds.) SBIA 2004. LNCS (LNAI), vol. 3171, pp. 506–515. Springer, Heidelberg (2004)
9. Huynh, T.D., Jennings, N.R., Shadbolt, N.R.: Developing an integrated trust and reputation model for open multi-agent systems. In: Falcone, R., Barber, S., Sabater, J., Singh, M. (eds.) AAMAS-04 Workshop on Trust in Agent Societies, pp. 62–77 (2004)
10. Huynh, T.D., Jennings, N.R., Shadbolt, N.R.: FIRE: An integrated trust and reputation model for open multi-agent systems. In: Proceedings of the 16th European Conference on Artificial Intelligence (ECAI), pp. 119–154 (2004)
11. Li, Y., Bandar, Z.A., McLean, D.: An approach for measuring semantic similarity between words using multiple information sources. IEEE Transactions on Knowledge and Data Engineering 15(4), 871–882 (2003)
12. Odell, J., Nodine, M.H., Levy, R.: A metamodel for agents, roles, and groups. In: Odell, J.J., Giorgini, P., Müller, J.P. (eds.) AOSE 2004. LNCS, vol. 3382, pp. 78–92. Springer, Heidelberg (2005)
13. Omicini, A., Ossowski, S., Ricci, A.: Coordination infrastructures in the engineering of multiagent systems. In: Bergenti, F., Gleizes, M.-P., Zambonelli, F. (eds.) Methodologies and Software Engineering for Agent Systems: The Agent-Oriented Software Engineering Handbook, ch. 14, pp. 273–296. Kluwer Academic Publishers, Boston, MA (2004)
14. Omicini, A., Ossowski, S.: Objective versus subjective coordination in the engineering of agent systems. In: Klusch, M., Bergamaschi, S., Edwards, P., Petta, P. (eds.) Intelligent Information Agents. LNCS (LNAI), vol. 2586, pp. 179–202. Springer, Heidelberg (2003)
15. Ramchurn, S.D., Sierra, C., Godó, L., Jennings, N.R.: A computational trust model for multi-agent interactions based on confidence and reputation. In: Proceedings of 6th International Workshop of Deception, Fraud and Trust in Agent Societies, pp. 69–75 (2003)
16. Sabater, J., Sierra, C.: REGRET: a reputation model for gregarious societies. In: Müller, J.P., Andre, E., Sen, S., Frasson, C. (eds.) Proceedings of the Fifth International Conference on Autonomous Agents, Montreal, Canada, pp. 194–195. ACM Press, New York (2001)
17. Sabater, J., Sierra, C.: Reputation and social network analysis in multi-agent systems. In: AAMAS '02: Proceedings of the first international joint conference on Autonomous agents and multiagent systems, pp. 475–482. ACM Press, New York (2002)

18. Schumacher, M., Ossowski, S.: The governing environment. In: Weyns, D., Parunak, H.V.D., Michel, F. (eds.) E4MAS 2005. LNCS (LNAI), vol. 3830, pp. 88–104. Springer, Heidelberg (2006)
19. Sensoy, M., Yolum, P.: A context-aware approach for service selection using ontologies. In: Proceedings of the Fifth International Joint Conference on Autonomous Agents and Multiagent Systems, AAMAS 2006, pp. 931–938 (2006)
20. Serrano, J.M., Ossowski, S., Fernández, A.: The pragmatics of software agents - analysis and design of agent communication languages. In: Klusch, M., Bergamaschi, S., Edwards, P., Petta, P. (eds.) Intelligent Information Agents. LNCS (LNAI), vol. 2586. Springer, Heidelberg (2003)
21. Serrano, J.M., Ossowski, S.: On the impact of agent communication languages on the implementation of agent systems. In: Klusch, M., Ossowski, S., Kashyap, V., Unland, R. (eds.) CIA 2004. LNCS (LNAI), vol. 3191, pp. 92–106. Springer, Heidelberg (2004)
22. Luke Teacy, W.T., Patel, J., Jennings, N.R., Luck, M.: Travos: Trust and reputation in the context of inaccurate information sources. Autonomous Agents and Multi-Agent Systems 12(2), 183–198 (2006)
23. Vázquez-Salceda, J., Dignum, V., Dignum, F.: Organizing multiagent systems. Autonomous Agents and Multi-Agent Systems 11(3), 307–360 (2005)
24. Yu, B., Singh, M.P.: An evidential model of distributed reputation management. In: AAMAS '02: Proceedings of the first international joint conference on Autonomous agents and multiagent systems, pp. 294–301. ACM Press, New York (2002)
25. Yu, B., Singh, M.P., Sycara, K.: Developing trust in large-scale peer-to-peer systems. In: Proceedings of First IEEE Symposium on Multi-Agent Security and Survivability, pp. 1–10 (2004)
26. Zambonelli, F., Jennings, N.R., Wooldridge, M.: Organizational abstractions for the analysis and design of multi-agent systems. In: Ciancarini, P., Wooldridge, M.J. (eds.) AOSE 2000. LNCS, vol. 1957, pp. 235–252. Springer, Heidelberg (2001)

Competence Checking for the Global E-Service Society Using Games

Kostas Stathis[1], George Lekeas[2], and Christos Kloukinas[2]

[1] Department of Computer Science, Royal Holloway, University London, UK
kostas.stathis@cs.rhul.ac.uk
[2] School of Informatics, The City University, London, UK
{g.k.lekeas,c.kloukinas}@soi.city.ac.uk

Abstract. We study the problem of checking the competence of communicative agents operating in a global society in order to receive and offer electronic services. Such a society will be composed of local sub-societies that will often be semi-open, viz., entrance of agents in a semi-open society is conditional to specific admission criteria. Assuming that a candidate agent provides an abstract description of their communicative skills, we present a test that a controller agent could perform in order to decide if a candidate agent should be admitted. We formulate this test by revisiting an existing knowledge representation framework based on games specified as extended logic programs. The resulting framework finds useful application in complex and inter-operable web-services construed as semi-open societies in support of the global vision known as the Semantic Web.

1 Introduction

The vision of the Semantic Web [2] has resulted in a tremendous effort aiming to build an open and distributed infrastructure of ubiquitous and semantic web-services available to both humans and software entities alike. If this effort carries on progressing with the current pace, it is only a matter of time before software components will be in a position to choose from a huge variety of globally available web-services when seeking to achieve their goals, just like humans. The problem then will not be simply how to describe services, publish them, and access them, but also how to organise them, compose them, and enact them, so that any software component can use them in the most effective and flexible manner.

To address the flexible organisation, composition and enactment of web-services, the position of this paper is that current web-services will need to be designed so that they will be part of actions mediated by software agents. Put another way, *agents can offer or receive a service by interacting with other agents*. Provided agents are a suitable abstraction for software components that access or offer web-services [3], the position of this paper goes on further to argue that artificial societies will act as a way of organising the complex interactions involved in composite and heterogeneous services. In this context, *agents can offer or receive a service if they are members of an artificial society.* The issue then becomes how an agent can be a member of a society [22] and interact with other member agents to receive or offer services.

For autonomous and heterogeneous interactions in artificial agent societies, however, we cannot always assume that (a) we have access to the action-selection strategy of the agent and (b) the protocols available in a society match perfectly with the action-selection strategies of the member agents. In [5] we relaxed (a) so that the action-selection strategy of the agent is kept private but the space of communicative responses is made public [6]. In this way, the agent revealed only the actions it could perform abstractly (e.g. query or refuse in Fig. 1), without giving the conditions under which it would select these acts. Then to address (b) we checked if an agent is *competent*, by checking that the agent is able to *reach* specific states of the interaction (e.g. states s3 and s4 in Fig. 1).

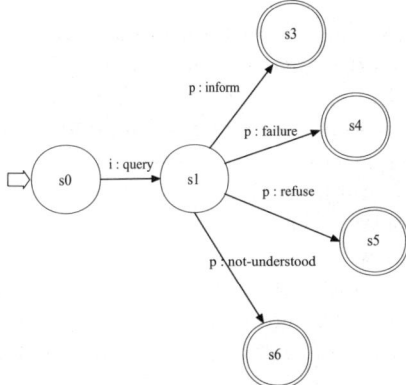

Fig. 1. A simple interaction protocol allowing agents to query other agents about the truth value of a proposition. The protocol starts at state s0 where an agent playing the role of the *initiator* i asks a query, giving rise to state s1. From s1 an agent playing the role of the *participant* p can then reply with: an inform, giving rise to final state s3; a failure, giving rise to state s4; a refuse, giving rise to final state s5; or a not-understood, giving rise to final state s6.

Competence as *reachability* allows us to check whether agents that wish to join a society have the *potential* to terminate the interactions in which they might participate, *provided the other participants allow them to do so*. However, in [5] we did not present the computational part of the competence checking procedure but referred the reader to the games framework in [19]. Here we extend [5] by linking the representation of competence checking using games as the methodology to support the structuring of e-service applications as artificial societies. We look at these issues by concentrating on competence checking of e-services only, i.e. other related issues such as *trust* or *workflow management* are beyond the scope of this work.

After this introduction, we discuss in Section 2 how to move from the current web-service scenario to one where e-services are mediated by artificial societies, including a social organisation stating how competent agents can become members of societies. In Section 3 we illustrate how interactions in artificial societies can be represented as gaming situations, by providing a concrete computational framework specified in terms of normal logic programs that have a direct Prolog implementation. The resulting computational framework is then extended in Section 4 where we show how to test competence of agents in interactions that require time. Section 5 summarises our contributions, evaluates it, and discusses related and future work.

2 Web-Services, Agents, and the Global E-Service Society

2.1 From Web-Services to Agents

A large part of the Semantic Web effort is currently being directed to *web-services*, software systems designed to support machine-to-machine interaction over a network. One of the advantages of the approach is interoperability, i.e., applications written in various programming languages and running on various platforms can use web-services to exchange data over the Internet in a way similar to inter-process communication on a single computer.

Fig. 2 depicts the typical service provision context, where a *service requester* identifies how to access a web-service by contacting a *service broker*, who holds information about services and how these can be obtained from *service providers*. One issue of Fig. 2 is that although conceptually the participating components are being thought of as roles of artificial or human agents, the figure focuses on

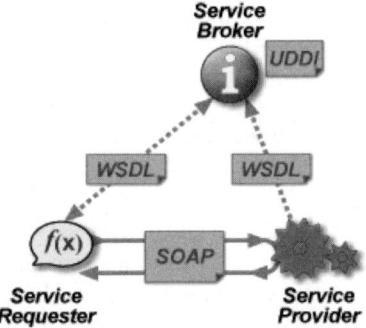

Fig. 2. The diagram, taken from [25], shows how the public interface of a web-service is described using WSDL (*Web-Service Description Language*). Other software components interact with a web-service in a manner prescribed by its interface using messages, which may be enclosed in a SOAP (*Simple Object Access Protocol*) envelope. Such messages are typically conveyed using HTTP, and normally comprise XML in conjunction with other web-related standards. Discovery of a new web-service is achieved via the use of UDDI (*Universal Description, Discovery, and Integration*) protocol.

the low-level implementation of the communication between parties, further reducing it to web-based protocol standards for distributed programming. There is, obviously, a conceptual gap between the low-level implementation of distributed components and the high-level organisation of service requesters, providers and brokers, as web-services proliferate day by day.

To fill the conceptual gap of Fig. 2 we use the notion of agents as the extra-layer required for one or more web-services with related functionality to be composed into more complex services. These more complex services will be associated with action descriptions that the agent will be capable to perform, either alone, or through communication with other agents. For example, in this view, the web-service interfaces supporting the functionality of a search engine provider, will be designed as the actions of a search agent that is capable of indexing, searching, and presenting a set of documents as URIs. Under this view, a service requester agent will have to communicate with a broker agent to find the search agent and subsequently ask for any required services. Communication between interacting agents will be governed by communication protocols [13] build on top of on an *Agent Communication Language* (ACL) [17]; [7] presents a way of using ACL for agent-based web-services.

In addition (but unlike [7]), we assume that agents rely upon a logical process that allows them to reason about web-services. Such a process is separate from the way agents invoke web-services using low-level protocols such as SOAP. We achieve this separation by assuming that agents are build with a *mind* and a *body*. The mind of the agent allows us to describe the logical reasoning the agent needs to do, including the planning required to offer a complex service by composing basic services. Agents are competent in providing services, represented in the mind as complex terms. The logical term below:

```
order("Item":string, "Quantity":integer)
```

shows how an agent might represent a more realistic `order` in the context of the protocol defined in Fig. 3. On the other hand, the body situates the mind in the distributed infrastructure of the Semantic Web. Through the body's sensors and effectors the different low-level protocols such as WSDL, UDDI, or SOAP will be used to execute actions and observe the environment. For example, the body will perform an action about an order by translating them in an XML format as shown in the term below.

```
<message name = "order">
<part name = "Item" type="xsd:string"/>
<part name = "Quantity" type="xsd:integer"/>
</message>
```

This kind of mind-body organisation has already been tested successfully in [20], where the logical actions of agents are translated into physical XML documents that are in turn communicated using the P2P system JXTA [23].

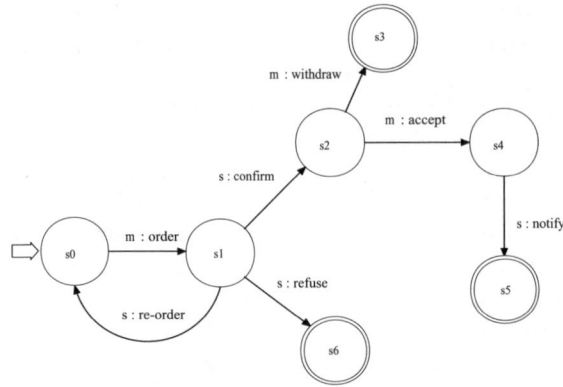

Fig. 3. A protocol where an agent in the role of a *manufacturer* m makes an order in s0, giving rise to situation s1. From s1 an agent in the role of a *supplier* s can reply with: a confirm, stating that the order must be confirmed by m, giving rise to state s2; a refuse, stating that the order cannot be carried out, giving rise to final state s6; and a reorder, stating that the order must be re-specified by m, returning to the initial state s0. If m is asked to confirm in s2, then it may reply either with a withdraw, giving rise to final state s3, or an accept, in which case the state s4 is reached. From s4 the supplier s needs to notify agent m on the details of the transaction, giving rise to final state s5.

2.2 Requirements for the Global E-Service Society

Although agents and the roles they play provide a first-level of semantic organisation of a set of web-services with related functionality, we argue that complex web-services can be best organised at another (higher) level as artificial agent societies. In this view we will use the notion of a *global society* structured in terms of *local sub-societies* as shown in Fig. 4. An agent will belong to a sub-society to start with and use the global society to communicate with agents from other sub-societies. To communicate in the global artificial society agents must be conversant in a global ACL (ACL_G in Fig. 4), possibly different to the local ACLs (such as ACL_K and ACL_N in Fig. 4) used in sub-societies. This choice of allowing different ACLs is not intended to ignore standards, but simply acknowledges heterogeneity, if it exists within an application.

The global society will be open in the sense of [14], while the local sub-societies might be in addition semi-open as in [4]. Members of a local sub-society will be individual agents acting as brokers, service requesters, and service providers, amongst other. To access a web-service within a particular sub-society, an agent must become a member if the sub-society is semi-open; we use semi-open societies to model the proliferation of web-sites that require registration for example. To join a sub-society we assume that candidate agents must reveal their *service needs*. A candidate agent will also need to make public to the sub-society it wishes to join the *service abilities* it can offer to the society. Service abilities are required so that a society can check whether the candidate agent can participate effectively in the service centric interactions within the society.

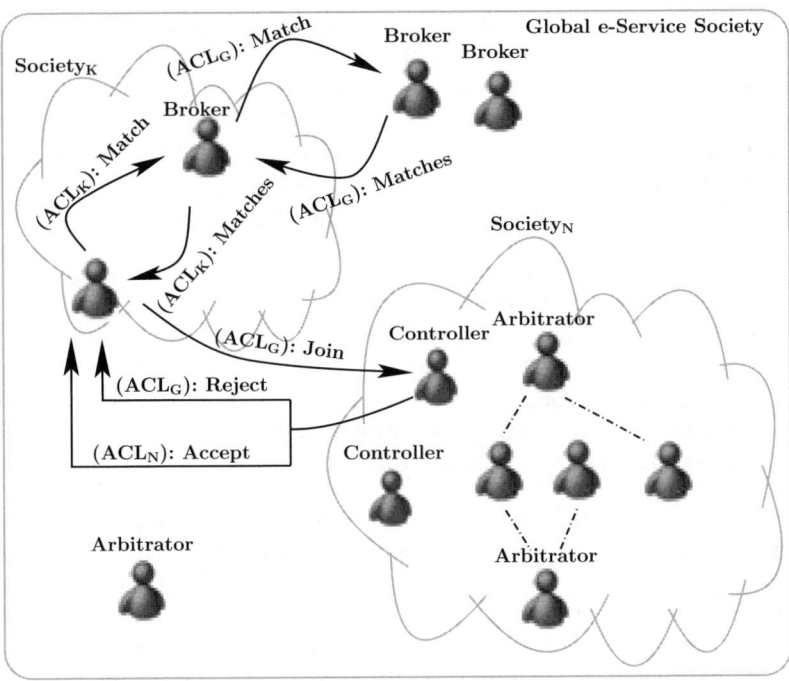

Fig. 4. Agent joining a society

Once the agent is allowed in a sub-society, the agent is given a particular *social position*, implying that the agent will be expected to play one or more *roles* associated with that position. The roles an agent plays condition the agent's participation in interaction *protocols* used in a society, thus regulating the interaction of the agent while receiving or offering a service. According to the protocols available in a society, we assume that a society will have a range of social positions on offer, with certain agents occupying some of these positions already, according to the way the society is organised. Apart from the usual positions encapsulating the roles of Fig.2, we anticipate to need (a) *controllers* to approve/disprove the entry of agents in a society and (b) *arbitrators* to observe the interactions between parties to enforce the social rules during the provision of a service.

The formal representation of a global society is beyond the scope of this work. Our primary concern is to use this notion informally to contextualise the knowledge representation framework that we propose in the next section, which is the main contribution of this work. This framework aims primarily to help with developing the functionality of controller agents, although from existing work it addresses well known issues with the development of arbitrators as game umpires [19].

3 Competence Checking Using Games

Following earlier work on the games metaphor [21], we view communicative interactions about web-services within an agent society abstractly as game interactions

[19]. More specifically, communicative acts about web-service execution and enactment are made according to protocols which are interpreted as moves made by players of a, possibly, complex game. We do not always look for a winner or a looser, but for the interaction to reach situations with a result, as in dialogue games [18]. We are motivated by communicative interactions that we envisage will play a role in web-services for e-commerce applications, see Fig. 3.

We represent the rules of a game as an normal logic program written in Prolog:

```
game(Situation, Result):-
        terminating(Situation, Result).
game(Situation, Result):-
        \+ terminating(Situation,_),
        valid(Situation, Move),
        effects(Situation, Move, NewSituation),
        game(NewSituation, Result).
```

To formulate a particular game we need to decide how to represent a game situation, its initial and terminating states, how players make valid moves, and how the effects of these moves change the current situation in to the next one until the terminating state is reached. In defining these details we specify what a controller agent needs to reason about how candidate agents can reach potentially terminating situations, by exploring the effects of valid moves for a social protocol. We will further extend this mechanism to plan for basic and complex interactions, involving many agents, to check the competency of such agents.

3.1 Game Situations

We represent game situations by terms of the form:

```
sit(Name, Id, Narrative).
```

Such a term labels a game with a Name represented by a constant, a unique Id also represented by a constant, and a Narrative of moves represented as a list. The Narrative can be empty, in which case the term represents the initial situation of a game. The term:

```
sit(order, s0, [])
```

can be thought of as representing the initial situation of the protocol depicted in Fig. 3 represented as a game. Game situations change by players making moves. The term:

```
select(Player, Action)
```

represents the fact that a Player has performed an Action as his move. For simplicity of presentation we will use here ground terms such as order or confirm to exemplify the discussion about action terms. Although such representation abstracts away from the content description of these actions, in practice terms will still be ground as we saw with the order term in section 2.1:

```
order("Item":string, "Quantity":integer).
```

We are now in a position to deal with what holds in the state of the game as a result of moves made by players. We combine our formulation with the *situation calculus* [11] expressed as a normal logic program:

```
holds(sit(Name, Id, []), F):-
    initially(sit(Name, Id, []), F).
holds(sit(Name, Id, [M | Ms]), F):-
    effect(F, M, sit(Name, Id, Ms)).
holds(sit(Id, [M | Ms]), F):-
    holds(sit(Name, Id, Ms), F),
    \+ abnormal(F, M, sit(Name, Id, Ms)).
```

Given this representation we need to express what holds initially, how effects of moves introduce new fluents, and how fluents that may hold abnormally can be excluded.

3.2 Initial and Terminating States

Consider a game where a manufacturer p1 and a supplier p2 want to communicate according to the protocol shown in Fig. 3. These roles will need to be specified when the game starts. Then the typical state of such a protocol will need to hold the roles of the players using a role_of/2 fluent. Another fluent last_move/1 will also be used to record the last move made. We can express the initial state of this protocol as:

```
initially(sit(order, s0, []), role_of(p1, manufacturer)).
initially(sit(order, s0, []), role_of(p2, supplier)).
```

The absence of last_move/1 from the initial situation allows our formulation of the situation calculus to capture that this does not hold, using negation as failure. Similarly, we can specify the terminating states of the protocol in Fig. 3 as:

```
terminating(Situation, Situation):-
        Situation = sit(order, Id, N),
        holds(Situation, last_move(select(P, Act))),
        on(Act, [refuse, withdraw, notify]).
```

In other words, in this instance we return as the result the whole of the situation term with all the moves selected so far.

3.3 Valid Moves

Differentiating between valid and invalid moves is of great importance in the analysis of interactive systems as games [18]. For social interactions using agents such as an auction this differentiation will allow the auctioneer to determine which bids are valid and therefore, which bids are eligible for winning the auction [1]. In our games framework, we represent valid moves as:

```
valid(S, Move) :- available(S, Move), legal(S, Move).
```

Available moves are all the moves afforded by the state of a game. To represent that order is an available move for the protocol of Fig.3 we write:

```
available(sit(order, Id, N), select(P, order)).
```

The specification of available moves should allow all specified moves to be selected by agents at any state. By adding conditions to `available/2` rules we can check the type preconditions of actions. As selecting an available move in a game does not always imply that this move is legal, we also need to specify legal moves separately. For example, to represent that it is legal to notify only after an accept as in the protocol of Fig. 3, we write:

```
legal(sit(order, Id, N), select(P1, notify)) :-
    holds(sit(order, Id, N), last_move(select(P2, accept))),
    holds(sit(order, Id, N), role_of(P1, supplier)).
```

`last_move/1` is what helps the definition of legal moves to ensure that communicative acts are ordered as expected by the protocol.

3.4 Representation of Effects

To represent the effects of a move on the game we distinguish between the effects of that move on the term representing a game situation and how these effects are brought about in the specific state represented by that situation. For example, to represent the effects on the state of the situation we simply extend the narrative of that situation with the move made:

```
effects(sit(Name, Id, Ms), Move, sit(Name, Id, [Move | Ms])).
```

The effects of such a move on the state representing a situation are obtained implicitly by the situation calculus effect and abnormality rules. To give an example of these predicates we consider again the protocol of Fig. 3. We write:

```
effect(last_move(M), M, sit(order, Id, N)).
```

to represent that when we apply a move on the state, it becomes the last move in the next situation. Note that with our formulation of the rules of a game we do not need to check for the preconditions of a move, as we have checked before the effects are carried out that the move is valid.

We also need to specify any abnormal situations where a fluent holds where it should not. For the protocol of Fig. 3, the assertion:

```
abnormal(last_move(M_old), M_new, sit(order, Id, N)).
```

will ensure that after a new move has been made it is abnormal to consider that the last move is the one made previously.

3.5 Competence Checking as Planning

Given the formulation of games so far we have a way of describing all valid situations that a set of agents can use according to the social rules of a protocol. In [19] we have shown how such rules can be used by an umpire (arbitrator) that checks conformance of the interactions or by a player who wants to play by the rules. However, [19] did not consider competence. To augment the applicability of the approach we view here competence checking as a particular instance of planning using the rules of the game. We will use the following program to plan according to the rules of a game:

```
plan(game(S, R), S, R):-
      achieved(terminating(S, R), S, R).
plan(game(S, R), S, R):-
      \+ terminating(S, _),
      assume(valid(S, M), S, M),
      apply(effects(S, M, NewS), S, M, NewS),
      plan(game(NewS, R), NewS, R).
```

That is, to plan for a game we need to stop when a terminating state has been achieved. Otherwise, in a non-terminating state, we need to assume a valid move, apply the effects of this move to get a new state, and then carry on planning in that new state.

We define `achieved/3` and `apply/4` simply by calling in Prolog the predicates that they take as their first argument (as they are instantiated in the `plan/3` program):

```
achieved(Terminating, Initial, Result):- call(Terminating).
```

```
apply(Effects, S, Move, NewS):- call(Effects).
```

To define `assume/3`, however, we need to rely on competence descriptions of players, which correspond to what we referred to in section 2 as the service abilities of agents. To represent such abilities for an agent we will assume rules of the form:

```
competent(Agent, do(Situation, Act)):- Conditions.
```

A controller agent will need to keep rules of this kind to test the competence of candidate agents. The controller must hold such models for all the members in the society too. For example for the protocol of Fig. 3, consider the models describing the competence of players p1 and p2:

```
competent(p1, do(sit(order, Id, N), order)).
competent(p1, do(sit(order, Id, N), accept)).

competent(p2, do(sit(order, Id, N), reorder)).
competent(p2, do(sit(order, Id, N), confirm)).
competent(p2, do(sit(order, Id, N), notify)).
```

We now define:

```
assume(Valid, Situation, select(Player, Act)):-
    call(Valid),
    competent(Player, do(Situation, Act)),
    acceptable(Situation, select(Player, Act)).
```

While planning, this definition allow us to generate a valid move, check that the agent is competent of performing it, and finally check that a move is acceptable. The definition of `acceptable/2` joins the assumed move with the rest of the narrative describing the current situation to filter unwanted loops. For the protocol of Fig. 3 such a loop is described by the sequence:

[select(A, order), select(B, reorder)]

which is allowed to be repeated only once. The implementation of acceptable moves for this example is not included here as it trivially checks for specific unwanted sub-lists of a list. We are now in a position to ask:

?- plan(Game, sit(order, s0, []), Result)

and get as part of the solution process all the valid states that can be planned for using the description of the protocol and the descriptions of the competence for individual players, with loops allowed only once, if they exist. What a controller agent can then do with the results is application specific.

4 Competence Checking in Timed Games

Combining our games framework with the normal logic programming formulation of the situation calculus allowed us to specify protocol-based interactions

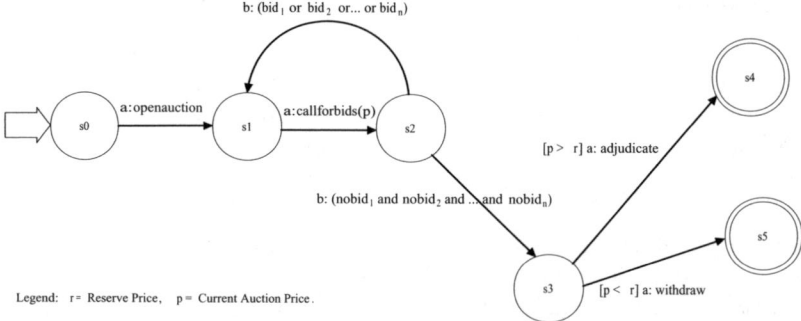

Fig. 5. The English auction protocol allowing an agent with the role of an auctioneer a and a set of agents with the role of bidder b to interact for the sale of a good. The auctioneer starts the auction and the calls for bids at a specific price. One or more bidders bid, in which case the auctioneer calls for new bids until no more bids are offered. At that point the auctioneer either adjudicates the good to the highest bidder or withdraws the good if the reserve price is not met.

and test for reachability of all the states of the protocol via planning. However, in many occasions social protocols do not assume strict turn-taking in that moves of players can occur at the same time. An example of such a protocol is that of an English auction, as shown in Fig. 5.

To allow for protocols of the kind describe in the above figure we introduce timed games, that is, games whose moves have also a representation of the time in which they happened.

4.1 Timed Games in the Event Calculus

In trying to formulate timed games we introduce timed situations of the form:

```
sit(Name, Id, Time, Narrative).
```

One difference from our earlier representation is that now we need to keep the current Time in the situation term. In addition, a narrative in timed games is represented in terms of *episodes*, that is collections of moves that can validly happen at the same time in a situation. We express episodes as:

```
at([select(Player1, Act1), ..., select(PlayerN, ActN)], T).
```

Using this representation, the term at(T, []) means that nothing happened at time T.

To reason about timed game situations, we use the *simple version* [15] of the *event calculus* [8] instead of the situation calculus, suitably adapted for our purposes as follows:

```
holds(sit(N,Id,Tn,Nn), P):-
        0 =< Tn,
        initially(sit(N,Id,Ti,Ni), P),
        \+ clipped(P, sit(N,Id,Ti,Ni), sit(N,Id,Tn,Nn)).
holds(sit(N,Id,Tn,Nn), P):-
        happens(E, Ti, Ni, Nn),
        Ti < Tn,
        initiates(E, P, sit(N,Id,Ti,Ni)),
        \+ clipped(P, sit(N,Id,Ti,Ni), sit(N,Id,Tn,Nn)).

clipped(P, sit(N,Id,Ti,Ni), sit(N,Id,Tn,Nn)):-
        happens(Estar, Tj, Nj, Nn),
        Ti < Tj, Tj < Tn,
        terminates(Estar, P, sit(N,Id,Tj,Nj)).
```

The important difference from the normal event calculus formulations is that narratives are held as lists in situation terms rather than as assertions in the knowledge base, in the spirit of the situation calculus. In this context, our representation of an event happening is re-specified as:

```
happens(E, Tn, [at(En, Tn)|Sn], [at(En, Tn)|Sn]):-
        member(E, En).
happens(E, Ti, [at(Ei,Ti)|Si], [at(En, Tn)|Sn]):-
        happens(E, Ti, [at(Ei,Ti)|Si], Sn).
happens(at(En,Tn), Tn, [at(En,Tn)|Sn], [at(En, Tn)|Sn]).
happens(at(Ei,Ti), Ti, [at(Ei,Ti)|Si], [at(En, Tn)|Sn]):-
        happens(at(Ei, Ti), Ti, [at(Ei,Ti)|Si], Sn).
```

The first two rules deal with individual events as in the simple event calculus, with the difference that now we need additional parameters to keep the narrative at intermediate times. Unlike the simple event calculus however, our formulation also requires additional rules (the last two) to deal with episodes that have happened in the narrative; like the events they contain, they too can affect the state of the game.

One implication of the use of episodes is that we need to change the way we update the narrative in a timed game. We write:

```
effects(sit(N,Id,T,Es), at(Ms, T), sit(N,Id,NewT,[at(Ms,T)|Es])):-
        T > 0,  NewT is T + 1.
```

The above definition makes the assumption that new episodes last for one unit of time. The rest of the generic representation for game remains the same, the only parts that change are the domain specific details. We give an example next.

4.2 Formulating an English Auction

To exemplify timed games we present briefly parts of our formulation for an auction as shown in Fig. 5. For simplicity, we will assume that there are two bidders and an auctioneer, and that in order to check the game we only need the last set of moves captured in the fluent (last_moves/1). We will represent the initial state as before, but now we will need to also specify the initial time, which we will assume it is 0. This gives rise to the initial state:

```
initially(sit(auction, s0,0,[]), role_of(p1, auctioneer)).
initially(sit(auction, s0,0,[]), role_of(p2, bidder)).
initially(sit(auction, s0,0,[]), role_of(p3, bidder)).
```

The terminating conditions are specified with holds axioms using the simple version of the event calculus presented in the previous section. For example, to define termination in the auction we write:

```
terminating(sit(auction,Id,T,N), sit(auction, Id, T, N)):-
        holds(sit(auction,Id,T,N), last_moves([select(P,X)])),
        member(X, [adjudicate,withdraw]).
```

The valid moves are specified as before, including available and legal moves, however now these need to be specific to the moves of the auction. For example, to specify a legal bid we write:

```
legal(sit(auction,Id,T,N), select(Player1, bid)):-
    holds(sit(auction,Id,T,N),role_of(Player1,bidder)),
    holds(sit(auction,Id,T,N),last_moves([select(Player2,cfp)])),
    holds(sit(auction,Id,T,N),role_of(Player2,auctioneer)).
```

The only aspect that really changes is the representation of effects, which are now expressed in terms of `initiates/3` and `terminates/3` instead of `effect/3` and `abnormal/3`.

```
initiates(at(Es, T),  last_moves(Es), sit(auction,Id,T,Ns)).

terminates(at(Es, T), last_moves(Old_M), sit(auction,Id,T,Ns)).
```

Notice that in this particular example `initiates/3` and `terminates/3` rules are written only for episodes, however, in general, these need to be specified also for individual events.

4.3 Competence Checking of an English Auction

To check the competence of a set of players for timed games we are going to assume, as before, that we have a set of statements regarding the competencies of individual players and the `plan/4` program. The main aspect that changes in timed games, however, is that instead of generating individual moves we need to generate individual episodes:

```
assume(Valid, Sit, at(Moves, T)):-
    Sit = sit(N,Id,T,Es),
    Valid = valid(Sit, select(P, M)),
    findall(M, (call(Valid, competent(P, do(Sit, M)))), All),
    sublist(Moves, All),
    acceptable(Sit, at(Moves, T)).
```

In other words, we need to change our definition of `assume/3` to deal with episodes, so that we get all the valid and acceptable subset of moves in the protocol. Running the query:

```
?- plan(Game, sit(auction, s0, 1, []), Result)
```

we will be in a position to find all the reachable states of the protocol, according to the description of the rules, and the competence of the players.

5 Concluding Remarks

We have investigated the issue of competence checking for agents operating in a global artificial society whose purpose is to organise complex services. Assuming that a candidate agent provides an abstract description of their communicative competence, we have formulated a test that a controller agent can perform to decide if the candidate agent should join a sub-society of the global society. We have formulated this test by revisiting an existing knowledge-based framework based on games represented in extensive form. Although [22,5,9] have motivated our framework, we have found no other related work that links agent competency with artificial societies using games.

In evaluating our approach we see that our formulation can integrate the situation and the event calculi according to the competence checking problem at hand. In this context we inherit from our original formulation of games the notion of *compound games*, viz., games built from *active* sub-games [18], thus allowing quite complex interactions to be checked for competency. Also, by using normal logic programs our approach can be implemented directly in Prolog, unlike other approaches that need to extend the proof-procedure e.g. agents based on abduction [22].

The current formulation of games and, as a result, the competence checking presented has the potential to build upon the methodology developed in [18]. One aspect of this is that it treats valid acts as an abstraction for different specification approaches of social action, as they may be required by different applications. We have for example assumed that valid acts must be available and legal. Nevertheless, not all applications need to be presented in this way. For example, in [1] valid acts are treated in a way that relies on a more elaborate representation of concepts such as those of obligation and permission. Investigation of these aspects will allow us to compare our framework with existing approaches that model web-services, e.g. see [12], but with an artificial societies approach.

By investigating how to best check the competency of agents in artificial societies for e-services we have identified the need to incorporate into our framework a mechanism that ensures that agents are not simply competent according to the acts of a protocol but also according to the expected order of acts described in it. In parallel, we also need to deal with the re-computation introduced from the use of event and situation calculi in more complex domains to the examples used here. An immediate remedy will be to run our games framework on a Prolog system that supports tabling [24], such XSB Prolog. How tabled execution compares with approaches based on model checking [10] and satisfiability [16] is another direction that we wish to investigate in this context.

Acknowledgements

We would like to thank the anonymous referees for their comments on a previous version of this paper. The first and third authors were partially supported by the EU IST6 *ArguGRID* and *SeCSE* projects respectively.

References

1. Artikis, A., Pitt, J., Sergot, M.: Animated specifications of computational societies. In: Castelfranchi, C., Lewis Johnson, W. (eds.) AAMAS 2002, pp. 1053–1061. ACM Press, New York (2002)
2. Burners-Lee, T., Hendler, J., Lassila, O.: The semantic web. Scientific American 284(5) (May 2001)

3. Curcin, V., Ghanem, M., Guo, Y., Stathis, K., Toni, F.: Building next generation Service-Oriented Architectures using Argumentation Agents. In: Polze, A., Kowalczyk, R. (eds.) 3rd International Conference on Grid Service Engineering and Management, Germany, pp. 249–263 (September 2006)
4. Davidsson, P.: Categories of artificial societies. In: Omicini, A., Petta, P., Tolksdorf, R. (eds.) ESAW 2001. LNCS (LNAI), vol. 2203, pp. 1–10. Springer, Heidelberg (2002)
5. Endriss, U., Lue, W., Maudet, N., Stathis, K.: Competent agents and customising protocols. In: Omicini, A., Petta, P., Pitt, J. (eds.) ESAW 2003. LNCS (LNAI), vol. 3071, pp. 168–181. Springer, Heidelberg (2004)
6. Endriss, U., Maudet, N., Sadri, F., Toni, F.: Protocol conformance for logic-based agents. In: Gottlob, G., Walsh, T. (eds.) Proceedings of the Eighteenth International Joint Conference on Artificial Intelligence, Acapulco, Mexico (IJCAI-03), Morgan Kaufmann, San Francisco (2003)
7. Kowalczyk, R., Yan, J., Yang, Y., Nguyen, X.T.: A service workflow management framework based on peer-to-peer and agent technologies. In: Proc. of International Workshop on Grid and Peer-to-Peer based Workflows, Melbourne, Australia (2005)
8. Kowalski, R.A., Sergot, M.: A logic-based calculus of events. New Generation Computing 4(1), 67–95 (1986)
9. Lekeas, G.K., Stathis, K.: Agents acquiring Resources through Social Positions: An Activity-based Approach. In: de Bruijn, O., Stathis, K. (eds.) Proceedings of the 1st International Workshop on Socio-Cognitive Grids, Santorini, Greece (June 2003)
10. Lomuscio, A., Raimondi, F.: Model checking knowledge, strategies, and games in multi-agent systems. In: Proceedings of the 5th International Conference on Autonomous Agents and Multi-Agent systems (AAMAS06), ACM Press, New York (2006)
11. McCarthy, J., Hayes, P.: Some philosophical problems from the standpoint of artificial intelligence. In: Meltzer, B., Michie, D. (eds.) Machine Intelligence 4, pp. 463–502. American Elsevier, New York (1969)
12. McIlraith, S., Cao Son, T., Zeng, H.: Semantic web services. IEEE Intelligent Systems 16(2), 46–53 (2001)
13. Pitt, J., Mamdani, A.: A Protocol-based Semantics for an Agent Communication Language. In: Proceedings of the Sixteenth International Joint Conference on Artificial Intelligence, Stockholm, Sweden (IJCAI-99), pp. 486–491. Morgan Kaufmann, San Francisco (1999)
14. Pitt, J.V.: The open agent society as a platform for the user-friendly information society. AI & Society 19(2), 123–158 (2005)
15. Shanahan, M.: The event calculus explained. In: Veloso, M.M., Wooldridge, M.J. (eds.) Artificial Intelligence Today. LNCS (LNAI), vol. 1600, pp. 409–430. Springer, Heidelberg (1999)
16. Shanahan, M., Witkowski, M.: Event Calculus Planning Through Satisfiability. Journal of Logic and Computation 14, 731–745 (2004)
17. Singh, M.P.: Agent communication languages: Rethinking the principles. In: Communication in Multiagent Systems, pp. 37–50 (2003)
18. Stathis, K.: Game–Based Development of Interactive Systems. PhD thesis, Department of Computing, Imperial College London (November 1996)
19. Stathis, K.: A Game-based Architecture for developing Interactive Components in Computational Logic. Functional and Logic Programming, Special Issue on Logical Formalisms for Program Composition, 2000(1) (March 2000)

20. Stathis, K., Kakas, A., Lu, W., Demetriou, N., Endriss, U., Bracciali, A.: PROSOCS: a platform for programming software agents in computational logic. In: Müller, J., Petta, P. (eds.) Proceedings of the Fourth International Symposium From Agent Theory to Agent Implementatio, Vienna, Austria (April 13-16, 2004)
21. Stathis, K., Sergot, M.J.: Games as a Metaphor for Interactive Systems. In: Sasse, M.A., Cunningham, R.J., Winder, R.L. (eds.) People and Computers XI (Proceedings of HCI'96), London, UK. BCS Conference Series, pp. 19–33. Springer-Verlag, Heidelberg (1996)
22. Toni, F., Stathis, K.: Access-as-you-need: a computational logic framework for flexible resource access in artificial societies. In: Petta, P., Tolksdorf, R., Zambonelli, F. (eds.) ESAW 2002. LNCS (LNAI), vol. 2577, Springer, Heidelberg (2003)
23. Traversat, B., Abdelaziz, M., Doolin, D., Duigou, M., Hugly, J.C., Pouyoul, E.: Project JXTA-C:Enabling a Web of Things. In: Proceedings of the 36th Hawaii International Conference on System Sciences (HICSS'03), pp. 282–287. IEEE Press, Los Alamitos (2003)
24. Warren, D.S.: Memoing for logic programs. Commun. ACM 35(3), 93–111 (1992)
25. Web-services. Home Page: http://en.wikipedia.org/wiki/Web_services

Author Index

Álvarez, Pedro   229
Artikis, Alexander   143

Bañares, José A.   229
Bernon, Carole   284
Billhardt, Holger   368
Boissier, Olivier   86
Bordini, Rafael H.   38
Bou, Eva   300
Bradshaw, Jeffrey M.   175

Caelen, Jean   354
Centeno, Roberto   368
Clancey, William J.   175

Dastani, Mehdi   38
Demazeau, Yves   354
Dikenelli, Oguz   209

Feltovich, Paul J.   175
Fuentes-Fernández, Rubén   126

García-Camino, Andrés   193
Gleizes, Marie-Pierre   284
Goknil, Arda   209
Gómez-Sanz, Jorge J.   25, 126

Hermoso, Ramón   368
Holvoet, Tom   62
Honiden, Shinichi   161
Hübner, Jomi Fred   86

Johnson, Matthew   175

Kamara, Lloyd   143
Kaponis, Dimosthenis   265
Kardas, Geylani   209
Kloukinas, Christos   384

Lekeas, George   384
López-Sánchez, Maite   300

Mata, Eloy J.   229
McBurney, Peter   245
Miller, Tim   245
Muldoon, C.   320

Noriega, Pablo   193

O'Grady, M.J.   320
O'Hare, G.M.P.   320
Ossowski, Sascha   368

Pavón, Juan   126
Picard, Gauthier   284
Piolle, Guillaume   354
Pitt, Jeremy   143, 265
Platon, Eric   161

Rodríguez-Aguilar,
    Juan-Antonio   193, 300
Rubio, Julio   229

Sabouret, Nicolas   161
Saunders, Rob   340
Sichman, Jaime Simão   86
Sierhuis, Maarten   1
Stathis, Kostas   384
Stuit, Marco   106
Szirbik, Nick B.   106

Topaloglu, N. Yasemin   209

Weyns, Danny   62
Winikoff, Michael   38

Printing: Mercedes-Druck, Berlin
Binding: Stein+Lehmann, Berlin

Lecture Notes in Artificial Intelligence (LNAI)

Vol. 4755: V. Corruble, M. Takeda, E. Suzuki (Eds.), Discovery Science. XI, 298 pages. 2007.

Vol. 4754: M. Hutter, R.A. Servedio, E. Takimoto (Eds.), Algorithmic Learning Theory. XI, 403 pages. 2007.

Vol. 4737: B. Berendt, A. Hotho, D. Mladenic, G. Semeraro (Eds.), From Web to Social Web: Discovering and Deploying User and Content Profiles. XI, 161 pages. 2007.

Vol. 4733: R. Basili, M.T. Pazienza (Eds.), AI*IA 2007: Artificial Intelligence and Human-Oriented Computing. XVII, 858 pages. 2007.

Vol. 4724: K. Mellouli (Ed.), Symbolic and Quantitative Approaches to Reasoning with Uncertainty. XV, 914 pages. 2007.

Vol. 4722: C. Pelachaud, J.-C. Martin, E. André, G. Chollet, K. Karpouzis, D. Pelé (Eds.), Intelligent Virtual Agents. XV, 425 pages. 2007.

Vol. 4720: B. Konev, F. Wolter (Eds.), Frontiers of Combining Systems. X, 283 pages. 2007.

Vol. 4702: J.N. Kok, J. Koronacki, R. Lopez de Mantaras, S. Matwin, D. Mladenič, A. Skowron (Eds.), Knowledge Discovery in Databases: PKDD 2007. XXIV, 640 pages. 2007.

Vol. 4701: J.N. Kok, J. Koronacki, R.L.d. Mantaras, S. Matwin, D. Mladenič, A. Skowron (Eds.), Machine Learning: ECML 2007. XXII, 809 pages. 2007.

Vol. 4696: H.-D. Burkhard, G. Lindemann, R. Verbrugge, L.Z. Varga (Eds.), Multi-Agent Systems and Applications V. XIII, 350 pages. 2007.

Vol. 4694: B. Apolloni, R.J. Howlett, L. Jain (Eds.), Knowledge-Based Intelligent Information and Engineering Systems, Part III. XXIX, 1126 pages. 2007.

Vol. 4693: B. Apolloni, R.J. Howlett, L. Jain (Eds.), Knowledge-Based Intelligent Information and Engineering Systems, Part II. XXXII, 1380 pages. 2007.

Vol. 4692: B. Apolloni, R.J. Howlett, L. Jain (Eds.), Knowledge-Based Intelligent Information and Engineering Systems, Part I. LV, 882 pages. 2007.

Vol. 4687: P. Petta, J.P. Müller, M. Klusch, M. Georgeff (Eds.), Multiagent System Technologies. X, 207 pages. 2007.

Vol. 4682: D.-S. Huang, L. Heutte, M. Loog (Eds.), Advanced Intelligent Computing Theories and Applications. XXVII, 1373 pages. 2007.

Vol. 4676: M. Klusch, K.V. Hindriks, M.P. Papazoglou, L. Sterling (Eds.), Cooperative Information Agents XI. XI, 361 pages. 2007.

Vol. 4667: J. Hertzberg, M. Beetz, R. Englert (Eds.), KI 2007: Advances in Artificial Intelligence. IX, 516 pages. 2007.

Vol. 4660: S. Džeroski, J. Todorovski (Eds.), Computational Discovery of Scientific Knowledge. X, 327 pages. 2007.

Vol. 4659: V. Mařík, V. Vyatkin, A.W. Colombo (Eds.), Holonic and Multi-Agent Systems for Manufacturing. VIII, 456 pages. 2007.

Vol. 4651: F. Azevedo, P. Barahona, F. Fages, F. Rossi (Eds.), Recent Advances in Constraints. VIII, 185 pages. 2007.

Vol. 4648: F. Almeida e Costa, L.M. Rocha, E. Costa, I. Harvey, A. Coutinho (Eds.), Advances in Artificial Life. XVIII, 1215 pages. 2007.

Vol. 4635: B. Kokinov, D.C. Richardson, T.R. Roth-Berghofer, L. Vieu (Eds.), Modeling and Using Context. XIV, 574 pages. 2007.

Vol. 4632: R. Alhajj, H. Gao, X. Li, J. Li, O.R. Zaïane (Eds.), Advanced Data Mining and Applications. XV, 634 pages. 2007.

Vol. 4629: V. Matoušek, P. Mautner (Eds.), Text, Speech and Dialogue. XVII, 663 pages. 2007.

Vol. 4626: R.O. Weber, M.M. Richter (Eds.), Case-Based Reasoning Research and Development. XIII, 534 pages. 2007.

Vol. 4617: V. Torra, Y. Narukawa, Y. Yoshida (Eds.), Modeling Decisions for Artificial Intelligence. XII, 502 pages. 2007.

Vol. 4612: I. Miguel, W. Ruml (Eds.), Abstraction, Reformulation, and Approximation. XI, 418 pages. 2007.

Vol. 4604: U. Priss, S. Polovina, R. Hill (Eds.), Conceptual Structures: Knowledge Architectures for Smart Applications. XII, 514 pages. 2007.

Vol. 4603: F. Pfenning (Ed.), Automated Deduction – CADE-21. XII, 522 pages. 2007.

Vol. 4597: P. Perner (Ed.), Advances in Data Mining. XI, 353 pages. 2007.

Vol. 4594: R. Bellazzi, A. Abu-Hanna, J. Hunter (Eds.), Artificial Intelligence in Medicine. XVI, 509 pages. 2007.

Vol. 4585: M. Kryszkiewicz, J.F. Peters, H. Rybinski, A. Skowron (Eds.), Rough Sets and Intelligent Systems Paradigms. XIX, 836 pages. 2007.

Vol. 4578: F. Masulli, S. Mitra, G. Pasi (Eds.), Applications of Fuzzy Sets Theory. XVIII, 693 pages. 2007.

Vol. 4573: M. Kauers, M. Kerber, R. Miner, W. Windsteiger (Eds.), Towards Mechanized Mathematical Assistants. XIII, 407 pages. 2007.

Vol. 4571: P. Perner (Ed.), Machine Learning and Data Mining in Pattern Recognition. XIV, 913 pages. 2007.

Vol. 4570: H.G. Okuno, M. Ali (Eds.), New Trends in Applied Artificial Intelligence. XXI, 1194 pages. 2007.

Vol. 4565: D.D. Schmorrow, L.M. Reeves (Eds.), Foundations of Augmented Cognition. XIX, 450 pages. 2007.

Vol. 4562: D. Harris (Ed.), Engineering Psychology and Cognitive Ergonomics. XXIII, 879 pages. 2007.

Vol. 4548: N. Olivetti (Ed.), Automated Reasoning with Analytic Tableaux and Related Methods. X, 245 pages. 2007.

Vol. 4539: N.H. Bshouty, C. Gentile (Eds.), Learning Theory. XII, 634 pages. 2007.

Vol. 4529: P. Melin, O. Castillo, L.T. Aguilar, J. Kacprzyk, W. Pedrycz (Eds.), Foundations of Fuzzy Logic and Soft Computing. XIX, 830 pages. 2007.

Vol. 4520: M.V. Butz, O. Sigaud, G. Pezzulo, G. Baldassarre (Eds.), Anticipatory Behavior in Adaptive Learning Systems. X, 379 pages. 2007.

Vol. 4511: C. Conati, K. McCoy, G. Paliouras (Eds.), User Modeling 2007. XVI, 487 pages. 2007.

Vol. 4509: Z. Kobti, D. Wu (Eds.), Advances in Artificial Intelligence. XII, 552 pages. 2007.

Vol. 4496: N.T. Nguyen, A. Grzech, R.J. Howlett, L.C. Jain (Eds.), Agent and Multi-Agent Systems: Technologies and Applications. XXI, 1046 pages. 2007.

Vol. 4483: C. Baral, G. Brewka, J. Schlipf (Eds.), Logic Programming and Nonmonotonic Reasoning. IX, 327 pages. 2007.

Vol. 4482: A. An, J. Stefanowski, S. Ramanna, C.J. Butz, W. Pedrycz, G. Wang (Eds.), Rough Sets, Fuzzy Sets, Data Mining and Granular Computing. XIV, 585 pages. 2007.

Vol. 4481: J. Yao, P. Lingras, W.-Z. Wu, M. Szczuka, N.J. Cercone, D. Ślęzak (Eds.), Rough Sets and Knowledge Technology. XIV, 576 pages. 2007.

Vol. 4476: V. Gorodetsky, C. Zhang, V.A. Skormin, L. Cao (Eds.), Autonomous Intelligent Systems: Multi-Agents and Data Mining. XIII, 323 pages. 2007.

Vol. 4457: G.M.P. O'Hare, A. Ricci, M.J. O'Grady, O. Dikenelli (Eds.), Engineering Societies in the Agents World VII. XI, 401 pages. 2007.

Vol. 4456: Y. Wang, Y.-m. Cheung, H. Liu (Eds.), Computational Intelligence and Security. XXIII, 1118 pages. 2007.

Vol. 4455: S. Muggleton, R. Otero, A. Tamaddoni-Nezhad (Eds.), Inductive Logic Programming. XII, 456 pages. 2007.

Vol. 4452: M. Fasli, O. Shehory (Eds.), Agent-Mediated Electronic Commerce. VIII, 249 pages. 2007.

Vol. 4451: T.S. Huang, A. Nijholt, M. Pantic, A. Pentland (Eds.), Artifical Intelligence for Human Computing. XVI, 359 pages. 2007.

Vol. 4441: C. Müller (Ed.), Speaker Classification. X, 309 pages. 2007.

Vol. 4438: L. Maicher, A. Sigel, L.M. Garshol (Eds.), Leveraging the Semantics of Topic Maps. X, 257 pages. 2007.

Vol. 4434: G. Lakemeyer, E. Sklar, D.G. Sorrenti, T. Takahashi (Eds.), RoboCup 2006: Robot Soccer World Cup X. XIII, 566 pages. 2007.

Vol. 4429: R. Lu, J.H. Siekmann, C. Ullrich (Eds.), Cognitive Systems. X, 161 pages. 2007.

Vol. 4428: S. Edelkamp, A. Lomuscio (Eds.), Model Checking and Artificial Intelligence. IX, 185 pages. 2007.

Vol. 4426: Z.-H. Zhou, H. Li, Q. Yang (Eds.), Advances in Knowledge Discovery and Data Mining. XXV, 1161 pages. 2007.

Vol. 4411: R.H. Bordini, M. Dastani, J. Dix, A.E.F. Seghrouchni (Eds.), Programming Multi-Agent Systems. XIV, 249 pages. 2007.

Vol. 4410: A. Branco (Ed.), Anaphora: Analysis, Algorithms and Applications. X, 191 pages. 2007.

Vol. 4399: T. Kovacs, X. Llorà, K. Takadama, P.L. Lanzi, W. Stolzmann, S.W. Wilson (Eds.), Learning Classifier Systems. XII, 345 pages. 2007.

Vol. 4390: S.O. Kuznetsov, S. Schmidt (Eds.), Formal Concept Analysis. X, 329 pages. 2007.

Vol. 4389: D. Weyns, H. Van Dyke Parunak, F. Michel (Eds.), Environments for Multi-Agent Systems III. X, 273 pages. 2007.

Vol. 4386: P. Noriega, J. Vázquez-Salceda, G. Boella, O. Boissier, V. Dignum, N. Fornara, E. Matson (Eds.), Coordination, Organizations, Institutions, and Norms in Agent Systems II. XI, 373 pages. 2007.

Vol. 4384: T. Washio, K. Satoh, H. Takeda, A. Inokuchi (Eds.), New Frontiers in Artificial Intelligence. IX, 401 pages. 2007.

Vol. 4371: K. Inoue, K. Satoh, F. Toni (Eds.), Computational Logic in Multi-Agent Systems. X, 315 pages. 2007.

Vol. 4369: M. Umeda, A. Wolf, O. Bartenstein, U. Geske, D. Seipel, O. Takata (Eds.), Declarative Programming for Knowledge Management. X, 229 pages. 2006.

Vol. 4363: B.D. ten Cate, H.W. Zeevat (Eds.), Logic, Language, and Computation. XII, 281 pages. 2007.

Vol. 4343: C. Müller (Ed.), Speaker Classification I. X, 355 pages. 2007.

Vol. 4342: H. de Swart, E. Orłowska, G. Schmidt, M. Roubens (Eds.), Theory and Applications of Relational Structures as Knowledge Instruments II. X, 373 pages. 2006.

Vol. 4335: S.A. Brueckner, S. Hassas, M. Jelasity, D. Yamins (Eds.), Engineering Self-Organising Systems. XII, 212 pages. 2007.

Vol. 4334: B. Beckert, R. Hähnle, P.H. Schmitt (Eds.), Verification of Object-Oriented Software. XXIX, 658 pages. 2007.

Vol. 4333: U. Reimer, D. Karagiannis (Eds.), Practical Aspects of Knowledge Management. XII, 338 pages. 2006.

Vol. 4327: M. Baldoni, U. Endriss (Eds.), Declarative Agent Languages and Technologies IV. VIII, 257 pages. 2006.